S0-CAS-548

KINETIC METHODS IN ANALYTICAL CHEMISTRY

ELLIS HORWOOD SERIES IN ANALYTICAL CHEMISTRY

Series Editors: Dr R. A. CHALMERS and Dr MARY MASSON, University of Aberdeen

Consultant Editor: Prof. J. N. MILLER, Loughborough University of Technology

Author(s)	Title
S. Allenmark	**Chromatographic Enantioseparation — Methods and Applications**
G.E. Baiulescu, P. Dumitrescu & P.Gh. Zugravescu	**Sampling**
H. Barańska, A. Łabudzińska & J. Terpiński	**Laser Raman Spectrometry**
G.I. Bekov & V.S. Letokhov	**Laser Resonant Photoionization Spectroscopy for Trace Analysis**
K. Beyermann	**Organic Trace Analysis**
O. Budevsky	**Foundations of Chemical Analysis**
J. Buffle	**Complexation Reactions in Aquatic Systems: An Analytical Approach**
D.T. Burns, A. Townshend & A.G. Catchpole	**Inorganic Reaction Chemistry Volume 1: Systematic Chemical Separation**
D.T. Burns, A. Townshend & A.H. Carter	**Inorganic Reaction Chemistry: Volume 2: Reactions of the Elements and their Compounds: Part A: Alkali Metals to Nitrogen, Part B: Osmium to Zirconium**
E. Casassas & S. Alegret	**Solvent Extraction**
J. Churáček	**New Trends in the Theory & Instrumentation of Selected Analytical Methods**
E. Constantin, A. Schnell & M. Mruzek	**Mass Spectrometry**
R. Czoch & A. Francik	**Instrumental Effects in Homodyne Electron Paramagnetic Resonance Spectrometers**
T.E. Edmonds	**Interfacing Analytical Instrumentation with Microcomputers**
Z. Galus	**Fundamentals of Electrochemical Analysis, Second Edition**
S. Görög	**Steroid Analysis in the Pharmaceutical Industry**
T. S. Harrison	**Handbook of Analytical Control of Iron and Steel Production**
J.P. Hart	**Electroanalysis of Biologically Important Compounds**
T.F. Hartley	**Computerized Quality Control: Programs for the Analytical Laboratory**
Saad S.M. Hassan	**Organic Analysis using Atomic Absorption Spectrometry**
M.H. Ho	**Analytical Methods in Forensic Chemistry**
Z. Holzbecher, L. Diviš, M. Král, L. Šůcha & F. Vláčil	**Handbook of Organic Reagents in Inorganic Chemistry**
A. Hulanicki	**Reactions of Acids and Bases in Analytical Chemistry**
David Huskins	**Electrical and Magnetic Methods in On-line Process Analysis**
David Huskins	**Optical Methods in On-line Process Analysis**
David Huskins	**Quality Measuring Instruments in On-line Process Analysis**
J. Inczédy	**Analytical Applications of Complex Equilibria**
M. Kaljurand & E. Küllik	**Computerized Multiple Input Chromatography**
S. Kotrlý & L. Šůcha	**Handbook of Chemical Equilibria in Analytical Chemistry**
J. Kragten	**Atlas of Metal-ligand Equilibria in Aqueous Solution**
A.M. Krstulović	**Quantitative Analysis of Catecholamines and Related Compounds**
F.J. Krug & E.A.G. Zagotto	**Flow Injection Analysis in Agriculture & Environmental Science**
V. Linek, V. Vacek, J. Sinkule & P. Beneš	**Measurement of Oxygen by Membrane-covered Probes**
C. Liteanu, E. Hopîrtean & R. A. Chalmers	**Titrimetric Analytical Chemistry**
C. Liteanu & I. Rîcă	**Statistical Theory and Methodology of Trace Analysis**
Z. Marczenko	**Separation and Spectrophotometric Determination of Elements**
M. Meloun, J. Havel & E. Högfeldt	**Computation of Solution Equilibria**
M. Meloun, J. Militky & M. Forina	**Chemometrics in Instrumental Analysis: Solved Problems for IBM PC**
O. Mikeš	**Laboratory Handbook of Chromatographic and Allied Methods**
J.C. Miller & J.N. Miller	**Statistics for Analytical Chemistry, Second Edition**
J.N. Miller	**Fluorescence Spectroscopy**
J.N. Miller	**Modern Analytical Chemistry**
J. Minczewski, J. Chwastowska & R. Dybczyński	**Separation and Preconcentration Methods in Inorganic Trace Analysis**
T.T. Orlovsky	**Chromatographic Adsorption Analysis**
D. Pérez-Bendito & M. Silva	**Kinetic Methods in Analytical Chemistry**
B. Ravindranath	**Principles and Practice of Chromotography**
V. Sediveč & J. Flek	**Handbook of Analysis of Organic Solvents**
O. Shpigun & Yu. A. Zolotov	**Ion Chromatography in Water Analysis**
R.M. Smith	**Derivatization for High Pressure Liquid Chromatography**
R.V. Smith	**Handbook of Biopharmaceutic Analysis**
K.R. Spurny	**Physical and Chemical Characterization of Individual Airborne Particles**
K. Štulík & V. Pacáková	**Electroanalytical Measurements in Flowing Liquids**
J. Tölgyessy & E.H. Klehr	**Nuclear Environmental Chemical Analysis**
J. Tölgyessy & M. Kyrš	**Radioanalytical Chemistry, Volumes I & II**
J. Urbanski, *et al.*	**Handbook of Analysis of Synthetic Polymers and Plastics**
M. Valcárcel & M.D. Luque de Castro	**Flow-Injection Analysis: Principles and Applications**
C. Vandecasteele	**Activation Analysis with Charged Particles**
F. Vydra, K. Štulík & E. Juláková	**Electrochemical Stripping Analysis**
N. G. West	**Practical Environmental Analysis using x-ray Fluorescence Spectrometry**
J. Zupan	**Computer-supported Spectroscopic Databases**
J. Zýka	**Instrumentation in Analytical Chemistry**

KINETIC METHODS IN ANALYTICAL CHEMISTRY

D. PEREZ-BENDITO, Ph.D.
Professor of Analytical Chemistry
and
M. SILVA, Ph.D.
Assistant Professor of Analytical Chemistry
both of Faculty of Sciences, University of Córdoba, Spain

Translation Editors:
J. R. MAJER
formerly Department of Chemistry, University of Birmingham
R. A. CHALMERS
Department of Chemistry, University of Aberdeen

ELLIS HORWOOD LIMITED
Publishers · Chichester

Halsted Press: a division of
JOHN WILEY & SONS
New York · Chichester · Brisbane · Toronto

First published in 1988
ELLIS HORWOOD LIMITED
Market Cross House, Cooper Street,
Chichester, West Sussex, PO19 1EB, England
The publisher's colophon is reproduced from James Gillison's drawing of the ancient Market Cross, Chichester.

Distributors:
Australia and New Zealand:
JACARANDA WILEY LIMITED
GPO Box 859, Brisbane, Queensland 4001, Australia
Canada:
JOHN WILEY & SONS CANADA LIMITED
22 Worcester Road, Rexdale, Ontario, Canada
Europe and Africa:
JOHN WILEY & SONS LIMITED
Baffins Lane, Chichester, West Sussex, England
North and South America and the rest of the world:
Halsted Press: a division of
JOHN WILEY & SONS
605 Third Avenue, New York, NY 10158, USA
South-East Asia
JOHN WILEY & SONS (SEA) PTE LIMITED
37 Jalan Pemimpin # 05–04
Block B, Union Industrial Building, Singapore 2057
Indian Subcontinent
WILEY EASTERN LIMITED
4835/24 Ansari Road
Daryaganj, New Delhi 110002, India

British Library Cataloguing in Publication Data
Pérez-Bendito, D. *1943–*
Kinetic methods in analytical chemistry.
1. Chemical reactions. Kinetics
I. Title II. Silva, M. *1951–*
541.3′94

Library of Congress CIP availabe

ISBN 0–7458–0105–6 (Ellis Horwood Limited)
ISBN 0–470–21181–4 (Halsted Press)

Typeset in Times by Ellis Horwood Limited
Printed in Great Britain by Unwin Bros., Woking

Table of contents

Preface

Reaction-rate methods are becoming increasingly important in analytical chemistry. Their present degree of development relies heavily on recent breakthroughs in instrumental design and, especially, on the incorporation of microcomputers into analytical chemical configurations.

Surprisingly enough, neither the evolution of these methods nor the innovations introduced in the last two decades have to date been dealt with comprehensively in a book in English. Filling this gap was our chief goal in writing this monograph on reaction-rate methods.

This book is concerned with every relevant aspect of kinetic methods used in analytical chemistry and presents the different methodologies on the basis of appropriate mathematical support. The methods described involve the use of both homogeneous catalysed (non-enzymatic) and uncatalysed reactions for the determination of single species or analysis of mixtures (differential rate methods), or even of their analytical features (sensitivity, selectivity, precision). The comprehensive description of the instrumentation typically employed in these methods is complemented by a number of practical examples and applications, in addition to a wealth of literature references.

We hope that this monograph, intended both as textbook and as an up-to-date reference book (it deals with contributions to the field up to 1987), will be of help both to chemistry graduates and to analytical chemical teachers and researchers with an interest in reaction-rate methods. Our endeavour will be much more than repaid if these basic aims are fulfilled.

Finally, we wish to express our gratitude to those who in one way or another have aided this venture. Thus, we are grateful to Dr Robert A. Chalmers for his invaluable suggestions and support as an experienced editor. We are also indebted to our disciple Antonio Losada for translation of the manuscript, and to Francisco J. Doctor for the artwork.

Córdoba
December 1986

THE AUTHORS

1

Introduction to kinetic methods

1.1 INTRODUCTION

Modern analytical chemistry comprises a vast range of methods based on the physical, chemical or physicochemical changes exhibited by substances on chemical reaction, with or without prior separation of the analyte.

Thus, every process, whatever its nature, takes place at a finite rate, tending to an equilibrium position, and therefore comprises two regions: the kinetic (dynamic) region, in which the system approaches equilibrium, and the equilibrium (static) region, which occurs once all the processes involved in the system have attained equilibrium (Fig. 1.1). Both regions are of high informative potential in analytical

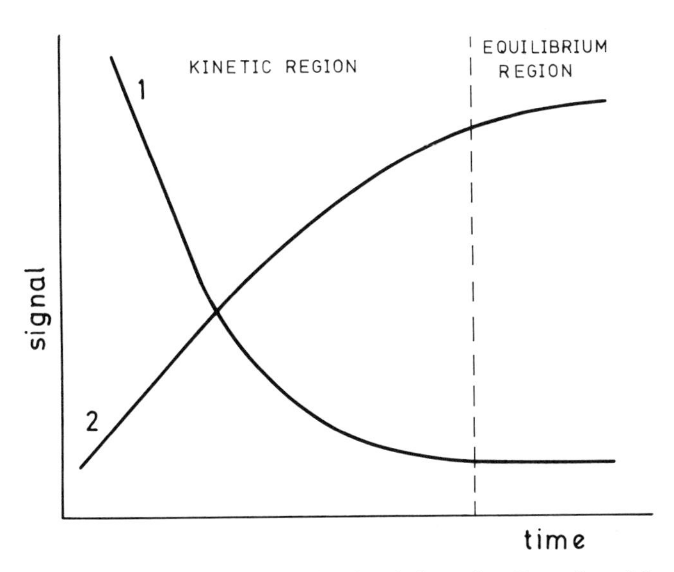

Fig. 1.1 — Kinetic and equilibrium regions of a chemical reaction. Recording of the analytical signal as a function of time, for (1) disappearance of a reactant; (2) formation of a product.

chemistry.

From a physicochemical viewpoint, the concept of 'equilibrium method' is not very accurate as it may lead to the conclusion that measurements are always made under equilibrium conditions and hence that only reversible reactions can be dealt with. In fact, equilibrium methods, whether classical (gravimetry, titrimetry) or physicochemical (instrumental), are all applied under steady-state conditions (i.e. both at equilibrium and after completion of the reaction).

Signal measurements made under dynamic conditions in systems approaching equilibrium compete in efficiency with static or equilibrium measurements. The choice is usually dictated by the particular problem addressed, as instrumental limitations are currently no solid argument for avoiding the use of kinetic data.

Not all reactions meet the requirements made by static measurements. Some of them involve side-processes; in others, equilibrium is reached very slowly or the final product is not formed quantitatively. Under these restrictive conditions, equilibrium methods fail to reach the standards set by kinetic methods, which are thus to be preferred.

Kinetic determinations, with catalytic and differential reaction-rate methods established as the main areas of kinetic analysis, are not the sole objective of use of kinetics in analytical chemistry. There are other aspects of increasing relevance [1], and since 1982, the annual reviews published by *Analytical Chemistry* are headed "Kinetic determinations and some kinetic aspects of analytical chemistry". Such aspects should not be ignored as they are the key to a large variety of analytical methods not always directly related to the measurement process. The development of many of these methods depends to a great extent on a better knowledge of the kinetics of the processes on which they are based or on the kinetic components involved.

Dynamic physical processes decisively influence both the quality of the analytical signal and the information obtained from a chemical system, so much so that the combination of physical and chemical processes in a dynamic system has given rise to one of the most recent and fruitful techniques of continuous-flow analysis, viz. (unsegmented) **flow-injection analysis** (FIA) [2,3]. Some authors claim that continuous-flow sample processing may change the potential use of kinetic methods. Such a statement is not valid insofar as there are a number of other sample-handling procedures based on kinetics (e.g. the **stopped-flow** mixing technique, highly competitive as regards sample throughput, though less flexible and somewhat more complicated). A combination of continuous and stopped flow implemented by Malmstadt [4] with the aid of an automated system offers interesting possibilities in this sense. The **resolved-time fluorescence** technique is a more recent alternative (which has so far provided fewer practical results) founded on the principles of **fluorescence half-life methods** [5]. These quantitative techniques can be regarded as kinetic methods, although they are not considered in this monograph, which deals with kinetic methods from a traditional viewpoint.

Kinetics plays a significant role in separation techniques, as was acknowledged well before the earliest kinetic measurement methods were developed. In precipitation assays, the rate of crystal growth affects the purity and ease of collection of the precipitate. The kinetic approach to the mechanisms of crystal formation and growth has contributed considerable improvements to the classic technique [6,7].

As a rule, kinetics has little influence on analytical methods based on liquid–liquid extraction [8], though its effect is sometimes useful. Conversely, in ion-exchange processes (whether carried out in columns or by the batch technique) the time elapsed until equilibrium is attained is always one of the factors to be considered [9]. Chromatographic processes are generally based on the establishment of multi-equilibria (one equilibrium per theoretical plate) and are thermodynamically-based methods. However, the kinetic foundation of chromatographic processes is also undeniable: the solutes (analytes) traverse the column (thin layer, paper, etc.) at different velocities and it is this dynamic behaviour which makes their separation possible. A distinction should be made between the kinetics of distribution throughout each theoretical plate and that related to the movement of the solute over the chromatographic substrates, which is influenced by the kinetics and thermodynamics of the distribution process [10].

Other instrumental techniques to which kinetics is of relevance are differential scanning calorimetry [11] and atomic-absorption spectrometry. In the latter, kinetics is of decisive influence on the fast processes (duration of the order of a few milliseconds) taking place in the flame, which are thus dynamic in nature [12–14]. Electroanalytical techniques such as chronoamperometry, chronocoulometry and chronopotentiometry (somewhat more recent), which have by now become traditional, can also be considered kinetic methods as they are based on time-dependent measurements [15, 16].

The arguments above testify to the present impact of kinetics on analytical chemistry and to the very promising outlook in this field.

1.1.1 Classification of kinetic methods

The rate at which a chemical reaction develops depends on factors such as temperature, pressure or reactant concentration, as well as on the presence of catalysts, activators or inhibitors. Some reactions (e.g. the neutralization of a strong acid with a strong base) are so fast that they attain equilibrium practically instantaneously whereas others, despite their thermodynamic feasibility, are so slow that they cannot be detected even after a long period of time, as is the case with the well-known reaction between As(III) and Ce(IV) at room temperature.

The study of the reaction rates and the influence of experimental conditions on them is of major importance as a knowledge of the kinetics of a given system allows the rate of the reaction involved to be matched to the experimenter's requirements. Thus, by careful choice of the conditions, the rate of reaction of any given species in a mixture can be made sufficiently different from the rates for the other species present to allow it to be readily resolved. Unforeseeably slow reactions under ordinary conditions can be satisfactorily accelerated by selecting a suitable temperature or by adding a catalyst to lower the activation energy barrier.

Kinetic methods have been classified according to different criteria. Probably, the best classification is that distinguishing between catalytic and non-catalytic methods (Table 1.1). In turn, catalytic methods have been further divided according to the type of reaction involved, and non-catalytic methods have been classified according to whether they are applied to the determination of a single species or of several components in a mixture (differential reaction-rate methods). Other classifications based on the methodology applied or the instrumentation used are dealt with in the corresponding chapters.

Kinetic methods, despite the difficulty involved in making measurements in a

Table 1.1 — General classification of kinetic methods

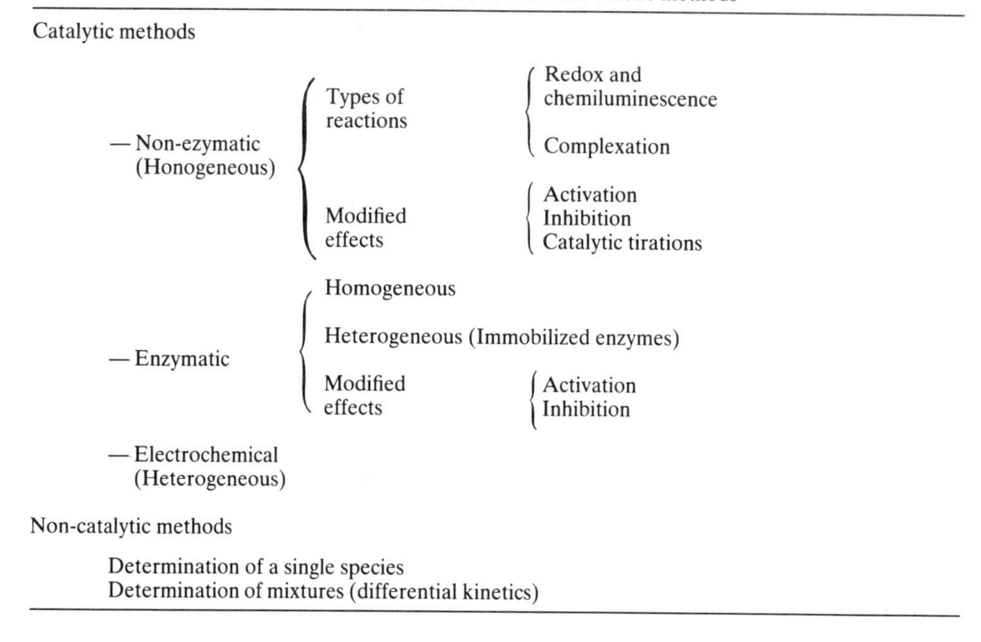

Catalytic methods

- — Non-ezymatic (Honogeneous)
  - Types of reactions
    - Redox and chemiluminescence
    - Complexation
  - Modified effects
    - Activation
    - Inhibition
    - Catalytic tirations
- — Enzymatic
  - Homogeneous
  - Heterogeneous (Immobilized enzymes)
  - Modified effects
    - Activation
    - Inhibition
- — Electrochemical (Heterogeneous)

Non-catalytic methods

- Determination of a single species
- Determination of mixtures (differential kinetics)

dynamic system, possess some advantages over equilibrium methods, as follows.

(a) They allow the use of a variety of chemical reactions to which equilibrium measurements cannot be applied (e.g. because of their slowness).
(b) Slowly reacting systems may give rise to side-reactions as the process of interest approaches completion, but this generally poses no problem with kinetic methods.
(c) Non-quantitative reactions, which cannot be dealt with by equilibrium methods, are not forbidden to kinetic methods.
(d) Kinetic methods have a greater potential than equilibrium methods in trace analysis on account of their greater sensitivity and, especially when catalytic reactions are used.
(e) Kinetic methods permit the resolution of mixtures of closely related compound by reaction at different rates with a common reagent (differential reaction-rate methods).

On the other hand, the most serious disadvantage of kinetic methods is the need for accurate reproduction of the reaction conditions in each experimental determination. Reproducibility in the experimental conditions is thus more critical in kinetic methods than in equilibrium methods insofar as time is always a variable of decisive importance in the former.

The increasing development of kinetic methods can be attributed to: (a) the need to analyse very small amounts (of the order of a few nanograms or even less) of substances contained in high-purity materials or in environmental or biological samples, (b) the better knowledge of reaction mechanisms, and especially (c) the great advances in instrumental techniques, particularly in the field of automation and computerization.

Therefore, it is not surprising that analytical chemistry manuals have for some time included chapters devoted to kinetic methods of analysis. However, there is a scarcity of bibliographical material available on the subject, apart from the monographs by Yatsimirskii [17], and Mark and Rechnitz [18], published in 1966 and 1968, respectively, and now outdated. More topical material is included in reviews published on the subject over the past two decades [19–22], and in the contribution of Kopanica and Stará [23].

1.2 REACTION RATE AND KINETIC EQUATION

The determination of a given species by a kinetic method is based on the direct or indirect measurement of its reaction rate, which in turn involves measuring the change in the reactant or product concentration as a function of time. Thus, the reaction rate is defined as the number of moles of material consumed or formed per unit volume per unit time.

Consider the reaction

$$\mathrm{A+B \rightarrow P} \qquad \mathrm{P = Products} \tag{I}$$

The reaction rate at time t will be given by the derivative of the concentration of any of the species involved, with respect to time:

$$\text{rate} = \frac{\mathrm{d[P]}}{\mathrm{d}t} = -\frac{\mathrm{d[A]}}{\mathrm{d}t} = -\frac{\mathrm{d[B]}}{\mathrm{d}t} \tag{1.1}$$

The derivatives of [A] and [B] with respect to time are negative since these species disappear as the reaction develops. Thus, Eq. (1.1) can be considered the mathematical definition of the reaction rate.

However, the reaction rate is proportional to the concentrations of all the species taking part in the reaction. If it is the change in the concentration of one of the reaction products, P, which is measured, then

$$\text{rate} = \frac{\mathrm{d[P]}}{\mathrm{d}t} = k\mathrm{[A][B]} \tag{1.2}$$

where k is the so-called **rate constant** and represents the reaction rate per unit concentration of reactants.

The sum of the exponents of [A] and [B] in Eq. (1.2) is known as the **reaction order**; thus, the reaction represented by this equation is of second order. The reaction order is an empirical parameter; it can be a fraction, though not necessarily a fixed one.

If a large excess of one of the reactants involved in reaction (I) (e.g. B) is used, so that its concentration changes are negligible, parameter [B] can be included in the constant k in Eq. (1.2) and the reaction considered to be **pseudo first-order** in A or **pseudo zero-order** with respect to B, i.e.

$$\frac{d[P]}{dt} = k'[A]$$

The reaction order is therefore an experimental result obtained for a chemical reaction under a given set of conditions and should not be confused with the **molecularity**, viz. the number of solute species reacting in an elementary process — elementary reactions are generally unimolecular or bimolecular and occasionally termolecular, and their equation rates are given by their stoichiometries. Thus, the partial orders in the rate equation of the well-known reaction

$$5Br^- + BrO_3^- + 6H^+ \rightarrow 3Br_2 + 3H_2O \tag{II}$$

do not coincide with its stoichiometric coefficients:

$$\frac{-d[BrO_3^-]}{dt} = k[BrO_3^-][Br^-][H^+]^2$$

In fact, the reaction is first-order in BrO_3^- and Br^-, and second-order in H^+, while the overall order is 4.

Table 1.2 lists the mathematical equations, both in differential and in integral form, representing straightforward irreversible reactions of pseudo-zero, first, second, fractional and *n*th order, together with their corresponding half-reaction times and the dimensions of the rate constant.

The concentrations in Eq. (1.1) can be replaced by any measurable quantity provided it is directly proportional to the concentration. The changes in the reactant or product concentration can be followed as a function of time by physical or chemical analytical techniques. Chemical monitoring techniques are only applicable to rather slow reactions and do not permit the continuous measurement of the monitored species. Conversely, physical techniques, based on the measurement of solution properties (proportional to the concentration), allow faster reactions to be monitored and continuous measurements to be made. The rate of the reaction in question is limited only by the response time of the instrument. Hence physical techniques are more frequently used than chemical ones for these purposes, the commonest properties measured being absorbance, potential, temperature, luminescence and conductivity.

If the concentration of the species to be determined and hence the measured property, R, decrease as a function of time, the plot representing the evolution of such a species with time is a decreasing exponential curve such as that shown in Fig. 1.2, which is representative of any of the differential kinetic equations listed in Table 1.2. It is also possible to plot the change in concentration of a product, P, or a property directly related to it, as its concentration is related to that of A through the expression $[A] = [A]_0 - [P]$. In this case, the kinetic plot is an increasing exponential curve. In each case, the reaction rate is given by the slope of the curve at each point ($\tan \alpha$).

Table 1.2 — Kinetic equations corresponding to simple irreversible reactions. $\upsilon = k[A]^a[B]^b$; $(n = a + b)$

Order			Differential form	Integral form	Half-life	Dimensions of k
n	a	b				
0	0	0	$-\frac{d[A]}{dt} = k$	$[A]_0 - [A]_t = kt$	$[A]_0/2k$	$mole.l^{-1}.sec^{-1}$
$\frac{1}{2}$	$\frac{1}{2}$	0	$-\frac{d[A]}{dt} = k[A]^{1/2}$	$2([A]_0^{1/2} - [A]_t^{1/2}) = kt$	$\frac{1}{k}(2-\sqrt{2})[A]_0^{1/2}$	$mole^{1/2}.l^{-1/2}.sec^{-1}$
1	1	0	$-\frac{d[A]}{dt} = k[A]$	$\ln\frac{[A]_0}{[A]_t} = kt^{(*)}$	$\ln 2/k$	sec^{-1}
2	1	1	$-\frac{d[A]}{dt} = k[A][B]$	$\frac{1}{[B]_0 - [A]_0}\ln\frac{[A]_0[B]_t}{[B]_0[A]_t} = kt$	$\frac{1}{k([B]_0 - [A]_0)}\ln\left(2 - \frac{[A]_0}{[B]_0}\right)$	$l.mole^{-1}.sec^{-1}$
2	2	0	$-\frac{d[A]}{dt} = k[A]^2$	$\frac{1}{[A]_t} - \frac{1}{[A]_0} = kt$	$1/k[A]_0$	$l.mole^{-1}.sec^{-1}$
$-\frac{1}{2}$	$-\frac{1}{2}$	0	$-\frac{d[A]}{dt} = k[A]^{-1/2}$	$\frac{2}{3}([A]_0^{3/2} - [A]_t^{3/2}) = kt$	$\frac{1}{6k}(4-\sqrt{2})[A]_0^{3/2}$	$mole^{3/2}.l^{-3/2}.sec^{-1}$
-1	-1	0	$-\frac{d[A]}{dt} = k[A]^{-1}$	$\frac{1}{2}([A]_0^2 - [A]_t^2) = kt$	$3[A]_0^2/8k$	$mole^2.l^{-2}.sec^{-1}$
-2	-2	0	$-\frac{d[A]}{dt} = k[A]^{-2}$	$\frac{1}{3}([A]_0^3 - [A]_t^3) = kt$	$7[A]_0^3/24k$	$mole^3.l^{-3}.sec^{-1}$
n	n	0	$-\frac{d[A]}{dt} = k[A]^n$	$\frac{1}{n-1}\left(\frac{1}{[A]_t^{n-1}} - \frac{1}{[A]_0^{n-1}}\right) = kt$	$\frac{2^{n-1}-1}{(n-1)k[A]_0^{n-1}}$	$mole^{-(n-1)}.l^{(n-1)}.sec^{-1}$

(*) Or, in exponential form: $[A]_t = [A]_0 \exp(-kt)$

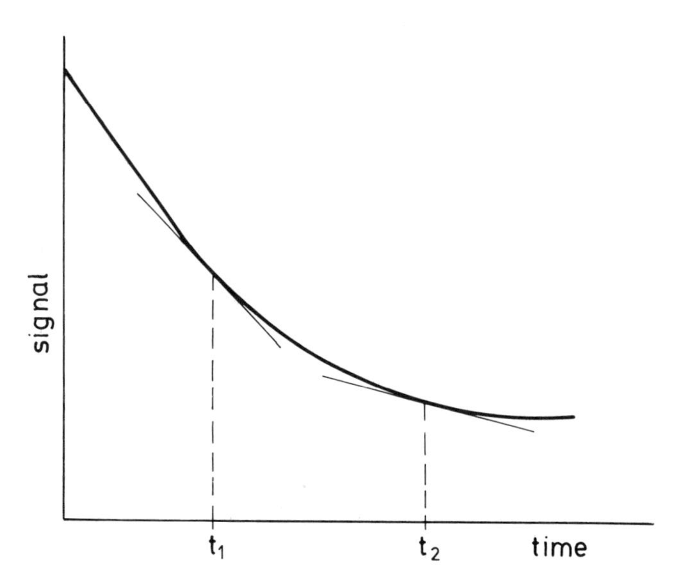

Fig. 1.2 — Typical kinetic curve showing the initial straight portion.

The commonest kinetic methods are applied to the initial portion of the curve, i.e. when the reaction is only 1–3% developed. This portion of the curve is normally linear and its slope is proportional to the concentration of the species measured (initial-rate method).

1.2.1 Determination of partial orders and rate constants

To propose a kinetic or rate equation representative of a given system it is necessary to determine the partial reaction orders with respect to the different variables influencing the process.

Such a determination can be carried out in two ways, according to whether the differential or the integral form of the rate equation is considered. As a rule, differential methods (based on the measurement of the initial rate) are chiefly used for determination of partial orders, whereas integral methods are normally applied to the determination of rate constants.

Experimentally, the determination of the partial order with respect to a given species involves introducing suitable changes in its concentration while keeping essentially constant (and hence in large excess) those of the other species present, and plotting as a function of time the values of the measured property obtained at different times and concentrations, and subsequently determining the initial rates (tangents of the initial straight segments) in the case of differential methods, or the value of the measured property at different time intervals in the case of integral methods. In this way the variations in the measured parameter are rendered exclusively due to the reactant of which the concentration is changed, and the kinetic order determined corresponds to the partial order with respect to that reactant.

1.2.1.1 Differential methods

For a partial reaction order n, the initial rate ($\tan \alpha$) is related to the concentration through the rate equation

$$\text{rate} = \tan \alpha = -\frac{d[A]}{dt} = k_A[A]^n \tag{1.3}$$

where A is the species for which the partial order is to be determined and k_A is the 'pseudo-nth-order' rate constant for such a species. Taking logarithms of Eq. (1.3) gives

$$\log(\tan \alpha) = \log k_A + n \log[A] \tag{1.4}$$

i.e. the equation of a straight line of slope n and intercept $\log k_A$. Thus, from the slope and the intercept of the plot of $\log(\tan \alpha)$ *vs.* $\log[A]$ it is possible to obtain the partial reaction order with respect to this reactant, and the corresponding rate constant.

An aspect of singular interest is the determination of partial orders with respect to species involved both in the reaction of interest and in another equilibrium, as in such cases it is impossible to keep constant (and in excess) the concentration of the remaining species. This is usually the case when using weak acid or base buffers, as the concentration of their conjugate forms is also affected by the pH. Thus, for a weak acid HA dissociating according to the reaction

$$HA \rightleftharpoons A^- + H^+ \qquad K_a = \frac{[A^-][H^+]}{[HA]}$$

which will influence the overall reaction kinetics, the rate equation will be given by

$$v = \tan \alpha = K[HA]^n[A^-]^m[H^+]^x \tag{1.5}$$

where n, m and x are the partial orders with respect to HA, A^- and H^+, respectively. A recently reported procedure [24] entails studying the influence of the pH by using two sets of solutions in which either $[A^-]$ or [HA] is kept constant. Under these conditions, the reaction rate will be given by

$$v = \tan \alpha = k_1[HA]^n[H^+]^x$$

or, in logarithmic form,

$$\log(\tan \alpha/[HA]^n) = \log k_1 - x\text{pH} \tag{1.6}$$

for $[A^-]$ = const. and, similarly,

$$\log(\tan\alpha/[A^-]^m) = \log k_2 - x\text{pH} \tag{1.7}$$

for [HA] = const.

The plot corresponding to Eq. (1.6) or (1.7) is a polygonal curve if the influence of the H^+ ion varies over the pH-range considered, or a straight line if it does not. The slopes of these linear sections of the plots allow the calculation of the corresponding partial orders with respect to the proton, x. From Eqs. (1.6) and (1.7) it is obvious that the x values obtained will depend on the particular n and m values used, so every partial order found for the hydrogen ion will correspond to a given pair of partial orders in HA (n) and A^- (m), which can thus be determined simultaneously.

Once the influence of each variable and concentration on the reaction rate has been determined, the corresponding kinetic equation can be formulated, and will have the general form

$$v = \frac{d[P]}{dt} = k[A]^a[B]^b[C]^c \ldots$$

where P is the reaction product by means of which the reaction is monitored, k is the reaction rate constant, and the indices denote the partial reaction orders with respect to the species to which they refer.

The partial order with respect to a given reactant will be unity if the plot of the initial rate against the reactant concentration is a straight line, since if $n = 1$, Eq. (1.3) represents a straight line of zero intercept.

1.2.1.2 Integral methods

The determination of partial order by integration methods involves plotting the integral form of the kinetic equation for a previously assumed reaction order. Thus, according to Table 1.2, the integrated rate equation for a hypothetical first-order reaction is

$$\ln[A]_t = \ln[A]_0 - k_A t \tag{1.8}$$

where $[A]_t$ is the concentration of species A at reaction time t, $[A]_0$ is the initial concentration of the same species and k_A is its rate constant. If the reaction in question is followed by means of its product, P, monitored photometrically, then

$$\ln(D_\infty - D_t) = \ln D_\infty - k_A t \tag{1.9}$$

since $[A]_t = [P]_\infty - [P]_t$ and $[A]_0 = [P]_\infty$, i.e. [P] and D (the absorbance) are proportional to each other.

Thus, if the plot of $\ln(D_\infty - D_t)$ *vs*. time for different initial concentrations of A consists solely of parallel straight lines, the initial reaction can be said to be of first order in A.

Equation (1.9) allows the rate constant k_A to be determined by a non-graphical method. Taking logarithms gives

$$\log (D_\infty - D_t) = \log D_\infty - \frac{k_A t}{2.303}$$

and hence,

$$k_\mathrm{A} = \frac{2.303}{t} \log \frac{D_\infty}{D_\infty - D_t}$$

Thus, the determination of the rate constant involves first obtaining D_∞ and then several D_t values corresponding to different time intervals on the kinetic curve, and introducing these into the expression above to obtain an average value for k_A.

If it is the disappearance of a species (e.g. A) with time, rather than the appearance of the reaction product, which is monitored photometrically, then Eq. (1.8) takes the form:

$$\log D_t = \log D_0 - \frac{k_\mathrm{A} t}{2.303} = \log \varepsilon_\mathrm{A} l [\mathrm{A}]_0 - \frac{k_\mathrm{A} t}{2.303}$$

where D_0 and D_t are the absorbances corresponding to $[\mathrm{A}]_0$ and $[\mathrm{A}]_t$, respectively, ε_A is the molar absorptivity of A, and l is the path-length of the photometric cell.

From this equation

$$k_\mathrm{A} = \frac{2.303}{t} \log \frac{D_0}{D_t}$$

Comparison of the two expressions derived for calculation of the rate constant shows the method involving monitoring of species A is the faster, as it does not require previous knowledge of D_∞, which can be rather cumbersome to obtain in very slow reactions.

The simplest case of second-order dependence is represented by the equation

$$-\frac{\mathrm{d}[\mathrm{A}]}{\mathrm{d}t} = k_\mathrm{A}[\mathrm{A}]^2$$

or, in integrated form,

$$\frac{1}{[\mathrm{A}]_t} = \frac{1}{[\mathrm{A}]_0} + k_\mathrm{A} t$$

If species A is monitored photometrically, then

$$\frac{1}{D_t} = \frac{1}{D_0} + \frac{k_A t}{\varepsilon_A}$$

The experimental procedure is similar to that for first-order reactions.

The calculation of any partial order by the differential or integral method can be illustrated by the following example. Applying the differential method to the kinetic plots of $[A]_t$ as a function of time (Fig. 1.3), gives the curves shown in Figs. 1.4a and

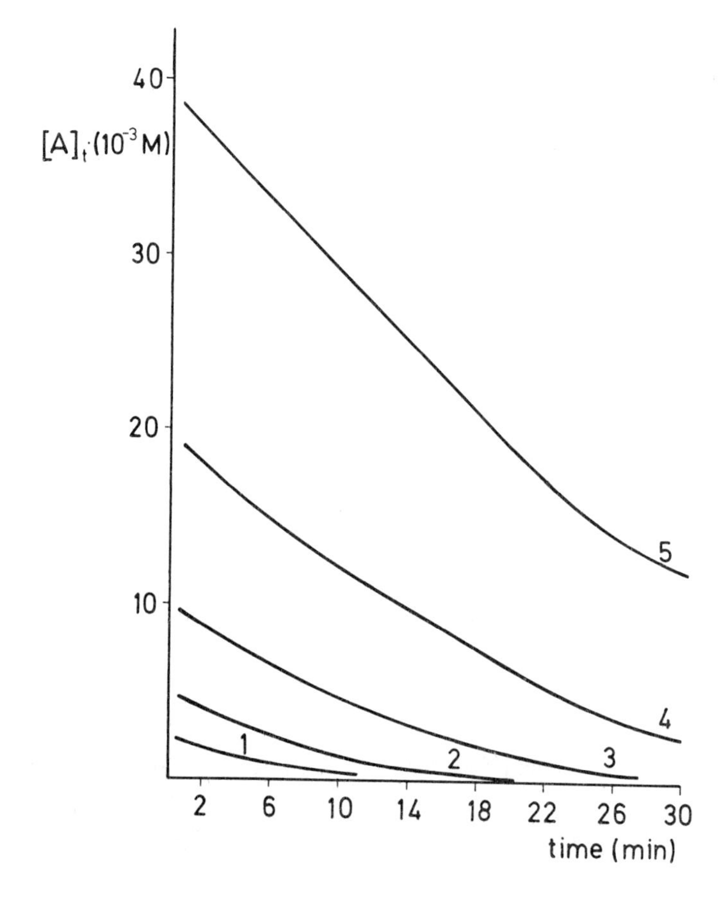

Fig. 1.3 — Kinetic plots of $[A]_t$ as a function of time for different concentrations of species A. Curves 1–5 correspond to 0.25, 0.5, 1.0, 2.0 and 4.0 × 10^{-2}M concentrations of A.

b. From the latter a partial order of 1/2 is obtained. Plotting $[A]_t^{1/2}$ as a function of time (Fig. 1.4c), gives a set of parallel straight lines, which confirms the partial order of 1/2 calculated by the differential method. The rate constant can be readily calculated from the slope of any of the lines.

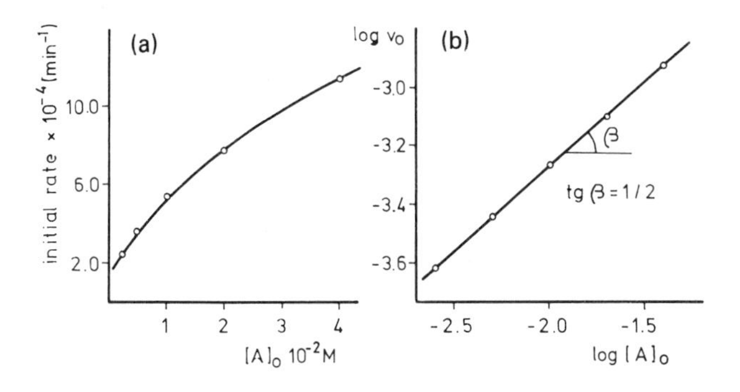

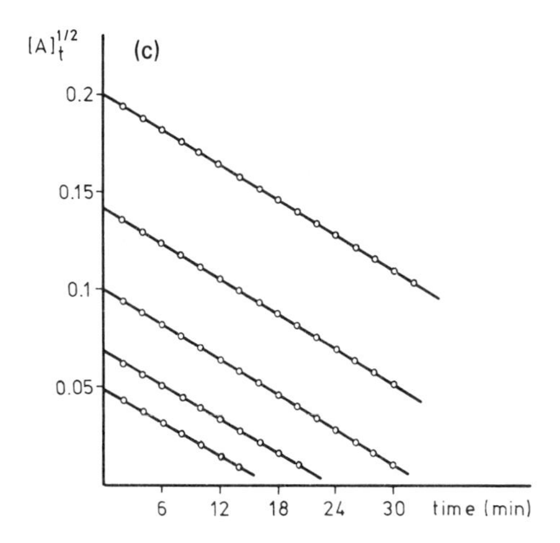

Fig. 1.4 — Determination of partial reaction orders by differential (a, b) and integral (c) methods.

1.3 KINETIC DETERMINATION OF A SINGLE SPECIES

As with the reaction order, the kinetic determination of a single species can be approached by two types of method: differential and integral.

In the differential methods, the initial rate of a chemical reaction $A + B \rightarrow P$ (i.e. the rate for the first 1–3% of reaction, where pseudo zero-order conditions hold, so the rate is not influenced by the concentration of A or B) is given by

$$v_0 = \frac{d[P]}{dt} \sim k[A][B] \sim \text{const.}$$

If the reaction is made first order in A, then the initial rate will be given by

$$v_0 = \frac{d[P]}{dt} = k_A[A]_0 \tag{1.10}$$

where k_A is the pseudo first-order rate constant and $[A]_0$ the initial concentration of A.

From Eq. (1.10) it follows that the plot of the initial rate as a function of $[A]_0$ will be a straight line which can be used as a calibration curve in determining species A.

If the integrated form of the rate equation, i.e. Eq. (1.8), is used, the plot of ln $[A]_t$ as a function of time gives a straight line with an intercept which allows the initial concentration of A, $[A]_0$, to be calculated.

These methods are described in greater detail in the section devoted to the methods used for both catalysed and uncatalysed reactions.

1.4 FACTORS INFLUENCING THE REACTION RATE

There are two general types of reactions associated with kinetic methods: slow reactions, viz. those with half-lives of 10 sec or longer, and fast reactions, i.e. those reaching half-completion in less than 10 sec. The latter are dealt with extensively in Chapter 7.

The rate of a slow reaction can be determined by straightforward conventional methods. The reaction is started by adding the reactants to a suitable vessel and is followed by means of performing titrations, or measurements of a physical property of the solution such as the absorbance, potential, or diffusion current, at preselected intervals. As the reaction mixture is a closed system and none of the species involved leaves it at any time during the measurement period, this monitoring technique is usually regarded as a *static* or *closed* procedure.

The rate of mixing of the reactants within the vessel determines the maximum measurable half-life. Thus, if the mixture is stirred magnetically, the mixing time is a few seconds, so reactions with half-lives less than 10 sec are difficult to monitor accurately.

On the other hand, determining the kinetics of slow reactions takes some time and is considered impracticable for reactions with half-lives longer than 2 hr. If the reaction takes more than 2 hr or less than 10 sec to develop, its rate can be adjusted so that its half-life may fall within the accepted practical range of slow reactions, by (a) selecting a more favourable temperature for the reaction system, (b) changing the reactant concentration, (c) using another solvent or (d) adjusting the ionic strength of the reaction medium to a suitable value.

1.4.1 Effect of temperature

Most of the slow systems dealt with in kinetic analysis are influenced by temperature. The change in the reaction rate constant as a function of temperature is given by the Arrhenius equation:

$$\frac{d(\ln k)}{dT} = \frac{E}{RT^2}$$

or, in integrated form

$$k = Ae^{-E/RT} \tag{1.11}$$

where E is the activation energy, R is the universal gas constant, A is the frequency factor (including the steric factor and the overall number of collisions), and T is the absolute temperature.

Taking logarithms of Eq. (1.11) gives

$$\log k = \log A - \frac{E}{2.303RT} \quad (1.12)$$

On introduction of the value of R (8.31 $J.K^{-1}.mole^{-1}$), Eq. (1.12) becomes:

$$\log k = \log A - \frac{E}{19.14T} \quad (1.13)$$

from which the activation energy ($J.K^{-1}.mole^{-1}$) can be readily calculated. In fact, it suffices to plot $\log k$ (or indeed any proportional factor such as $\log v_0$) as a function of $1/T$ to obtain a straight line of slope $E/19.14$.

For many homogeneous reactions, in the neighbourhood of room temperature the rate constant increases by a factor of 2–3 per 10°C increase in temperature. Thus, reactions developing rapidly at room temperature can be slowed down by cooling. On the other hand, slow reactions can be accelerated by raising the temperature. This behaviour can be utilized kinetically for the resolution of mixtures of several species. Thus, when a fructose/glucose mixture is reacted at room temperature with anthrone (a compound resulting from the reduction of anthraquinone), fructose is rapidly consumed (within the first 10 min of reaction), yielding an intensely coloured product, whereas glucose scarcely reacts. Once all of the fructose has disappeared, the glucose is determined by raising the temperature to 100°C, at which it reacts completely in a few minutes.

1.4.2 Effect of reactant concentration

Reactions with high rate constants can be readily monitored by using low reactant concentrations, thanks to the sensitivity of some analytical techniques which allow the measurement of very small changes in concentration.

Spectrophotometric techniques are of great use for measuring extremely low concentrations of highly coloured compounds. A typical reaction with a high rate constant is that taking place between iron(II) and trioxalatocobalt(III) ions

$$Fe^{2+} + [Co(C_2O_4)_3]^{3-} \rightarrow Fe^{3+} + [Co(C_2O_4)_3]^{4-}$$

which can be slowed down by merely using small reactant concentrations.

On the other hand, high reactant concentrations can be used to accelerate reactions with low rate constants.

1.4.3 Effect of the solvent

The rate of a chemical reaction can be modified by changing the solvent used, or its concentration. The effect of the solvent on the rate of a hypothetical reaction between species A and B to yield an activated complex AB* is defined by the

Kirkwood equation [25]:

$$\ln k = \ln k_0 + \frac{e^2}{2kT}\left(1-\frac{1}{\varepsilon}\right)M - \frac{3}{8kT}\left(\frac{1-\varepsilon}{1+\varepsilon}\right)N \tag{1.14}$$

where

$$M = \frac{(Z_A+Z_B)^2}{r_{AB^*}} - \frac{Z_A^2}{r_A} - \frac{Z_B^2}{r_B}$$

and

$$N = \frac{\mu_A^2}{r_A^3} + \frac{\mu_B^2}{r_B^3} - \frac{\mu_{AB^*}^2}{r_{AB^*}^3}$$

where k_0 is the rate constant of the reaction in a medium of infinite relative permittivity; ε is the relative permittivity of the solvent; μ_A, μ_B, and μ_{AB^*} are the dipole moments of species A and B and their activated complex, respectively; Z_A, Z_B are the charges on A and B; r_A, r_B and r_{AB^*} are the ionic radii of A, B and AB*; e is the electric charge of the proton (1.602×10^{-19} C).

Term M in Eq. (1.14) considerably exceeds N in the same expression for any reaction between ionic species (even if these are dipolar) so N can safely be neglected. The negligible term in the case of reactions between neutral, though dipolar, species, or between an ion and a dipolar molecule, is M, which becomes zero or very small.

For reactions between ions, an increase in the dielectric constant of the solvent results in an increase (decrease) in the rate of reaction between ions of the same (opposite) charge.

The rate of reactions between neutral molecules yielding very polar activated complexes will increase with increasing polarity of the solvent, since $\mu_{AB^*} \gg \mu_B, \mu_B$. If the reaction takes place between an ion and a neutral molecule to yield a weakly polar activated complex, its rate will not change appreciably as N approaches zero.

1.4.4 Influence of ionic strength

The influence of ionic strength on reaction rate has been extensively studied from a variety of viewpoints. Of all the treatments made so far, the simplest considers the formation of an activated complex between the reactants according to:

$$A + B \rightarrow AB^* \rightarrow \text{Products}$$

The reaction rate of this system is given by

$$\upsilon = k_1\,[AB^*] \tag{1.15}$$

On the other hand, the constant of the equilibrium between the reactants and the activated complex can be expressed as:

$$K = \frac{[AB^*]}{[A][B]} = K_0\frac{f_A f_B}{f_{AB^*}} \tag{1.16}$$

where K_0 is the thermodynamic equilibrium constant and f denotes the activity coefficient of a given species.

Solving Eq. (1.16) for [AB*] and substituting this value into Eq. (1.15) yields

$$\upsilon = k_1 K_0 \frac{f_A f_B}{f_{AB^*}}[A][B] = k_0 \frac{f_A f_B}{f_{AB^*}}[A][B]$$

k_0 being the rate constant corresponding to $f = 1$. Thus, the rate constant, k, for $f \neq 1$ will be given by:

$$k = k_0 \frac{f_A f_B}{f_{AB^*}} \tag{1.17}$$

After taking of logarithms and replacement of the activity coefficient by the expressions derived from the Debye–Hückel law, Eq. (1.17) becomes:

$$\log k = \log k_0 + Z_A Z_B I^{1/2} \tag{1.18}$$

where I is the ionic strength. According to this equation, the greater the ionic strength, the higher the reaction rate between ions of the same charge and the lower the rate between ions with opposite charges.

However, Eq. (1.18) does not predict the variation of the rate constant when the reaction involves two neutral species or a neutral molecule and an ion. Such cases entail using the complete form of the definition of the activity coefficient, which, for more concentrated solutions, is given by

$$\log f = -Z^2 I^{1/2} + bI$$

where the last term (including the proportionality constant b) becomes significant in the case of neutral species. Thus, Eq. (1.18) becomes:

$$\log k = \log k_0 + (b_A + b_B + b_{AB^*})I$$

or, in simplified form:

$$\log k = \log k_0 + b'I \tag{1.19}$$

As b' is usually small, changes in the ionic strength scarcely affect the rate of reaction between two neutral species or an ion and a neutral molecule.

Both Eq. (1.18) and Eq. (1.19) allow the calculation of k_0 from the intercept of the straight line obtained by plotting log k as a function of $I^{1/2}$ or I, respectively. Nevertheless, the k_0 value thus calculated is only of qualitative use, as both equations are derived by introducing some approximations.

REFERENCES

[1] H. A. Mottola, *J. Chem. Ed.*, 1981, **58**, 399.
[2] J. Růžička and E. H. Hansen, *Flow Injection Analysis*, 2nd Ed., Wiley, New York, 1988.
[3] M. Valcárcel and M. D. Luque de Castro, *Flow Injection Analysis: Principles and Applications*, Ellis Horwood, Chichester, 1987.
[4] H. V. Malmstadt, K. M. Walczack and M. A. Kouparis, *Am. Lab.*, 1980, **12**, No. 9, 27.
[5] G. M. Hiefte and E. E. Vogelstein, *A linear response theory approach to time-resolved fluorimetry*, in *Modern Fluorescence Spectroscopy*, Vol. 4, E. Wehry (ed.), Plenum Press, New York, 1981.
[6] I. M. Kolthoff, E. B. Sandell, E. J. Meehan and S. Bruckenstein, *Quantitative Chemical Analysis*, 4th Ed., Macmillan, Toronto, 1969, Chapter 10.
[7] A. E. Nielsen, *The Kinetics of Precipitation*, Pergamon Press, Oxford, 1964.
[8] M. Valcárcel and M. Silva, *Teoría y práctica de la extracción líquido-líquido*, Alhambra, Madrid, 1984, Chapter 3.
[9] F. Helfferich, *Ion-exchange Kinetics*, Vol. 1 in the series *Ion-Exchange and Solvent Extraction*, J. A. Marinsky and Y. Marcus (eds.), Dekker, New York, 1966.
[10] G. H. Weiss, *Sep. Sci. Technol.*, 1982, **17**, 1609.
[11] J. L. McNaughton and C. T. Mortimer, *IRS*; *Physical Chemistry Series* 2, Butterworths, London, Vol. 10, 1975.
[12] B. V. L'vov and P. A. Bayunov, *Zh. Analit. Khim.*, 1981, **36**, 837, 1877.
[13] Kuang-Pang Li, *Anal. Chem.*, 1981, **53**, 317.
[14] Kuang-Pang Li and Yue-Yue Li, *Anal. Chem.*, 1981, **53**, 2217.
[15] A. J. Bard and L. R. Faulkner, *Electrochemical Methods: Fundamentals and Applications*, Wiley, New York, 1980, Chapter 3.
[16] J. A. Plambeck, *Electroanalytical Chemistry: Basic Principles and Applications*, Wiley, New York, 1980, Chapter 17.
[17] K. B. Yatsimirskii, *Kinetic Methods of Analysis*, Pergamon Press, Oxford, 1966.
[18] H. B. Mark, Jr. and G. A. Rechnitz, *Kinetics in Analytical Chemistry*, Interscience, New York, 1968.
[19] G. A. Rechnitz, *Anal. Chem.*, 1964, **36**, 453R; 1966, **38**, 513R; 1968, **40**, 455R.
[20] G. G. Guilbault, *Anal. Chem.*, 1970, **42**, 334R.
[21] R. A. Greinke and H. B. Mark, Jr., *Anal. Chem.*, 1972, **44**, 295R; 1974, **46**, 413R; 1976, **48**, 87R; 1978, **50**, 70R.
[22] H. A. Mottola and H. B. Mark, Jr., *Anal. Chem.*, 1980, **52**, 31R; 1982; **54**, 62R; 1984, **56**, 96, 96R; 1986, **58**, 264R.
[23] M. Kopanica and V. Stará, in *Wilson and Wilson's Comprehensive Analytical Chemistry*, Gy. Svehla (ed.), Elsevier, Amsterdam, 1983, Vol. XVII.
[24] F. Salinas López, J. J. Berzas Nevado and A. Espinosa Mansilla, *Talanta*, 1984, **31**, 325.
[25] K. J. Laidler, *Chemical Kinetics*, 2nd Ed., McGraw-Hill, New York, 1965, p. 227.

2

Catalysed reactions

2.1 INTRODUCTION

The earliest applications of catalysed reactions to quantitative analysis were reported about five decades ago, but the use of catalysts for analytical purposes dates from much earlier.

It is well known that the presence of certain substances (metals, non-metals and even ligands) accelerates some slow reactions through their catalytic effect. Insofar as the rate of a catalysed reaction is directly proportional to the catalyst concentration, such a reaction can be used for determination of the catalyst. 'Catalytic methods' of this kind are extremely sensitive, as the catalyst is not consumed in the reaction, but takes part in it in a cyclic manner. The fact that "small causes have great effects" (on colour development, luminescence, voltammetric currents, etc.) is the basis for trace and ultratrace analysis by means of instrumental techniques.

Historically, the earliest kinetic determination of a species on the basis of its catalytic effect was reported by Guyard, who in 1876 proposed a method for determining vanadium through its catalytic effect on the oxidation of aniline by potassium chlorate [1,2]. Nearly 50 years later, Sandell and Kolthoff discovered the catalytic effect of iodide on the oxidation of As(III) by Ce(IV) [3]. Later, these authors reported the specific conditions required for the catalytic determination of iodide [4] described in detail below.

2.2 DEFINITION OF A CATALYST

Broadly speaking, a catalyst can be defined as a substance modifying the rate of a chemical reaction without altering its equilibrium state. Both the concept of catalysis and the term itself (from the greek for 'loosen') are due to Berzelius (1835–36). According to his definition: (a) the catalyst remains chemically unchanged at the end of the reaction; (b) a small amount is often sufficient to cause the reaction to develop to a considerable extent and (c) it does not affect the equilibrium position of a reversible reaction.

According to Mottola [5], these concepts allow a more precise definition of a catalyst, which can be considered "a substance lowering the free activation energy without altering the equilibrium position".

The lowering of the free activation energy is accomplished thanks to the continuous regeneration of the catalyst (catalytic cycle) in one of the reaction steps, so that, for practical purposes, its initial concentration remains constant. Since the equilibrium constant of a given reaction, K, is equal to the ratio between the rate constants of the forward and reverse reactions, a catalyst will influence both reactions to the same extent. Hence a catalyst accelerates attainment of the equilibrium, but does not alter its position, so for analytical purposes nothing is gained by increasing the rate of non-spontaneous processes. It may be remarked that the term 'negative catalysis', applied to substances which raise the free activation energy, is not technically correct.

The reaction catalysed by the substance to be determined is known as the 'indicator reaction'. In order for an indicator reaction to be useful for analytical purposes, its rate must be very low or negligible compared to that of the catalysed reaction.

2.3 KINETIC EQUATIONS AND REACTION MECHANISMS

The kinetic equations representing a catalysed reaction can be formulated on the basis of a mechanism involving formation of a transient intermediate complex, according to the following scheme:

$$C + B \underset{k_{-1}}{\overset{k_1}{\rightleftharpoons}} CB + Y \tag{I}$$

$$CB + A \xrightarrow{k_2} P + C \tag{II}$$

where C denotes the catalyst, B an excess reactant forming the intermediate complex (CB) with the catalyst, A the monitored reactant, and P and Y are reaction products.

Two different cases can be considered, according to whether the rate-determining step (rds) corresponds to reaction (I) or reaction (II), the overall rate of the process depending on the relative values of k_1, k_{-1} and k_2.

If (II) is the rds (i.e. $k_2 \ll k_{-1}$), the complex dissociation reaction will be much faster than the reaction between the complex and reactant A. Under such conditions, the process is said to be in 'pre-equilibrium'. If equilibrium (I), represented by the constant

$$K_e = \frac{[CB][Y]}{[C][B]} = \frac{k_1}{k_{-1}} \tag{2.1}$$

is assumed to occur first, then the reaction rate will be given by that of the limiting step (II):

$$\frac{d[P]}{dt} = -\frac{d[A]}{dt} = k_2\,[A][CB] \tag{2.2}$$

The concentrations of C and B at equilibrium will be given by:

$$[B] = [B]_0 - [CB] \tag{2.3}$$

$$[C] = [C]_0 - [CB] \tag{2.4}$$

$[B]_0$ and $[C]_0$ being the initial concentrations of B and C, respectively.

If $[B]_0 \gg [C]_0$, as is usually the case, then since [CB] cannot exceed $[C]_0$ at any time, [B] can be assumed equal to $[B]_0$ at equilibrium. Substituting Eqs. (2.3) and (2.4) into (2.1) then yields

$$K_e = \frac{[CB][Y]}{[B]_0\,([C]_0 - [CB])}$$

from which [CB] can be readily derived:

$$[CB] = \frac{K_e\,[C]_0\,[B]_0}{K_e\,[B]_0 + [Y]}$$

Substitution of this relation into Eq. (2.2) gives

$$v = \frac{d[P]}{dt} = -\frac{d[A]}{dt} = \frac{k_2 K_e [C]_0 [B]_0 [A]}{K_e [B]_0 + [Y]} \tag{2.5}$$

[A] being the concentration of reactant A at time t.

In the other case (i.e. $k_2 \gg k_{-1}$ and $k_2 \gg k_1$), the reaction of the intermediate complex with reactant A is much faster than the complex formation or dissociation, and reaction (II) takes over from reaction (I). Under these conditions, the concentration of CB in the system is always constant and very small, hence the term 'steady state' is used.

The steady-state principle establishes that

$$\frac{d[CB]}{dt} = 0 = k_1[C][B] - k_{-1}[CB][Y] - k_2[CB][A] \tag{2.6}$$

from which the concentration of complex CB can be expressed as:

$$[CB] = \frac{k_1[C][B]}{k_{-1}[Y] + k_2[A]}$$

From substituting the concentrations of C and B derived from Eqs. (2.3) and (2.4),

$$[\mathrm{CB}] = \frac{k_1([\mathrm{C}]_0 - [\mathrm{CB}])\,([\mathrm{B}]_0 - [\mathrm{CB}])}{k_{-1}[\mathrm{Y}] + k_2[\mathrm{A}]}$$

and on rearrangement:

$$[\mathrm{CB}] = \frac{k_1[\mathrm{C}]_0[\mathrm{B}]_0}{k_1\,([\mathrm{C}]_0 + [\mathrm{B}]_0 - [\mathrm{CB}]^2) + k_{-1}[\mathrm{Y}] + k_2[\mathrm{A}]}$$

From this, after neglecting $[\mathrm{CB}]^2$ which is small compared to $[\mathrm{C}]_0 + [\mathrm{B}]_0$ in the denominator, and taking into account that, according to the steady-state condition

$$\upsilon = \frac{\mathrm{d}[\mathrm{P}]}{\mathrm{d}t} = -\frac{\mathrm{d}[\mathrm{A}]}{\mathrm{d}t} = k_1[\mathrm{C}][\mathrm{B}] - k_{-1}[\mathrm{CB}][\mathrm{Y}] = k_2[\mathrm{CB}][\mathrm{A}]$$

substitution for [CB] into the right-hand side of this expression gives:

$$\upsilon = \frac{\mathrm{d}[\mathrm{P}]}{\mathrm{d}t} = -\frac{\mathrm{d}[\mathrm{A}]}{\mathrm{d}t} = \frac{k_1\,k_2[\mathrm{C}]_0[\mathrm{B}]_0[\mathrm{A}]}{k_1\,([\mathrm{C}]_0 + [\mathrm{B}]_0) + k_{-1}[\mathrm{Y}] + k_2[\mathrm{A}]} \tag{2.7}$$

Both Eq. (2.5) and Eq. (2.7) can be simplified for practical purposes as follows: replacing [A] by $[\mathrm{A}]_0 - [\mathrm{P}]$ (where [P] is the decrease in the concentration of reactant A or increase in the concentration of product P, measured during the reaction), results in

$$\upsilon = \frac{\mathrm{d}[\mathrm{P}]}{\mathrm{d}t} = \boldsymbol{K}\alpha_c[\mathrm{C}]_0\,([\mathrm{A}]_0 - [\mathrm{P}]) \tag{2.8}$$

where α_c is a function containing the remainder of the concentration terms appearing in Eqs. (2.5) and (2.7), except for $[\mathrm{C}]_0$, $[\mathrm{A}]_0$ and [P], and $\boldsymbol{K}$ is a term including all the rate or equilibrium constants involved.

On the other hand, if the indicator reaction takes place at a certain rate reflected in the constant k_3 for the reaction in the absence of a catalyst, then

$$\left(\frac{\mathrm{d}[\mathrm{P}]}{\mathrm{d}t}\right)_{\mathrm{uncat}} = k_3[\mathrm{B}]_0\,([\mathrm{A}]_0 - [\mathrm{P}]) \tag{2.9}$$

provided that $[\mathrm{B}]_0$ is greater than $[\mathrm{A}]_0$.

Therefore, the overall reaction rate will be given by the sum of Eqs. (2.8) and (2.9):

$$\left(\frac{d[P]}{dt}\right)_{obs} = (\boldsymbol{K}\alpha_c[C]_0 + k_3[B]_0)([A]_0 - [P]) \tag{2.10}$$

This equation is the basis for the analytical application of this type of reaction to the determination of a catalyst concentration.

At times close to the start of the reaction, [P] can be neglected relative to $[A]_0$ and Eq. (2.10) can be rewritten as:

$$v_0 = \left(\frac{d[P]}{dt}\right)_{obs,\, t=0} = \boldsymbol{K}'[C]_0 + \boldsymbol{K}'' \tag{2.11}$$

where $\boldsymbol{K}'$ and $\boldsymbol{K}''$ are constants provided that $[A]_0$ and $[B]_0$ are sufficiently greater than $[C]_0$ and that $[B]_0 \gg [A]_0$. This equation is valid for times close to zero and, in general, as long as neither the reverse reaction nor any side-reactions affect the main process. However, it is not applicable to mechanisms involving chain reactions.

2.3.1 General mechanisms

Consider the indicator reaction

$$A + B \rightarrow P + Y \tag{III}$$

the rate of which can be increased by the action of a catalyst, C, which can act in two different ways [6]: by forming a complex with one of the reactants (e.g. B) or by reaction with one of the reactants (e.g. A) to yield the reaction product (P) and an activated form of the catalyst.

In the first case, the catalyst–reactant complex (BC) interacts with the other reactant (A) to yield the reaction product while regenerating the catalyst:

$$B + C \overset{\text{fast}}{\rightleftharpoons} BC$$

$$BC + A \rightarrow P + Y + C$$

As a rule, the first of these reactions develops at a significantly greater rate than the second, which will thus be the rds. However, in some processes (chiefly enzymatic reactions), the opposite holds and the first step is the rds (steady state).

This mechanism in which the catalyst undergoes no change in its oxidation state in the course of the process generally comprises oxidation by hydrogen peroxide in an acid medium, as well as metal-catalysed decomposition, substitution and hydrolysis reactions involving both organic and inorganic substances.

The reactions in which hydrogen peroxide is involved can be catalysed by ions in a high oxidation state, capable of forming unstable peroxo complexes that readily decompose to radicals such as $HO\cdot$ and $HO_2\cdot$, which act as oxidants in the reaction.

Decomposition, substitution and hydrolysis reactions are of great interest since they are often catalysed by metals lacking vacant *d*-orbitals, which are inefficient catalysts in redox reactions. The catalytic action of such metals lies in the polarization effects that they cause on the substrate molecules or to an 'orientation' effect on the catalyst itself that makes the reaction possible. The formation of a chelate or a complex between the catalyst and the substrate is of major importance in such catalytic cycles.

As stated above, the second possible way in which the catalyst may interact with the components of the indicator reaction involves its reaction with A to yield P and its own activated form, C*, from which it is regenerated upon subsequent reaction with B to yield Y:

$$A + C = P + C^*$$

$$C^* + B \xrightarrow{\text{fast}} Y + C$$

Therefore, the first step will be the rds.

Most catalytic methods are based on reactions of this type in which the formation of species C* involves a change in the oxidation state of the catalyst. As a rule, these reactions are more sensitive than the other types for the determination of the catalyst.

This mechanism is valid only if two conditions are fulfilled, namely: (i) the oxidation potential of the catalytic system E_c, must be more positive than that of the P/A system and more negative than that of the B/Y couple (i.e. $E_{B/Y} > E_C > E_{P/A}$); (ii) direct interaction between A and B, though thermodynamically permitted, must be kinetically forbidden. In addition, the reaction between the catalyst and species B should be very fast.

In general, these redox reactions involve both organic and inorganic substrates, although the former have been dealt with more extensively in the literature.

The interaction between the catalyst and an organic substrate (e.g. amine or phenol) generally involves the prior formation of a charge-transfer complex between both species, the stability of which is directly related to the pH of the medium, hence the great influence of this variable on the catalytic activity of the metal ion. On the other hand, inorganic substrates interact through electron transfers with the catalyst and may give rise to unstable intermediates in other steps [e.g. the well-known As(III)–Ce(IV) reaction, catalysed by iodide].

The sum of the reactions proposed for either of the mechanisms would yield the overall indicator reaction (III). As stated above, these mechanisms are not so simple as they appear to be; in fact, each of their steps consists of a series of stages, one of which can be the actual rds.

Finally it should be pointed out that, whatever the mechanism, Eqs. (2.7) and (2.9) are equally valid for the kinetic determination of the catalyst, the nature of the rds being the decisive factor in their applicability.

2.4 HOMOGENEOUS CATALYTIC REACTIONS

Apart from enzymatic and voltammetric reactions, which are not dealt with in this monograph, homogeneous catalytic reactions can be classified into three major categories: (a) ordinary catalytic reactions; (b) Landolt reactions and (c) oscillating reactions.

2.4.1 Ordinary catalytic reactions

This group includes redox and chemiluminescence reactions and ligand-exchange reactions. .

The commonest indicator reactions are redox in nature and involve oxidants such as H_2O_2, O_2 (usually atmospheric), BrO_3^-, ClO_3^-, IO_3^-, IO_4^-, $S_2O_8^{2-}$, Fe(III), Ce(IV), Ag(I), etc., and reductants such as Sn(II), Fe(II), As(III), I^-, $S_2O_3^{2-}$ (inorganic) and amines, phenols, azo dyes, etc, (organic).

The catalysts determined are generally metals in a high oxidation state, possessing vacant *d*-orbitals and capable of forming co-ordination compounds with one of the components of the indicator reaction. They are generally transition metals of one of the following types: (a) quadrivalent (Zr, Hf, Th); (b) quinquevalent (Nb, Ta, V); (c) sexivalent (Mo, W); (d) bi- and tervalent (Fe^{3+}, Mn^{2+}, Cu^{2+}, Co^{2+}, Ni^{2+}, etc); (e) the platinum family (Pt, Os, Pd, Ru, Rh, Ir, Ag). Anionic catalysts such as iodide, nitrite and a few others, are also, though less frequently, determined by these types of reaction.

Chemiluminescence reactions are closely related to redox processes. The catalytic effect is reflected in the release of radiant energy. The commonest of all catalytic chemiluminescence reactions is the oxidation of luminol by hydrogen peroxide in an aqueous medium,

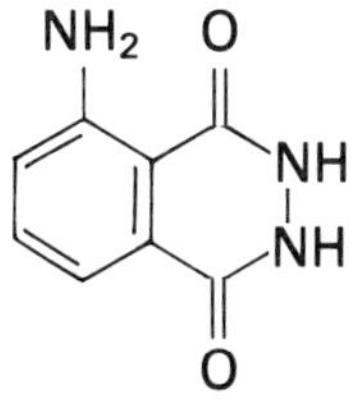

NH_2 O NH NH O

$$+ OH^- + H_2O_2 \rightarrow \text{Products} + h\nu$$

the mechanism of which is not well understood. It is catalysed by metal ions such as Co(II), Cu(II), Ni(II), Cr(III) and Mn(II) and is usually performed at pH 10–11. Detection limits between 10^{-11} and $10^{-8}M$ are readily achieved.

Catalytic ligand-exchange and complex-formation reactions are more recent and open new prospects for catalytic methods as they allow the determination of non-transition metals such as the alkaline-earth metals (individually and in binary or ternary mixtures) and of species such as ammonia, lanthanides, etc.

All three types of reaction are described in greater detail in the applications section (Section 2.6).

2.4.2 Landolt reactions

Some catalysed reactions involve a longer or shorter induction period, defined as the interval between addition of the last reactant (i.e. the start of the catalysed reaction) and the appearance of the reaction product, and throughout which the reaction does not seem to develop.

This phenomenon was first observed by Landolt [7] in the reaction between iodate and sulphite in an acid medium, which releases iodine after an induction period. This effect, known as the 'Landolt effect', is undergone by redox reactions involving halogens in various oxidation states, and (less often) by acid–base and complexation reactions.

The reactions representing the Landolt effect are:

$$A + B \underset{\text{slow}}{\overset{k_1}{\rightarrow}} P \qquad \text{(IV)}$$

$$P + L \underset{\text{fast}}{\overset{k_2}{\rightarrow}} Y \qquad \text{(V)}$$

where $k_2 > k_1$. Thus, in Landolt processes, a slow reaction (IV) is linked to a fast one (V) by the reaction product of the former. Since the second reaction is faster than the first, its product (P) can only be detected once L (the 'Landolt reagent') has disappeared completely as a result of the second reaction. If reaction (IV) is accelerated by a catalyst (Fig. 2.1), then the time elapsed until the appearance of the

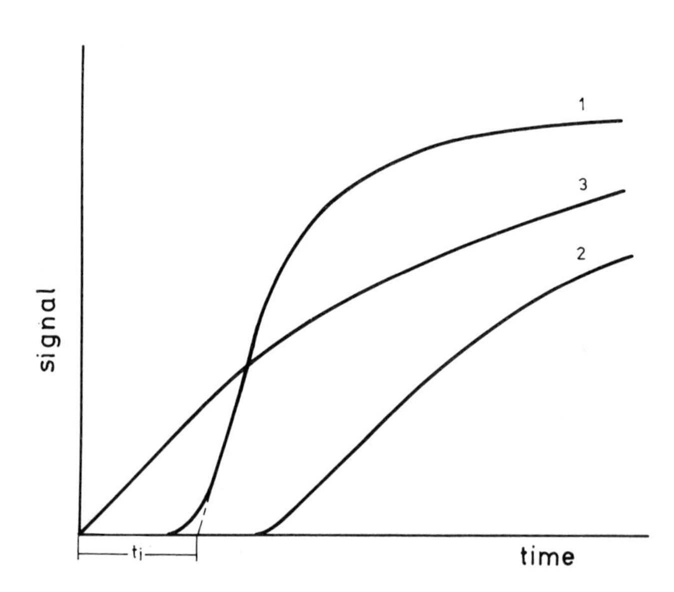

Fig. 2.1 — Kinetic curves corresponding to Landolt reactions: (1) in the presence of a catalyst (C); (2) in the absence of a catalyst and (3) indicator reaction ($A + B \rightarrow P$).

product (P) will be a measure of the catalyst concentration, which is directly related to the induction period, t_i, through empirical equations [8] of the form $[C]_0 = K_i/t_i$ or $[C]_0 = K_i/t_i^2$, depending on the particular system. The validity of the first expression has been verified by Svehla [9], although its exact solution indicates slight deviations from linearity (errors of the order of 6%). Some slow reactions can be turned into Landolt processes with the aid of a retardant as the Landolt reagent.

Landolt reactions have frequently been employed for analytical purposes because of their operational and instrumental simplicity and high sensitivity (detection limits between 0.1 and 1.0 μg/ml are easily achieved). The experimental procedure involves measurement of the time elapsed between the addition of the last reactant (A, B or the catalyst), which starts reaction (IV), and the first appearance of the signal corresponding to the product, (determined by consecutive measurements) and extrapolation of the measured signal to zero.

This modification of kinetic methods has been named the 'chronometric' or 'tempometric' method and can be regarded either as a fixed-concentration or a variable-time kinetic method.

An advantageous variant of the Landolt reaction has been proposed by Weisz and Pantel [10] for the determination of molybdenum, copper and vanadium. Instead of adding the reactants (A and B), the catalyst and the Landolt reagent (L) simultaneously, these authors add only L and one of the components of reaction (IV); the catalyst and the other component (the oxidant in this case) are dispensed at constant speed from a burette. The rate of addition must be sufficiently high that even at high catalyst concentrations (short induction periods) the oxidant is not consumed in the reaction and may build up in the solution. One such reaction is the oxidation of I^- by H_2O_2, BrO_3^- or $S_2O_8^{2-}$, with ascorbic acid or thiosulphate as the Landolt reagent. The catalyst concentration is related to the volume of oxidant added (or to time, since the rate of addition is kept constant) up to the appearance of free I_2, which is detected biamperometrically. In this manner it is possible to determine these catalysts in the following ranges: 0.4–4.0 μg/ml Mo, 0.2–1.8 μg/ml Cu and 0.02–0.3 μg/ml V.

2.4.3 Oscillating reactions

This type of reaction is of purely theoretical interest and is usually applied in interpreting biological processes such as muscle contraction or cell division. This interpretation is based on the cyclic periodic variation of the concentration of some intermediate products of the catalysed reactions affecting the reaction rate to the same extent.

Reactions catalysed by metal ions which can exchange a single electron at potentials between 0.9 and 1.6 V [e.g. the Ce(III)/Ce(IV) system in the presence of ferroin as indicator] are typical representatives of oscillating reactions, as is the Ce(IV)-catalysed oxidation of succinic, citric or malonic acid by bromate [11]. Thus, the oxidation of malonic acid is believed to take place according to the following scheme [10]:

$$\text{Ce(III)} + \text{HBrO}_3 \rightarrow \text{Ce(IV)} + \text{Products} \qquad \text{(VI)}$$

$$\text{Ce(IV)} + \text{Malonic acid} \rightarrow \text{Ce(III)} + \text{Products} \qquad \text{(VII)}$$

The process is monitored either by photometric measurement of the changes in the Ce(IV) concentration or by non-equilibrium measurements of the redox potential of the Ce(IV)/Ce(III) pair, which show the periodic variation of the Ce(IV) concentration. Reaction (VI) is autocatalytic, as an intermediate generated in the reduction of BrO_3^- acts as an autocatalyst; on the other hand, the regeneration of the catalyst through reaction (VI) is inhibited by bromide ions released in the decomposition of the different bromo derivatives of malonic acid formed in reaction (VII).

The periodic nature of the process (assuming the system to contain a certain amount of Ce^{4+}) can be accounted for as follows. Reaction (VII) produces bromide ions which slow down reaction (VI). However, the concentration of bromide in the system depends on the rates of reaction (VII) and of the reaction in which bromide is consumed through interaction with bromate ($BrO_3^- + Br^- \rightarrow Br_2$). If the bromide concentration is sufficiently high, then reaction (VI) is halted as no Ce(IV) is regenerated by oxidation of Ce(III) by BrO_3^- and the catalytic cycle is interrupted as a result. When the Ce(IV) concentration, which is decreased by reaction (VII), reaches its minimum possible value, the Br^- concentration starts to decrease sharply. Then reaction (VI) is accelerated markedly and the Ce(IV) concentration increases up to a given value of which the bromide concentration starts to increase rapidly, thus retarding reaction (VI). The complete cycle is then repeated.

The number of cycles per unit time (frequency) or the reciprocal of the time elapsed to complete a given number of cycles is proportional to the initial reactant or catalyst concentration. However, such a dependence is only valid for frequency measurements made over a short interval, since the frequency decreases with decreasing concentration of the reactants (bromate and malonic acid). The frequency is also proportional to the concentration of the primary catalyst, Ce(IV), although this catalytic effect is not sufficiently powerful to be utilized for the determination of the catalyst. On the other hand, Ru(III) and Ru(IV) increase the frequency of the catalytic cycle of this reaction and can be determined at concentrations below 0.01 μg/ml with a relative standard deviation (rsd) of 0.03% in the presence of small amounts of Pt, Rh or Ir [12], which act as secondary catalysts. Figure 2.2 shows the typical curves obtained for two different concentrations of a secondary catalyst.

2.5 METHODS OF DETERMINATION

The application of any kinetic method for the determination of a catalyst entails plotting (usually in an automatic or semiautomatic manner) the variation of the measured property as a function of time. Such curves, as can be seen in Fig. 2.3, can be rising or falling according to whether the monitored species is a reactant or a product.

The catalyst is then determined by means of calibration curves obtained by a method suited to the particular case.

Kinetic methods can be classified according to the kinetic order of the indicator reaction into: (a) *differential* or *pseudo zero-order* methods; and (b) *integral*, *first-order* or *pseudo first-order*, and *second-order* methods. With both types of method the tangent, variable-time and fixed-time techniques can be used. The differential variant of the tangent method is more correctly known as the 'initial-rate method'.

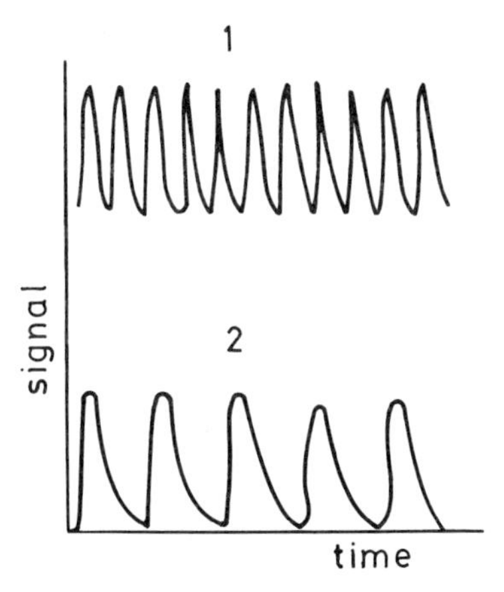

Fig. 2.2 — Signal *vs.* time curves for an oscillating reaction recorded at two different concentrations of secondary catalyst. Curve 1 corresponds to the higher concentration.

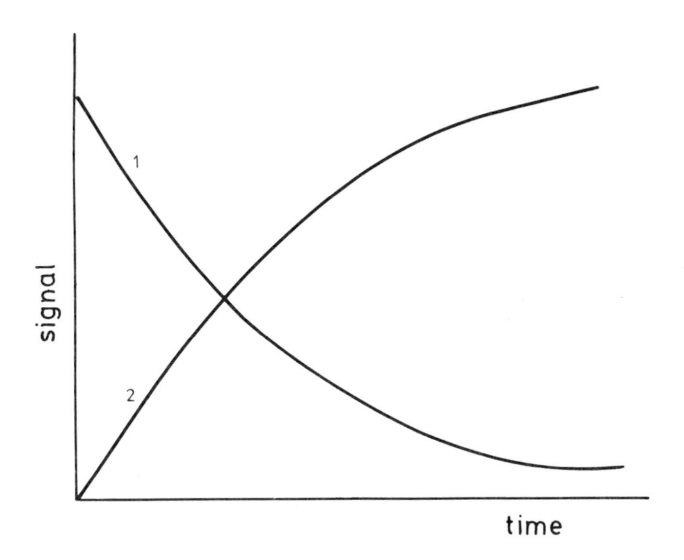

Fig. 2.3 — Curve representing the variation of the analytical signal as a function of time and corresponding to one of the reactants (1) or products (2).

In this classification those methods based on kinetic or induction period measurements can also be included (Table 2.1). The methods commonly used to measure induction periods have been described above in dealing with Landolt reactions. Methods based on kinetic graphs are described in detail in the chapters devoted to uncatalysed reactions and differential reaction-rate methods.

Table 2.1 — Methods of determination

Differential
— Initial-rate
— Fixed-time
— Variable-time
Integral
— Tangent
— Fixed-time
— Variable-time
Based on kinetic plots
Based on induction period measurements

Differential and integral methods are most frequently used for catalysed reactions and are the subject matter of the discussion below.

The rate equation corresponding to the catalysed reaction

$$A + B \xrightarrow{C} P$$

is

$$v = -\frac{d[A]}{dt} = k[A][C]_0 + k_1[A] \tag{2.12}$$

or

$$v = \frac{d[P]}{dt} = k([A]_0 - [P])[C]_0 + k_1([A]_0 - [P]) \tag{2.13}$$

according to whether reactant A or product P is monitored, $[A]_0$ being the initial concentration of A, and [P] the concentration of product formed and k and k_1 denoting the rate constants of the catalysed and uncatalysed reactions.

In either case, the reaction must be made pseudo first-order with respect to the monitored reactant since the catalyst concentration, by definition, does not change during the reaction.

The different methods applicable on the basis of Eqs. (2.12) and (2.13) are commented on below.

2.5.1 Differential methods

Differential methods involve pseudo zero-order reactions, as they make use of measurements made at the start of the process, i.e. when the changes in the concentration of reactants or products are virtually negligible.

2.5.1.1 Initial-rate method

If initial-rate measurements are made, [P] can be neglected with respect to $[A]_0$ in Eq. (2.13) and, since readings are normally taken from $t = 0$, then

$$[\mathrm{P}] = (k[\mathrm{A}]_0[\mathrm{C}]_0 + k_1[\mathrm{A}]_0)t = k'[\mathrm{C}]_0 t + k_1' t \tag{2.14}$$

The variation of [P] for the measured property as a function of time gives a straight segment (Fig. 2.4) which is the ultimate basis for this method, also known as the 'tangent method' because the slope of the straight line is a function of only the catalyst concentration and obeys the general equation

$$v_0 = \frac{\mathrm{d}[\mathrm{P}]}{\mathrm{d}t} = \tan\alpha \simeq \frac{\Delta[\mathrm{P}]}{\Delta t} = k'[\mathrm{C}]_0 + k_1' \tag{2.15}$$

Thus, plotting v_0 as a function of the catalyst concentration, $[\mathrm{C}]_0$, for a series of samples containing known catalyst concentrations, will give a straight line of zero or non-zero intercept, according to whether the uncatalysed reaction develops to a negligible or appreciable extent (curves A and B in Fig. 2.4). Such a line is used as a calibration graph.

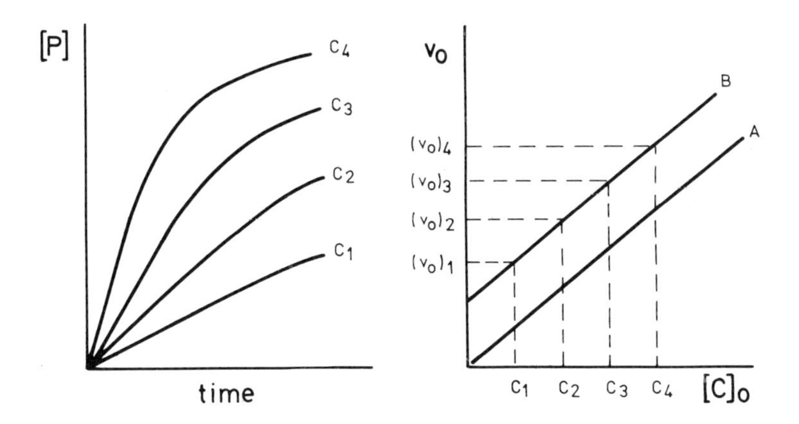

Fig. 2.4 — Initial-rate method

The initial rate is linearly related to the catalyst concentration, whether or not the initial portion of the kinetic curve is linear, provided (if this portion is non-linear) that all the different v_0 measurements are made after the same time interval for the different catalyst concentrations, in order to ensure constancy of the proportionality factor; hence the name 'single-point method'.

The chief difficulty of this method lies in accurate measurement of the initial slope, which is subject to the errors inherent in every graphical method. This shortcoming is currently circumvented by the use of microcomputers which provide the slope of the curve directly, in a more precise manner.

This method features some advantages, namely: (a) the reaction obeys pseudo zero-order kinetics, as the changes in the concentrations of the reactants are very small; (b) since the amount of product formed during the measurement period is very small, the reverse reaction virtually does not take place at all, so the overall rate is not modified; (c) complications arising from possible side-reactions are negligible during the initial rate period; (d) when the reaction rate is in the usual 'useful range', the initial rate measurements will be more precise than those made at later stages, since the faster rate at the start of the reaction (greater slope) results in better signals; (e) initial-rate methods are applicable to reactions with rate constants too small to be used in equilibrium methods.

As the application of the initial-rate method entails plotting the kinetic curve, it is sometimes preferable to use the fixed-time or the variable-time methods, which require no kinetic curves to be drawn.

2.5.1.2 Fixed-time method

This method involves measuring the concentration of a reactant or product at a predetermined time from the start of the reaction.

Equation (2.15) can be written in incremental form as:

$$\Delta[\mathrm{P}] = (k'[\mathrm{C}]_0 + k_1')\Delta t = k'[\mathrm{C}]_0\Delta t + k_1'\Delta t \qquad (2.16)$$

If Δt is constant (i.e. if [P] is measured at a fixed time), the catalyst concentration will be directly proportional to the change in the concentration of the indicator reaction product, P, over a preselected time interval, Δt. Thus, plotting $\Delta[\mathrm{P}]$ as a function of $[\mathrm{C}]_0$ yields a straight line of slope $k'\Delta t$ and intercept $k_1'\Delta t$. Moreover, if $k_1' \approx 0$, then $\Delta[\mathrm{P}] = k'\Delta t\,[\mathrm{C}]_0 = k''[\mathrm{C}]_0$ and the curve will have zero intercept. The application of this method is illustrated in Fig. 2.5.

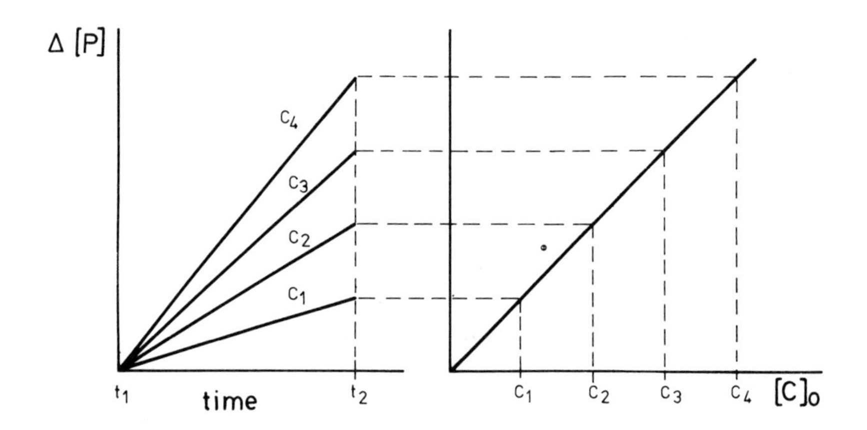

Fig. 2.5 — Fixed-time method.

So far, Δt has been taken from the start of the reaction, i.e. $t_1 = 0$ ($t_2 - t_1 = t_2$) and hence $\Delta t = t_2$. Thus, in the example in Fig. 2.5, this time falls within the straight portion of the kinetic curve. This method is equally valid for measurements made over any interval Δt for which $t_1 = 0$, provided that they are made not too far from the start of the reaction.

Ingle and Crouch [13] have shown that $\Delta[P]$ and $[C]_0$ are proportional even in the case of reversible reactions, and that the fixed-time procedure is theoretically and practically superior for pseudo first-order reactions or processes involving measurements of the concentration of enzymatic substrates.

2.5.1.3 Variable-time method

Also known as the *fixed-* or *constant-concentration method*, this technique entails measuring the time required for a predetermined change in the solution to take place.

Solving Eq. (2.16) for $1/\Delta t$ gives

$$\frac{1}{\Delta t} = \frac{k'[C]_0 + k_1'}{\Delta[P]} \tag{2.17}$$

Therefore, since $\Delta[P] = \text{const.}$, the plot of $1/\Delta t$ *vs.* $[C]_0$ will be a straight line of slope $k'/\Delta[P]$ and intercept $k_1'/\Delta[P]$. If the uncatalysed reaction does not develop to an appreciable extent, then Eq. (2.17) is reduced to $1/\Delta t = k''[C]_0$, with $k'' = k'/\Delta[P]$, and the curve has a zero intercept. The application of this method is illustrated in Fig. 2.6.

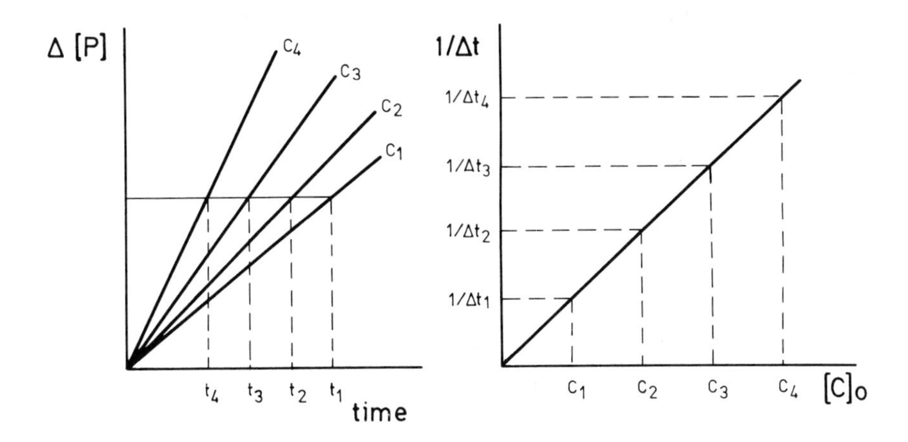

Fig. 2.6 — Variable-time method.

In practice, the method involves fixing a limit for the measured parameter (e.g. absorbance) and measuring the time required by a series of samples with known catalyst concentrations to reach such a value. The plot of the reciprocal of time *vs.* catalyst concentration constitutes the calibration graph (Fig. 2.6).

This method is more limited than the previous one, as securing valid results requires the reaction to develop to an extent proportional to time over an interval Δt that does not start from $t_1 = 0$.

2.5.2 Integration methods

When it is impossible to neglect [P] relative to $[A]_0$ in Eq. (2.13), this differential equation has to be integrated over a finite, though not necessarily short, time interval $\Delta t = t_2 - t_1$. Integration of Eq. (2.12) over the limits $[A]_1$ to $[A]_2$ (i.e. the $[A]_t$ values corresponding to $t = t_1$ and $t = t_2$) yields:

$$\ln([A]_1/[A]_2) = (k[C]_0 + k_1)(t_2 - t_1) \quad (2.18)$$

which, assuming $t_1 = 0$ (start of the reaction), reduces to:

$$\ln \frac{[A]_0}{[A]_0 - [P]} = (k[C]_0 + k_1)t \quad (2.19)$$

When the reaction is monitored through one of the reactants (which is not always possible), t_1 can be non-zero, whereas if the monitored species is a product, it is advisable to take $t_1 = 0$ so that the simplified expression represented by Eq. (2.19) can be used.

These equations are the basis for the tangent, fixed-time and variable-time methods described below.

2.5.2.1 Tangent method

From Eq. (2.19) the following expression can be derived:

$$\log([A]_0 - [P]) = \log [A]_0 - (1/2.303)(k[C]_0 + k_1)t \quad (2.20)$$

Obviously, the plot of $\log([A]_0 - [P])$ as a function of time will be a straight line, with a slope which will be a function of the catalyst concentration, and an intercept which will be constant if a fixed initial concentration of A, $[A]_0$, is taken for every assay (Fig. 2.7).

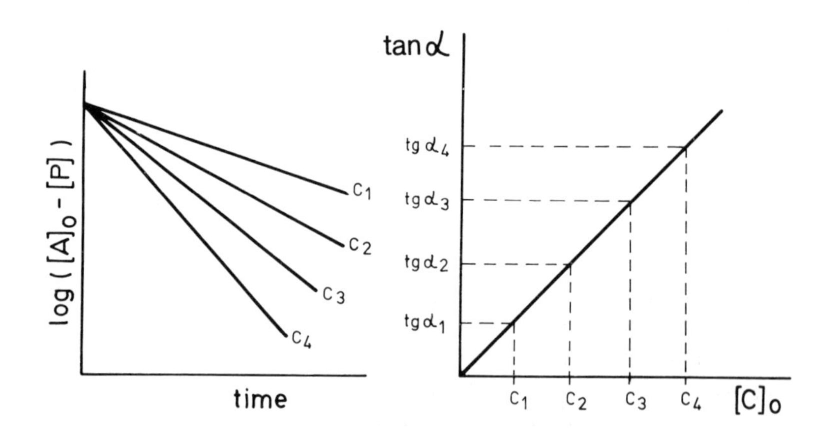

Fig. 2.7 — Tangent method.

Thus, the plot of the slope of these lines against the catalyst concentration will constitute the calibration graph.

In practice it is possible to plot log [P] (or the logarithm of the absorbance, for example) as a function of time. The result is a series of straight lines with positive slopes from which the calibration graph can be constructed. This approach is correct as $[A]_0$ remains constant in every experiment.

2.5.2.2 Fixed-time method

From Eq. (2.18) and assuming $\Delta t = t_2 - t_1 = \text{const.}$,

$$\ln \frac{[A]_1}{[A]_2} = \Delta(\ln [A]) = k'[C]_0 + k'_1$$

with $k' = k\Delta t$ and $k'_1 = k_1\Delta t$. The plot of $\Delta(\ln [A])$ *vs.* $[C]_0$ for a fixed concentration of A, $[A]_0$, constitutes the calibration graph in this case.

Since the $[A]_1/[A]_2$ ratio is constant for any catalyst concentration over a preselected time interval, if all reaction conditions are kept constant, then it will be possible to calculate the overall pseudo first-order rate constant, k_T, from

$$k_T = \frac{\Delta(\ln [A])}{\Delta t} = k[C]_0 + k_1$$

which, if plotted against the catalyst concentration, will give a straight line that can be used as the calibration plot.

If the reaction is monitored by means of the formation of the product, P, rather than through the disappearance of reactant A, and if $t_1 = 0$, then $[A]_1 = [A]_0$ and $[A]_2 = [A]_0 - [P]$, so

$$\ln \frac{[A]_0}{[A]_0 - [P]} = k'[C]_0 + k'_1$$

and the plot of $\ln \{[A]_0/([A]_0 - [P])\}$ against the catalyst concentration will be the calibration line to be used.

2.5.2.3 Variable-time method

From Eq. (2.18) and assuming $\Delta[A]$ to be constant

$$\frac{1}{\Delta t} = k''[C]_0 + k''_1$$

where $k'' = k/\Delta(\ln [A])$ and $k''_1 = k_1/\Delta(\ln [A])$. This expression is equally valid whether the reaction is monitored through one of the reactants or the product.

The plot of $1/\Delta t$ as a function of the concentration of a series of catalyst standards constitutes the calibration graph sought. In fact, Δt can be any time interval along the kinetic curve, though the best results are obtained when it falls within the straight portion of the curve. More often than not, t_1 corresponds to the start of the reaction. It has been shown [14] that the error made in the determination of the catalyst is at a minimum if the $[A]_1/[A]_2$ ratio in Eq. (2.18) is equal to e, whereby the expression is simplified to

$$1/t_e = k[C] + k_1$$

which is the basis for the determination of the catalyst.

As integration methods are applied under conditions where the reaction has proceeded to a greater extent, the use of the variable-time method is more limited here than in its differential form, as such conditions may also result in a loss of linearity in the variation of [P] with time. However, the fixed-time method is more general, since it is valid even when applied to the non-linear portion of the plot of the measured property as a function of time. These considerations are supported mathematically in Chapter 5, which is devoted to uncatalysed reactions.

2.6 APPLICATIONS OF CATALYSED REACTIONS

The vast development of kinetic methods in the last few years is a result of the improvements in sensitivity and selectivity brought about by the use of catalysed reactions. These have facilitated the determination of microconstituents in various types of samples by means of straightforward and inexpensive instrumentation. However, it should be borne in mind that such determinations are usually limited to inorganic species (metal ions and anions). The determination of some organic species is also possible by the exploitation of modified catalytic effects (Chapters 3 and 4). The resolution of mixtures has been rendered feasible by the development of differential reaction-rate methods (Chapter 6).

Catalytic kinetic methods have attracted attention from various authors in a number of reviews and chapters in books of a general character, mostly written in the 1970s. Thus, it is worth mentioning the reviews of Bontchev [6], Svehla [15] and Mottola [5], as well as Guilbault's chapter in Kolthoff and Elving's *Treatise on Analytical Chemistry* [16]. Recently, Müller [17] published a comprehensive review on this topic, and Valcárcel and Grases [18] re-examined catalytic methods (both enzymatic and non-enzymatic) involving monitoring of the reaction rate by fluorescence measurements. Fluorimetric detection further lowers the detection limits achievable in kinetic determinations based on catalysed reactions.

As stated above, the description of the analytical applications of catalysed reactions has been divided into four broad categories according to the chemical nature of the indicator reaction, namely: redox reactions, chemiluminescence reactions, reactions involving complexes and other reactions.

2.6.1 Redox reactions

This is undoubtedly the widest and most extensively studied group of kinetic methods based on catalysed reactions. Some of these determinations, because of their acceptance by many authors, the wide range of instrumentation used for measuring the reaction rate, and their applicability to real problems, have become the standards for determination of a variety of species. Noteworthy examples are the determinations of iodide by the Sandell and Kolthoff reaction, sulphur-containing compounds with the iodine/sodium azide system, copper with hydroquinone/H_2O_2 and manganese with Malachite Green/periodate. Most redox reactions are catalysed by metal ions (only about a fifth of them are catalysed by inorganic anions).

2.6.1.1 Determination of metal ions

Because of the large number of literature references available on this aspect of catalytic kinetic methods, the entries in Tables 2.2–2.9 correspond chiefly to the decade 1975–1985; earlier references can be found in the above-mentioned reviews.

Obviously, not all metal ions can be used as efficient catalysts for redox reactions. Figure 2.8 shows the numbers of methods based on different indicator reactions for a large number of metal ions over the period 1975–1985. As can be seen, Co(II) and (III), Fe(II) and (III), V(IV) and (V), Mn(II) and Cu(II) are by far the most frequently investigated (especially the last).

Among indicator reactions, oxidation processes (involving either organic or inorganic compounds) are the more common. In fact, reduction reactions are rarely used as the basis for kinetic catalytic determinations of metal ions. Among the few examples worth considering are those of copper, based on the catalytic effect of this metal on the reduction of 2,2′-diquinoxalyl with Sn(II) or Ti(III), bismuth and selenium with the PbO_2^{2-}/SnO_2^{2-} and Fe(III)/$SnCl_2$ systems, respectively, and those of palladium based on its catalytic effect on the reduction of various organic compounds by the hypophosphite ion.

As stated above, the field of oxidation reactions is comprised of two large groups: inorganic indicator reactions and reactions involving oxidation of organic compounds.

Kinetic catalytic methods make use of a number of inorganic oxidation indicator reactions, the most representative of which is probably the decomposition of hydrogen peroxide catalysed by copper at levels between 10 and 5000 ng/ml, iron, manganese and cobalt, which allows these species to be determined by the use of various detection systems.

The reaction between iodide and hydrogen peroxide,

$$H_2O_2 + 2I^- + 2H^+ \rightarrow I_2 + 2H_2O$$

has been studied in depth by Kataoka *et al.* [101]. Its kinetics is usually determined by means of an iodide-selective electrode. This reaction has been applied to the determination of iron, molybdenum, tungsten, zirconium and hafnium at the micromolar level, and some automated methods have been reported.

Other reactions, involving the oxidation of iodide with Ce(IV), peroxoborate or bromate, have also been used as indicators. Of these, the commonest are those using bromate and electrochemical detection (potentiometric and/or amperometric). It is also worth mentioning the fluorimetric determination of molybdenum proposed by Kataoka *et al.* [117], which has a significantly lower detection limit; the reaction is performed in the presence of ascorbic acid or thiosulphate, which transforms it into a

Table 2.2 — Kinetic catalytic methods for determination of copper.

Indicator reaction	Detection technique	Detection limit or range (rsd)	Observations	References
Decomposition of H_2O_2	Thermometric	0.6–6 μg	In the presence of cyanide	[20]
	'Biamperostat'			[21]
$Fe(CN)_6^{3-} + CN^-$	A ($\lambda = 422$ nm)	$< 4 \times 10^{-6} M$ (3%)	Constant-rate method	[22]
$IO_4^- + S_2O_3^{2-}$	A	10–100 ng/ml (2%)	Automated procedure	[23]
Fe(III)–$SCN^- + S_2O_3^{2-}$		10^{-8} μg/ml	Organic medium: 30% ethanol	[24]
Ce(IV) + I^-		$10^{-3} M$		[25]
Hydroquinone + H_2O_2	A ($\lambda = 453$ nm)	0.2–2.0 ng/ml	Pyridine as activator	[26,27]
Amidol + H_2O_2		5×10^{-4} μg/ml	In the presence of H_3BO_3	[28]
2,4-Diaminophenol + H_2O_2	A ($\lambda = 536$ nm)	0.1 ng/ml		[29]
1,3,5-Benzenetriol + H_2O_2		1 ng/ml		[30]
Pyrocatechol Violet + H_2O_2		3.3–13.2 ng/ml	In 30% acetonitrile	[31]
Sulphanilic acid + H_2O_2	A ($\lambda = 430$ nm)	5 ng/ml	In the presence of pyridine	[32]
1-Amino-2-naphthol-4-sulphonic acid + H_2O_2		0.5–4.0 ng/ml	Relative error 10–19%	[33]
R + H_2O_2	A ($\lambda = 533$ nm)	$3–27 \times 10^{-3}$ μg/ml	Tolerates Fe(III) in 100-fold excess	[34]
TNS + H_2O_2	A ($\lambda = 525$ nm)	0.25–5 ng/ml	In the presence of ascorbic acid	[35]
4,4′-DBPT + H_2O_2	A ($\lambda = 415$ nm)	10–90 ng/ml	Data on mechanism	[36]
p-Anisidine + *N*,*N*-dimethylaniline + H_2O_2	A ($\lambda = 740$ nm)	< 10 ng/ml	Fixed-time method (10 min)	[37]
p-Hydrazinobenzenesulphonic acid + H_2O_2 + *m*-phenylenediamine	A($\lambda = 454$ nm)	2 ng/ml	Fixed-time method (20 min)	[38]
N-Phenyl-*p*-phenylenediamine + *N*,*N*-dimethylaniline + H_2O_2	A ($\lambda = 728$ nm)	0–4 ng/ml (2%)	Fixed-time method (15 min)	[39]
Bindschedler Green + H_2O_2	A ($\lambda = 725$ nm)	0–20 ng/ml (1%)	Fixed-time method	[40]
2,2′-Dipyridyl ketone hydrazone + O_2	F($\lambda = 349$, 435 nm)	0.4–1.0 ng/ml (2%)	Data on the oxidation product	[41]
		8–30 ng/ml (2.3%)	FIA	[42]
	Thermometric	0.2–2.0 μg/ml (1.2%)		[43]
2,2′-Dipyridyl ketone hydrazone + H_2O_2	F ($\lambda = 339$, 435 nm)	0.2–2.0 ng/ml (1.7%)	Stopped-flow FIA	[44]
2,2′-Dipyridyl ketone azine + O_2	F ($\lambda = 347$, 430 nm)	0.2–0.5 μg/ml (3.5%)		[45]
Phenyl-2-pyridyl ketone hydrazone + O_2	F ($\lambda = 308$, 435 nm)	0.12–0.60 μg/ml (3%)		[45]
Cyclohexane-1,3-dione dithiosemicarbazone + O_2	A ($\lambda = 375$ nm)	0.05–0.65 μg/ml	Initial-rate method	[46]
Dimedone bisguanylhydrazone + O_2	A ($\lambda = 550$ nm)	24–330 ng/ml	In the presence of pyridine	[47]
Ascorbic acid + $S_2O_8^{2-}$		0.5–5.0 ng/ml	pH-stat method	[48]
2,2′-Diquinoxalyl + Sn(II) or Ti(III)		10^{-4}–10^{-5}%	Relative error 40%	[49]
Dibutyl phosphinodithionic acid + Fe(III)	A ($\lambda = 410$ nm)	0.5μM (1.9%)	Ni(II), Co(II) and Cr(III) are tolerated in 10000-fold excess	[50]

A = photometry; F = fluorimetry; R = 4-hydroxy-3-carboxyphenylazo-4′,4″-biphenylazo-2-chromotropic acid; TNS = sodium 1,2-naphthoquinone-4-sulphonate; 4,4′-DBT = 4,4′-dihydroxybenzophenone thiosemicarbazone; FIA = flow-injection analysis.

Table 2.3 — Kinetic methods for determination of manganese

Indicator reaction	Detection technique	Detection limit or range (rsd)	Observations	References
Decomposition of H_2O_2		5 μg/ml	In the presence of 1,10-phenanthroline	[51]
Malachite Green + IO_4^-	A	$0.1–1.0 \times 10^{-7} M$	In the presence of NTA	[52]
	A	0.02–2.0 μg/ml	Absorptiostat technique	[53]
	A	0.2–10 μg/ml	NaF masks Fe(III) and Al(III)	[54]
p-Fuchsine + IO_4^-	A ($\lambda = 540$ nm)	0.4 ng/ml	Reaction rate	[55]
Sulphanilic acid + IO_4^-		Submicrogram	Data on mechanism	[56]
Thymol Blue + IO_4^-				[57]
Acetylacetone + IO_4^-	Potentiometric	10–60 ng/ml (2–3%)	Variable-time method (10 mV)	[58]
Triethanolamine + IO_4^-	Potentiometric	0.01–0.6 μg/ml	NTA as activator	[59]
Acid Blue 45 + H_2O_2	A ($\lambda = 595$ nm)	4–25 ng/ml	Initial-rate method	[60]
Hydroxynaphthol Blue + H_2O_2	A ($\lambda = 645$ nm)	0.01–10 ng/ml		[61]
Tiron + H_2O_2		0.2 ng/ml	2,2′-Bipyridyl as activator	[62]
Azorubin S + H_2O_2	A($\lambda = 540$ nm)	5.5–33 ng/ml	Fixed-time method (10 min)	[63]
o-Dianisidine + H_2O_2	A ($\lambda = 440$ nm)	2–100 ng/ml (1.3%)	36% Dimethylformamide	[64]
Indigo sulphonates + H_2O_2		$2 \times 10^{-8}–2 \times 10^{-7} M$	Active species: $Mn(HCO_3)_2$	[65]
HBTS + H_2O_2	F ($\lambda = 365, 440$ nm)	2–9 ng/ml (1.2%)	Initial-rate method	[66]
HNTS + H_2O_2	F ($\lambda = 390, 450$ nm)	0.75–4.8 ng/ml (1.12%)	Kinetic study of HNTS	[67]
PPH + H_2O_2	F ($\lambda = 335, 425$ nm)	0.2–3.0 ng/ml (0.6%)	Data an oxidation products	[68]
OH–PDT + O_2	A ($\lambda = 594$ nm)	10–90 ng/ml (1.8%)	Data on mechanism	[69]
R + O_2		0.01–0.3μM	Cyanide masks heavy metals	[70]

HBTS = 2-hydroxybenzaldehyde thiosemicarbazone; HNTS = 2-hydroxynaphthaldehyde thiosemicarbazone; PPH = pyridoxal 2-pyridylhydrazone; OH–PDT = 1,4-dihydroxyphthalimide dithiosemicarbazone; R = 1,5-bis-(5-chloro-2-hydroxyphenyl)-3-cyanoformazan; NTA = nitrilotriacetic acid.

Table 2.4 — Kinetic catalytic methods for determination of vanadium

Indicator reaction	Detection technique	Detection limit or range (rsd)	Observations	References
$Br^- + BrO_3^-$	Potentiometric	0–0.6 μg/ml	Two reaction steps yield BrO^-	[71]
$I^- + BrO_3^-$	Potentiometric	0.1–1.5μ*M*	Iodide-selective electrode	[72]
Gallic acid + BrO_3^-	*A* (λ = 420 nm)	0.1 ng/ml (3.5%)	Only Fe(III) interferes	[73]
Chromotropic acid + BrO_3^-	*A* (λ = 430 nm)	0.5–2.5 ng/ml (2.5%)	Data on mechanism	[74]
Bindschedler Green + BrO_3^-	*A* (λ = 725 nm)	0.008 ng/ml	The reaction is accelerated by Fe(III) and Cu(II)	[75]
1,2-Phenylenediamine + BrO_3^-		5×10^{-5} μg/ml	Accelerated in an acetate buffer (pH 3.2) in the presence of Tiron	[76]
Pyrogallol Red + BrO_3^-	*A* (λ = 490 nm)	0–51 ng/ml	Fixed-time method (5 min)	[77]
Chlorpromazine + BrO_3^-	*A* (λ = 525 nm)	10–400 ng/ml (5%)	Initial-rate method	[78]
N-Phenyl-*p*-phenylenediamine + *N*,*N*-dimethylaniline + BrO_3^-	*A* (λ = 735 nm)	0–10 ng/ml (2.3%)	Fixed-time method (20 min)	[79]
p-Hydrazinobenzenesulphonic acid + *m*-phenylenediamine + ClO_3^-	*A* (λ = 454 nm)	0.008–0.04 ng/ml	Fixed-time method (25 min)	[80]
2-Aminophenol + ClO_3^-	*F*	0.1–5.0 μg/ml	No interference from Fe(II), Mn(VII), Cr(VI) or Mo(VI)	[81]
Tiron + $S_2O_8^{2-}$	*A* (λ = 364 nm)		Molybdenum is masked with citrate	[82]
Gallic acid + $S_2O_8^{2-}$	*A* (λ = 415 nm)	0.04–0.6 μg/ml	Modification of the Fishman and Skougstad method	[83]
I^- + peroxoborate	Potentiometric	0–40μ*M*	Iodide-selective electrode	[84]
o-Dianisidine + *t*-butyl peroxide		0.7 μg/ml	Uses the simplex method	[85]
4,8-Diamino-1,5-dihydroxyanthraquinone-2,6-disulphonate + O_2	*F*	0.04–0.05 μg/ml	Induction period measurements	[86]

Table 2.5 — Kinetic catalytic methods for determination of cobalt

Indicator reaction	Detection technique	Detection limit or range (rsd)	Observations	References
Decomposition of H_2O_2		10 ng/ml	In the presence of 1,10-phenanthroline	[51]
	Thermometric	50nM–10μM	In the presence of 2-aminopyridine	[87]
Gallomycin + H_2O_2		0.2–2.4 ng/ml		[88]
R + H_2O_2	A (λ = 549 nm)	5–50 ng/ml	Cu(II) is masked with citrate	[89]
1,4-Dihydroxyanthraquinone + H_2O_2		0.3 ng/ml	Borate buffer	[90]
Catechol + H_2O_2	A (λ = 365 nm)	3 ng/ml	Fixed time	[91]
Quinalizarin + H_2O_2	A	10^{-8}–$10^{-7}M$	Mechanism resembling enzymatic reaction	[92]
Salicylfluorone + H_2O_2		0.01 ng/ml	Fixed time	[93]
Tiron + H_2O_2		0.05 ng/ml	Differential and integral methods	[62]
SPADNS + H_2O_2	A (λ = 520 nm)	5–200 ng/ml (2%)	FIA stopped-flow	[94]
1-Naphthylamine + diperoxyadipic acid		10^{-8}–$2 \times 10^{-7}M$	Basic medium (pH 9.2)	[95]
Autoxidation of SO_3^{2-}	Thermometric	2×10^{-7}–$2 \times 10^{-5}M$ (3%)	Flow system	[96]
	Potentiometric	10^{-7}–$7 \times 10^{-7}M$	Oxygen electrode	[97]
	Thermometric	5–30 ng/ml	Few interferences	[98]
Gallic acid + O_2	Potentiometric	$<2\mu M$	Oxygen electrode	[99]
FPKH + O_2	F (λ = 308, 435 nm)	0.7–2.4 ng/ml (4.5%)	Contrast with non-kinetic methods	[100]
DPDKH + O_2	F (λ = 340, 450 nm)	0.08–0.24 μg/ml (4.1%)	Contrast with non-kinetic methods	[100]

R: sodium 1,2-naphthoquinone-4-sulphonate 2-thiosemicarbazone; DPDKH: dipyridylglyoxal hydrazone; SPADNS: 2-(4-sulphophenylazo)-1,8-dihydroxynaphthalene-3,6-disulphonic acid; FPKH: phenyl 2-pyridyl ketone hydrazone.

Table 2.6 — Kinetic catalytic methods for determination of iron

Indicator reaction	Detection technique	Detection limit or range (rsd)	Observations	References
Decomposition of H_2O_2		50 ng/ml	In the presence of 1,10-phenanthroline	[51]
$H_2O_2 + I^-$	Potentiometric	5–160μM	Iodide-selective electrode	[101]
Tris(oxalato)cobalt(III) + ascorbic acid	A ($\lambda = 600$ nm)	0.28–7.8 μg/ml	0.2M $HClO_4$ medium	[102]
p-Aminophenol + H_2O_2		5×10^{-2} μg/ml (1.6%)		[103]
p-Phenetidine + H_2O_2		0–2.0 mg/l.	1,10-Phenanthroline as activator	[104]
Metol + H_2O_2	A ($\lambda = 490$ nm)	0.1–0.6 μg/25 ml	Reaction-rate method	[105]
Anisole + H_2O_2	Gas chromatography	0.25–1000 ng/ml		[106]
HBTS + H_2O_2	F ($\lambda = 365$, 440 nm)	10–60 ng/ml	Data on mechanism	[107]
Pyridoxal 2-pyridylhydrazone + H_2O_2	F ($\lambda = 325$, 390 nm)	5–60 ng/ml	Fe(II) as inducer and Cr(VI) as inhibitor	[108]
p-Anisidine + *N*,*N*-dimethylaniline + H_2O_2		6–20 ng	Fixed time (60 min)	[109]
MDP + H_2O_2	A ($\lambda = 735$ nm)	0.002–0.014 μg/ml	1,10-phenanthroline as activator	[110]
N-Phenyl-*p*-phenylenediamine + H_2O_2 + *N*,*N*-dimethylaniline	A ($\lambda = 728$ nm)	0–6 ng/ml	Acetate as activator	[111]
Redoxin II + H_2O_2 + $h\nu$		10 ng/ml	30 min irradiation	[112]
Acridine + O_2 + $h\nu$	F	0.02–0.44 μg/ml	Irradiation at 360 nm	[113]

Metol: *p*-methylaminophenol sulphate; HBTS: 2-hydroxybenzaldehyde thiosemicarbazone; MDP: *N*-(*p*-methoxyphenyl)-*N*,*N*′-dimethyl-*p*-phenylenediamine.

Table 2.7 — Kinetic catalytic methods for determination of molybdenum and tungsten

	Indicator reaction	Detection technique	Detection limit or range (rsd)	Observations	References
MOLYBDENUM	$I^- + H_2O_2$		0.01 μg/ml	Automated method (35 samples/hr)	[114]
		Potentiometric	0.1–160μM	Landolt reaction	[115]
	$I^- + BrO_3^-$	Amperometric	0.5μM–0.2mM	Rotating Pt electrode	[116]
		F ($\lambda = 555, 573$ nm)	80–200nM	Landolt reaction	[117]
	I^- + peroxoborate	Potentiometric	0.09 μg/ml	Iodide-selective electrode	[118]
	1-Naphthylamine + BrO_3^-		10–500nM (11%)	Previous extraction of Mo with oxine	[119]
	2-Aminophenol + H_2O_2	A ($\lambda = 440$ nm)	10–2000 ng/ml		[120]
	Light-induced oxidation of 2-aminophenol	A ($\lambda = 430$ nm)		8 min irradiation at 380 nm. Sodium anthraquinone-2,6-disulphonate as activator	[121]
	Se(IV) + Sn(II)	A ($\lambda = 390$ nm)		Colloidal selenium is stabilized with gum arabic	[122]
TUNGSTEN	$I^- + H_2O_2$			Landolt reaction in the presence of ascorbic acid	[123]
	$I^- + BrO_3^-$	Amperometric	0.8μM–2mM	Rotating Pd electrode	[116]
	I^- + peroxoborate	Potentiometric	0–6μM	Iodide-selective electrode	[84]
	2-Aminophenol + H_2O_2	A ($\lambda = 440$ nm)	10–2000 ng/ml	Mo(VI) is masked with oxalate	[120]
	Victoria Blue + Ti(III)	A ($\lambda = 610$ nm)	1–20μM	Non-linear calibration curves resulting from the possible dimerization of W(V) formed	[124]
	Malachite Green + Ti(III)	A ($\lambda = 584$ nm)	1–20μM		[124]

Table 2.8 — Kinetic catalytic methods for determination of silver, gold and metals of the platinum family

	Indicator reaction	Detection technique	Detection limit or range (rsd)	Observations	References
SILVER	$Ce(IV) + Cl^-$	*A*	<2 µg/ml		[125]
	Oxine-5-sulphonic acid + $S_2O_8^{2-}$	*F* ($\lambda = 366, 480$ nm)	1.0–4.5 ng/ml	Prior interference removal by extraction with dithizone	[126]
	Indigo Carmine + $S_2O_8^{2-}$		0.3–4.0 µg/ml (2%)	Rate-constant method	[127]
	Antipyrine-8-hydroxyquinoline + $S_2O_8^{2-}$		5×10^{-4} µg/ml (4%)	2,2′-Bipyridyl as activator	[128]
	Ethylenediamine + $S_2O_8^{2-}$	Titrimetric	5×10^{-8}–$8 \times 10^{-6}M$	Titration of the oxidant at regular intervals	[129]
	Pyridoxal nicotinolhydrazone + $S_2O_8^{2-}$	*F* ($\lambda = 365, 450$ nm)	60–720 ng/ml (0.6%)	Data on the nature of the fluorescence	[130]
GOLD	$Hg(I) + Ce(IV)$	*A*	1.8 ng (9.7%)		[131]
	Variamine Blue + $S_2O_8^{2-}$	*A* ($\lambda = 540$ nm)	2 ng/ml (2.1%)	Fixed time (6 min)	[132]
	Molybdophosphoric acid + formic acid		0.1–0.8 ng/ml		[133]
	DPDKFH + O_2	*F* ($\lambda = 305, 437$ nm)	0.5–2.0 µg/ml (2.5%)	Data on the nature of the fluorescence	[134]
	2,2′-Dipyridyl ketone azine + O_2	*F* ($\lambda = 347, 435$ nm)	0.05–0.25 µg/ml	Data on the nature of the fluorescence	[135]
	$R + O_2$	*F* ($\lambda = 525, 585$ nm)	0.16–1.06 µg/ml	Interference by V(V), Fe(III) and Ce(IV)	[136]
PALLADIUM	Erioglaucine + hypophosphite		0.2–1 µg/ml	$Pd(II) + H_3PO_2^- \rightarrow Pd$; $Pd + H_2PO_2^- \rightarrow H_2$ (reductant)	[137]
	Phenol Red + hypophosphite	*A* ($\lambda = 420$ nm)	2–7 µg/ml	Segmented flow (AutoAnalyser)	[137]
	Tolidine Blue + hypophosphite	*A* ($\lambda = 638$ nm)	0.5–5.0 µg/ml	Inhibitors: Pt(IV) and Mo(VI); Activators: Br^- and SO_3^{2-}	[138]
	Phenosafrine + hypophosphite	*A* ($\lambda = 525$ nm)	0.1–1.1 µg/ml (2.6%)	Induction period measurements	[139]
	Pyronine G + hypophosphite	*A* ($\lambda = 548$ nm)	0.08–1.0 µg/ml (2.4%)	Iodide as inhibitor; Ag(I) results in induction period	[140]
OSMIUM	$AsO_3^{3-} + IO_4^-$		5×10^{-7} µg/ml	Tolerates 50-fold excess of Ru	[141]
	H_2O_2 + cyanocuprate(I)	Potentiometric or amperometric	0.03–1 ng/ml	Pt electrodes	[142]
	1-Naphthylamine + NO_3^-	*A* ($\lambda = 525$ nm)	1.6–40 ng/ml		[143]
	Neutral Red + BrO_3^-	*A* ($\lambda = 530$ nm)	50–600 pg/ml	Violet Red yields similar results	[144]

Element	Reaction	Detection	Range / limit	Remarks	Ref.
RUTHENIUM	Tropaeolin 00 + IO_4^-		4×10^{-6} μg/ml		[145]
	Diphenylamine + IO_4^-	*A* (λ = 530 nm)	0.05 μg/ml	Sulphate complexes as catalysts	[146]
	Direct Blue 6B + H_2O_2	*A*	0.0001–0.001%	Acid medium (pH 0.8–1.2)	[147]
	$R_1 + VO_3^-$		4×10^{-11} g/ml		[148]
	1-Naphthylamine + NO_3^-	*A* (λ = 525 nm)	10–60 ng/ml		[143]
RHODIUM	Mn(II) + BrO_3^-	*A*	0.01–0.06 μg/ml (10%)	Monitoring of MnO_4^- formation	[149]
	Cu(II) + IO_4^-	*A* (λ = 413 nm)	0.02–0.2 ng/ml	Fixed time (6 min)	[150]
	Ag(I) + Fe(II)		10^{-4} μg/ml	Catalyst: Rh	[151]
IRIDIUM	Hg(I) + Ce(IV)	*A* (λ = 459 nm)	0.1–2.0 ng/ml	Automatic colorimeter	[152]
	Photodecomposition of IO_4^-		80–800 ng (1.8%)		[153]
	$R_1 + VO_3^-$		2×10^{-10} g/ml		[154]
	Direct Blue 6B + IO_4^-	*A*	0.0001–0.01%	Alkaline medium (pH 8.5–9.5)	[147]
PLATINUM	$R_2 + O_2$	*F* (λ = 359, 435 nm)	0.2–0.6 μg/ml (3%)	Few interferences	[155]
RHENIUM	Malachite Green + $SnCl_2$	*A* (λ = 600 nm)	0.0019 μg/ml	Citric and tartaric acid as activators	[156]

DPDKHF: dipyridylglyoxal diphenylhydrazone; R: sodium 4,8-diamino-1,5-dihydroxyanthraquinone-2,6-disulphonate; R_1: 4(*N*-methylanilino)benzenesulphonic acid; R_2: 2,2′-dipridyl ketone hydrazone.

Table 2.9 — Kinetic catalytic methods for various metal ions

Ion	Indicator reaction	Detection technique	Detection limit or range (rsd)	Observations	References
Chromium(III)	$IO_4^- + AsO_2^-$	Potentiometric	40–300 ng (5%)	Perchlorate-selective electrode	[157]
	R + *N*,*N*-dimethylaniline + H_2O_2	*A* ($\lambda = 590$ nm)	0.4–10 ng/ml (3.5%)	EDTA as activator	[158]
Chromium(VI)	$I^- + BrO_3^-$	Amperometric	0.1μM–0.2mM	Rotating Pt electrode	[116]
	Aniline + ClO_3^-		0.05–0.5 μg/ml	No interference from Cr(III)	[159]
	o-Dianisidine + H_2O_2		30 ng/ml	Prior extraction into MIBK	[160]
Nickel	1-Naphthylamine + peroxycapric acid		0.1 μg/ml	Alkaline medium	[161]
	Diphenylcarbazone + H_2O_2		10^{-4} μg/ml	Reaction accelerated by organic solvents	[162]
	Pyridoxal hydrazone + $S_2O_8^{2-}$	*F* ($\lambda = 315, 370$ nm)	50–300 ng/ml (1.8%)	Data on the nature of the fluorescence	[163]
Lead	Pyrogallol Red + $S_2O_8^{2-}$		0.002–10 μg/ml	Prior extraction into dithizone	[164]
	PPH + H_2O_2	*F* ($\lambda = 355, 425$ nm)	0.05–0.40 μg/ml (1%)	Few interferences	[165]
Mercury	$AsO_2^- + Sn(II)$		4×10^{-8} g/ml	Gelatin as stabilizer	[166]
	DPKH + O_2	*F* ($\lambda = 359, 435$ nm)	80–320 ng/ml (2%)	Data on the nature of the fluorescence	[41]
Yttrium	MoO_4^{2-} + ethylhydrazine	Potentiometric	0.3 μg/ml	Various interferences	[167]
				Iodide-selective electrode	[101]
Zinc	Resorcinol + H_2O_2		0.9–8.0 μg/ml (10%)		[168]
Bismuth	$PbO_2^{2-} + SnO_2^{2-}$	Appearance of white precipitate	5–60 ng/ml	CN^- masks Cu, Hg and Ag	[169]
Tin	Isopolymolybdate + ascorbic acid	*A* ($\lambda = 650$ nm)	1–12μM (5.6%)	Tolerates Pb, Zn, Cr and Mg in 1000-fold excess	[170]
Selenium	Fe(III) + $SnCl_2$	*A* ($\lambda = 335$ nm)	0.2–1.2 μg/ml	In the presence of tartrate	[171]
	Picrate + S^{2-}	Potentiometric	3–30 μg (2%)	Picrate-selective electrode	[172]
	$R_1 + ClO_3^-$ + *m*-phenylenediamine	*A* ($\lambda = 454$ nm)	0.1μM	Many interferences	[173]
Tellurium	Fe(III) + $SnCl_2$	*A* ($\lambda = 335$ nm)	0.006–0.03 μg/ml	In the presence of tartrate	[171]
Germanium	MoO_4^{2-} + Sn(II)		3×10^{-2}–8×10^{-2} μg/ml		[174]
Titanium	PANH + O_2	*F* ($\lambda = 365, 445$ nm)	60–400 ng/ml	A non-kinetic method is also proposed	[175]
	o-Phenylenediamine + H_2O_2		5×10^{-4} μg/ml		[176]
Zirconium	$I^- + H_2O_2$	Amperometric	0.1–1μM (10%)	Variable-time method	[177]
Hafnium	$I^- + H_2O_2$	Amperometric	0.1–1μM (10%)	Variable-time method	[177]
Niobium	$S_2O_3^{2-} + H_2O_2$	Turbidimetric	0.5–100μM	Tolerates Ta in 10-fold excess	[178]
	o-Aminophenol + H_2O_2		1×10^{-7}–$3 \times 10^{-7}M$		[179]
Tantalum	*o*-Aminophenol + H_2O_2		0.5×10^{-7}–$2 \times 10^{-7}M$		[179]

R: 3-methyl-2-benzothiazole hydrazone or benzothiazolone; PPH: pyridoxal 2-pyridylhydrazone; R_1: *p*-hydrazinobenzenesulphonic acid; PANH: picolinaldehyde nicotinoylhydrazone.

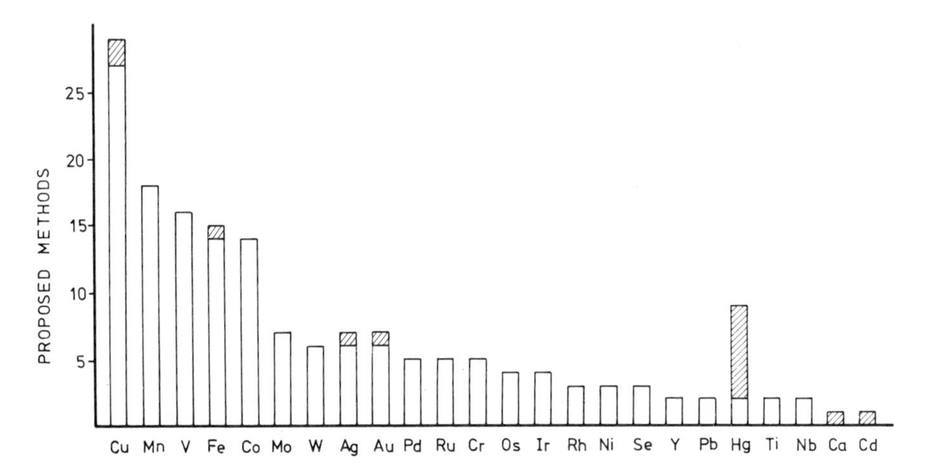

Fig. 2.8 — Diagram representing the number of indicator reactions proposed for the kinetic determination of catalysts in redox reactions (clear zones) and ligand-exchange and complex-formation reactions (hatched zones).

typical Landolt reaction. Under these conditions, the iodine formed reacts with the above-mentioned reductants and does not quench the natural fluorescence of Rhodamine B until the reductants have been fully depleted. Obviously, this procedure is equally suitable for the determination of ascorbic acid and/or thiosulphate.

Oxidants such as periodate, bromate or cerium(IV) have also been employed as inorganic indicators in the kinetic catalytic determination of metal ions. In this connection the study of the copper-catalysed oxidation of cyanide by ferricyanide, by López-Cueto and Casado-Riobó [19] is worthy of note. They proposed the following mechanism, both for the catalysed and the uncatalysed reaction:

$$Fe(CN)_6^{3-} + CN^- \underset{k_{-a}}{\overset{k_a}{\rightleftharpoons}} Fe(CN)_6^{4-} + CN\cdot$$

$$Fe(CN)_6^{3-} + CN\cdot \xrightarrow{k_b} Fe(CN)_6^{4-} + CN^+$$

$$CN^+ + 2OH^- \xrightarrow{\text{fast}} CNO^- + H_2O$$

which ultimately yields cyanate. Copper probably exerts its catalytic action through the cycle Cu(I)$\rightleftharpoons$Cu(II), in which cyano complexes are probably involved. After the first step,

$$Fe(CN)_6^{3-} + Cu(CN)_4^{3-} \underset{k_{-a}}{\overset{k_a}{\rightleftharpoons}} Fe(CN)_6^{4-} + Cu(CN)_4^{2-}$$

the cyano complex of copper(II), which is very unstable, reacts with the ferricyanide present to yield

$$Fe(CN)_6^{3-} + Cu(CN)_4^{3-} \xrightarrow{k_b} [(CN)_6FeCNCu(CN)_3]^{5-}$$

which rapidly decomposes in the alkaline medium, by

$$[(CN)_6FeCNCu(CN)_3]^{5-} + 2OH^- \xrightarrow{\text{fast}} Fe(CN)_3^{4-} + Cu(CN)_3^{2-} + CNO^- + H_2O$$

The catalytic cycle is closed by the formation of the original cyano complex of copper(I):

$$Cu(CN)_3^{2-} + CN^- \xrightarrow{\text{fast}} Cu(CN)_4^{3-}$$

This mechanism is experimentally supported by the kinetic data obtained by its proponents.

Processes involving the oxidation of organic compounds are those most frequently used as indicator reactions in the determination of metal species on the basis of their catalytic effect. Among the variety of oxidants used (Fig. 2.9), hydrogen

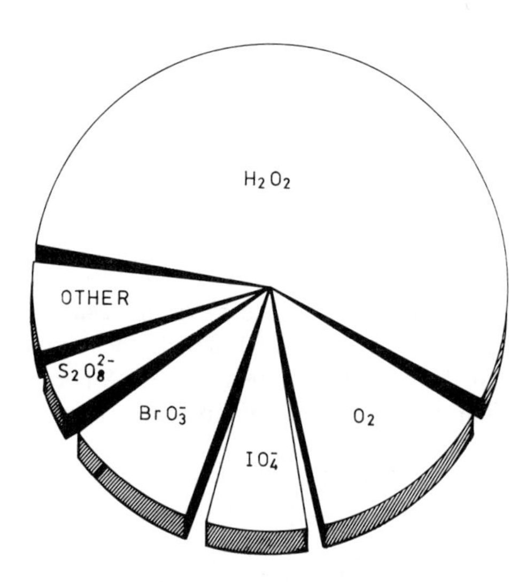

Fig. 2.9 — Relative use of different oxidants in the catalytic determination of Cu, Mn, Co and Fe with organic reagents.

peroxide is undoubtedly the commonest, followed by oxyanions such as BrO_3^-, IO_4^- or $S_2O_8^{2-}$ and by dissolved oxygen. Some of these oxidants are more or less 'specific' for certain metal ions. As shown in Fig. 2.10, which includes a number of ions of

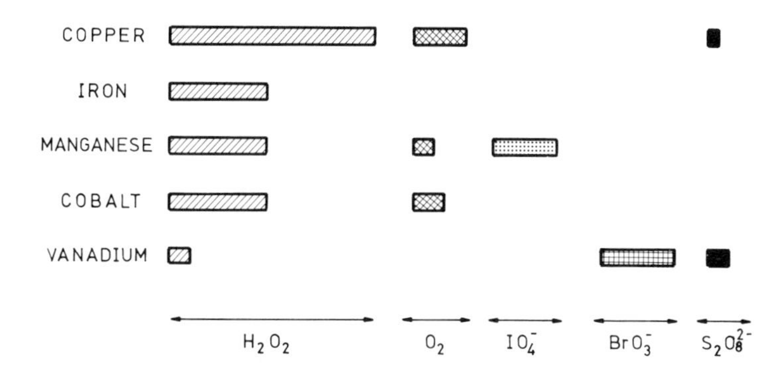

Fig. 2.10 — Diagram of the relative affinity (indicated by length of the rectangles) of the catalysts most commonly employed, for certain oxidants.

markedly catalytic character, periodate has so far only been used in manganese-catalysed reactions, and bromate is an efficient oxidant in catalytic processes involving vanadium. According to Bontchev [6], this behaviour is closely related to the relative redox potentials of the oxidant and the catalyst ($E_{M^{(n+1)}}/E_{M^{n+}}$). Thus, peroxodisulphate is capable of oxidizing silver to its bivalent state, so it is usually employed in Ag(I)-catalysed reactions. On the other hand, oxidants such as MnO_4^- or Ce(IV) are seldom utilized in kinetic catalytic methods owing to the relative slowness of their indicator reactions.

A study of Tables 2.2–2.9 allows some general observations to be made. Thus for detection, photometry occupies a prominent place as a result of the wide use of organic dyes as indicators. In some instances, neither the reactants nor the products of the indicator reaction have high enough absorptivity for use in monitoring the rate of the reaction. This drawback has been circumvented by Nakano *et al.* [38], who proposed combining the catalysed reaction with a second coupling reaction yielding a coloured compound. Thus, copper(II) catalyses the oxidation of *p*-hydrazinobenzenesulphonic acid by hydrogen peroxide to the *p*-sulphobenzenediazonium ion; the presence of *m*-phenylenediamine in the medium results in the formation of a yellow azo dye ($\lambda = 454$ nm) by means of which the course of the reaction is monitored. In a similar fashion the determination of vanadium or iron with *N*-phenyl-*p*-phenylenediamine/*N*,*N*-dimethylaniline is implemented by the use of potassium bromate or hydrogen peroxide, respectively, as oxidant [79,111].

As a rule, the organic compound participating in the indicator reaction is commercially available or has been synthesized. However, it is sometimes obtained *in situ* in the reaction medium. Thus, copper(II) catalyses the oxidation of Variamine Blue B by hydrogen peroxide [37], the leuco form of the dye being previously generated by reaction between *p*-anisidine and *N*,*N*-dimethylaniline. According to the proponents, the use of the commercially available base results in exceedingly high values of the blank reaction, probably arising from the oxidation products accumulating during the storage period.

Activators are frequently used in catalysed reactions in order to improve the detection limit. For example, pyridine is an efficient activator for copper(II), 1,10-phenanthroline for iron and nitrilotriacetic acid for manganese. The applications of this modifying effect are described in the next chapter, and its relevance to the selectivity of kinetic methods is discussed in Chapter 8.

Undoubtedly, the elucidation of the mechanism associated with catalysed reactions is a difficult task; sometimes it is only achieved by the use of very sophisticated instrumental techniques. The problem is not so serious when purely inorganic indicator reactions are concerned, thanks to the simplicity and knowledge of the species commonly involved. On the other hand, when organic compounds are involved in the indicator reaction, the researcher rarely proposes a mechanism, but concentrates instead on the identification of the reaction products. Thus, according to Yamane *et al.* [74], the vanadium-catalysed oxidation of chromotropic acid by bromate takes place through an intermediate complex formed between the acid and the metal. The participation of this hypothetical complex in the catalytic cycle is a credible supposition, since it has been shown that citrate and oxalate disfavour its formation and that these anions diminish the rate of the catalysed reaction.

As stated above, research in this field has been aimed at the identification of the reaction products. Thus, Valcárcel *et al.* [41] propose the following overall reaction for the oxidation of dipyridyl ketone hydrazone by dissolved oxygen, catalysed by Cu(II), Hg(II), Co(II) or Mn(II):

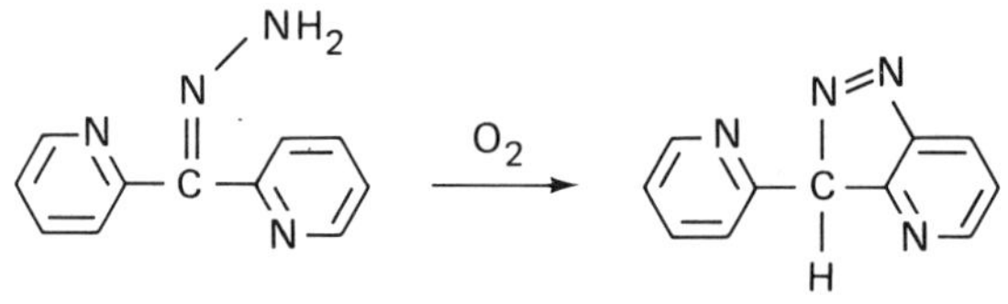

(λ_{exc} = 349 nm; λ_{em} = 435 nm)

The azo compound was isolated, and identified by mass spectrometry and infrared spectroscopy. According to these authors, the enhanced fluorescence of this compound in an acid medium can be attributed to the protection of the pyridine nitrogen atom and subsequent formation of an intramolecular hydrogen bond with one of the nitrogen atoms of the azo group, which would augment the planarity and rigidity (intrinsic characteristics of fluorescent substances) of the molecule.

2.6.1.2 *Determination of anions*

Determinations of inorganic ions by means of their catalytic effect on redox reactions are less common than those of metal ions. It can be safely stated that most such determinations involve the iodide ion, which shows a marked catalytic effect on a large number of reactions as a result of the redox characteristics of the iodine/iodide couple.

In this context three types of catalysed reactions are of note, namely: (a) the determination of iodide based on the reaction between Ce(IV) and As(III); (b) that of sulphur-containing compounds through their catalytic action on the iodine/sodium azide system, and (c) that of phosphates by their effect on the formation of molybdenum blue.

The catalytic effect of iodide on the redox reaction between Ce(IV) and As(III) was first shown and exploited for the determination of this halide and its parent halogen at the ng/ml level by Sandell and Kolthoff [3,4]. Several other authors later studied the effects of the reactant concentration ratio, the particular acid (nitric or sulphuric) used to adjust the pH, and the presence of chloride. Their conclusions are at variance, particularly as regards the acid to be used and the influence of chloride. In general, the reaction is performed in sulphuric acid medium, although Knapp and Spitzy [180] claim that the catalytic activity of iodide increases by a factor of twenty if nitric acid is used. As far as the effect of chloride ion is concerned, the findings of Sandell and Kolthoff [4] indicate that it inhibits the catalysed reaction at concentrations of the order of 0.34*M* or greater. Other authors, however, state that low concentrations of chloride increase the reaction rate [181,182] and even the linear range of the calibration curve. Rodríguez and Pardue [183] have published an interesting paper on the kinetics and mechanism of this reaction, based on initial rate data, which provide more comprehensive kinetic information.

The proposed mechanism, completely consistent with the experimental findings, is as follows:

$$I^- + Ce(IV) \xrightarrow{k_1} Ce(III) + I \qquad \text{(VIII)}$$

$$I + As(III) \underset{k_{-2}}{\overset{k_2}{\rightleftharpoons}} \text{Intermediate} \qquad \text{(IX)}$$

$$\text{Intermediate} + Ce(IV) \xrightarrow{k_3} As(V) + Ce(III) + I^- \qquad \text{(X)}$$

$$I + Ce(IV) \xrightarrow{k_4} Ce(III) + I^+ \qquad \text{(XI)}$$

$$I^+ + As(III) \underset{k_{-5}}{\overset{k_5}{\rightleftharpoons}} As(V) + I^- \qquad \text{(XII)}$$

It should be pointed out that reactions VIII–X and VIII, IX and XII present two different catalytic cycles which do not conform to the experimental data individually. As regards the reaction intermediate, this could be the species $I\text{–}As(OH)_2O$ detected in the reaction between iodine and arsenite at higher concentration levels, according to Schenk [184].

As can be seen in Table 2.10, the course of this reaction has been followed by different analytical techniques, photometry affording the lowest detection limit.

Iodide is not the only ion that can be determined on the basis of this reaction. In fact, various authors have proposed the determination of iodate [190,191] and even periodate [190], which requires only the prior conversion of the analyte into the catalytically active species, *viz*. iodide. The reduction is done by the arsenite itself on heating (for 30 min). This treatment allows the resolution of iodate–iodide mixtures, for which an automated procedure has been developed by Truesdale *et al*. [192,193]. They determined both species in water with the aid of a Technicon AutoAnalyzer. This reaction can also be used for the determination of iodine-containing biological

Table 2.10 — Determination of iodide with the Ce(IV)/As(III) system by different techniques

Detection technique	Determination range	Observations	References
Photometric (420 nm)	≈4–20 ng/ml	Range widened by the presence of Cl^-	[181]
Potentiometric	0.2–100 ng/ml	Measurement of the rate of change with a vitreous carbon electrode	[185]
Amperometric (dead-stop)	3.17–25.38 μg	Electrogeneration of Ce(IV)	[186]
Thermochemical	0.01–10μ*M*	Unsegmented flow method	[187]
Absorptiostat (415 nm)	11–110 ng/ml	Absorbance is kept constant by addition of Ce(IV)	[53]
Potentiostat	6–60 ng/ml	A fixed potential is kept constant by addition of Ce(IV)	[188]
Biamperostat	μg level		[189]

substances such as iodoproteins (thyroid hormones) [194–196] by simply subjecting the samples to a prior treatment aimed to release iodine.

As described in Chapter 9, the determination of iodide by the Sandell and Kolthoff method has been the basis of a large number of applications, to a variety of real samples.

The reaction between sodium azide and iodine

$$2NaN_3 + I_2 \rightarrow 3N_2 + 2NaI$$

is catalysed by substances containing sulphur in oxidation state −II and has been used for the determination of anions such as sulphide, thiocyanate and thiosulphate.

The determination of sulphide generally involves separating it by precipitation with zinc or cadmium and subsequently adding the ingredients of the indicator reaction. The process is usually monitored photometrically at 350 nm for 5 min, which allows sulphide to be determined in the range 2–25 ng/ml [197]. Alternatively, it can be followed by direct-injection enthalpimetry, on the basis of which sulphide can be determined in the range 0.02–0.5 μmole [198]. The dual photometric–thermometric indicator system proposed by Weisz *et al.* [199] is an interesting tool for monitoring this reaction, both in closed and in flow systems. In addition, it permits the determination of H_2S (5–100 μg/ml) in gases or when released from sodium sulphide solutions (0.1–10 μg/ml) with ascorbic acid.

Thiosulphate and thiocyanate are determined photometrically in the ranges 0.01–0.15 μg/ml [200] and 0.005–0.08 μg/ml [201], respectively. The above-mentioned dual detection system [199] allows thiosulphate to be determined in the range 32.4–324 μg/ml by observation of the turbidity caused by nitrogen formed in the reaction and the changes in temperature in a closed system, or in the range 112–1120 μg/ml in a flow system wherein sample and reagents are continuously mixed.

The analytical methods available for the determination of phosphate are usually based on the formation of 12-molybdophosphoric acid from phosphate and molybdate in an acid medium and the subsequent reduction to the blue heteropoly acid. In an interesting paper, Crouch and Malmstadt [202] reported an extensive study on the mechanism of this reaction. The first stage involves the formation of 12-molybdophosphoric acid (12-MPA), which (in nitric acid) can be formulated as:

$$H_3PO_4 + 6Mo(VI)_t \underset{k_{-1}}{\overset{k_1}{\rightleftharpoons}} 12\text{–}MPA + 9H^+$$

(note that this is not a balanced equation). The acid is subsequently reduced to the blue heteropoly acid (PMB),

$$12\text{-}MPA + n\ Red \overset{k_2}{\rightarrow} PMB + n\ Ox$$

where n is the number of moles of reductant (generally ascorbic acid) used. The rate of formation of this heteropoly acid, taking into account the steady-state condition for 12-MPA, can be expressed as:

$$\frac{d[PMB]}{dt} = \frac{k_1k_2[H_3PO_4][Mo(VI)]^6}{(k_{-1}[H^+]^9/k_2[Red]^n) + 1}$$

which is basically identical with that obtained when a sulphuric acid medium is used [202].

This reaction has been applied to the determination of phosphate and condensed phosphates by using ascorbic acid [203,204] or tin(II) [205] as reductant and making photometric measurements at 650 nm in the fixed-time kinetic mode. Phosphorus can thus be determined at the ng/ml level. The procedure has been automated by Crouch *et al.* [206–208], who applied it to the determination of phosphorus in samples of special interest, with the aid of various computer systems.

An interesting innovation in this regard lies in the work of Klockow *et al.* on the use of competitive reaction systems [209]. Essentially, the method involves coupling a competitive fast reaction to a slow catalysed reaction,

$$A + B \overset{k_I}{\rightarrow} P_I \qquad \text{(reaction of interest)}$$

$$B + R \overset{k_{II}}{\rightarrow} P_{II}$$

so that $k_{II} \gg k_I$. Thus, R (the 'competitor'), reacts with B as rapidly as it is introduced or generated in the system containing A and B. If the speed of addition of R is constant, then the time t_c required for the concentration of B to become zero (which should be detected with a suitable indicator system) will depend only on the rate of the slow reaction. The calibration graphs used are constructed by plotting t_c/t_R as a function of the catalyst concentration, where t_R is the time required for completion of the reaction between B and R in the absence of A.

For this particular case, the relevant reactions are as follows:

$$Mo(VI) + \text{ascorbic acid} \overset{H_2PO_4^-}{\rightarrow} Mo(V) + \text{dehydroascorbic acid}$$

$$I_2 + \text{ascorbic acid} \rightarrow 2I^- + \text{dehydroascorbic acid}$$

where molecular iodine acts as the competitor.

The joint use of this system and potentiometric detection allows phosphate to be determined in the range 0–1.2 μg/ml. In contrast to the usual kinetic methodology, this method does not require a fresh set of standards to be prepared for each series of samples, because any small change in the concentration of iodine or ascorbic acid will affect t_c and t_R to the same extent, so their ratio will remain virtually constant.

This methodology has also been applied to the determination of sulphur-containing compounds on the basis of the reaction between iodine and sodium azide and the same competitive reaction as above [209].

Tables 2.11 and 2.12 list other determinations of anions based on their catalytic effects on redox reactions. As can be seen, no less than twenty of these determinations involve iodide (Table 2.11).

The redox reactions on which these determinations rely, most often involve the oxidation of organic compounds by hydrogen peroxide or a chloramine derivative. Their mechanisms are rarely indicated, with the remarkable exception of the catalytic cycle for iodide in the redox reaction between hydrogen peroxide and chloramine-T. According to Koupparis and Hadjiioannou [222], the catalytic effect of iodide on this process arises from the formation of hypoiodite:

$$CH_3C_6H_4SO_2NCl^- + I^- + H_2O \rightarrow CH_3C_6H_4SO_2NH_2 + IO^- + Cl^-$$

$$H_2O_2 + IO^- \rightarrow H_2O + I^- + O_2$$

Of special interest in this context is the work of Hasty *et al.* on the elucidation of the behaviour of bromide in the decomposition of bromate in the presence of Methyl Orange [227].

At very low bromide concentrations the following reactions take place:

$$BrO_3^- + \text{Red} \xrightarrow{\text{slow}} Br^- + \text{Ox} \qquad \text{(XIII)}$$

$$BrO_3^- + 5Br^- + 6H^+ \xrightarrow{\text{fast}} 3Br_2 + 3H_2O \qquad \text{(XIV)}$$

$$Br_2 + \text{Red} \rightarrow \text{Ox} + 2Br^- \qquad \text{(XV)}$$

Thus, the initial attack of bromate on Methyl Orange (Red) gives rise to its colourless brominated product (Ox) and bromide, which in turn is rapidly converted into bromine, which also acts on the indicator. This is an autocatalytic process resulting in an induction period in the absorbance *vs.* time curves. Above a given bromide concentration only reactions (XIV) and (XV) take place.

Relatively high initial bromide concentrations result in the following reactions:

$$5Br^- + BrO_3^- + 6H^+ \rightarrow 3Br_2 + 3H_2O \qquad \text{(XVI)}$$

Table 2.11 — Kinetic catalytic methods for determination of iodide

Indicator reaction	Detection technique	Detection limit or range (rsd)	Observations	References
$FeSCN^{2+} + NO_2^-$	A ($\lambda = 460$ nm)	0.02–0.1 μg/ml	Different calibration curves depending on the I^- concentration	[210,211]
Ce(IV) + Sb(III)	A	< 1 μg/ml	Ferroin as indicator	[212]
As(III) + IO_3^-	Polarographic	5 ng/ml	Variable-time; polarographic monitoring of IO_3^-	[213]
$BrO_3^- + Br^-$	A	1–30 μg	Measurement of released iodine	[214]
	A ($\lambda = 490$ nm)	10 ng/ml		[215]
Catechol Violet + H_2O_2	A ($\lambda = 550$ nm)	< 0.1 μg/ml	Fixed time (8 min)	[216]
3,3′-Dimethylnaphthidine + H_2O_2		2–20 ng/ml	In the presence of formic acid	[217]
Triphenylmethane dyes + H_2O_2	A ($\lambda = 440$ nm)	10.9–1100 ng/ml	Sensitivity is increased by the presence of 0.2M Cl^-	[218]
Benzidine + H_2O_2	A	< 20 ng/ml	Fixed time (8 min)	[219]
o-Tolidine + H_2O_2	A	< 2 ng/ml	Fixed time (8 min)	[219]
o-Phenylenediamine + H_2O_2	A ($\lambda = 490$ nm)	20nM	Few interferences	[220]
Diphenylcarbazide + H_2O_2	A ($\lambda = 455$ nm)	90nM	Few interferences	[220]
Variamine Blue + H_2O_2	A ($\lambda = 580$ nm)	160nM	Few interferences	[220]
TNHS + H_2O_2	A ($\lambda = 549$ nm)	0.5–4 μg/ml	EDTA improves selectivity	[221]
Chloramine-T + H_2O_2	Potentiometric	1.5–9.0 μg	Chloramine-T-selective electrode	[222]
Chloramine-T + AsO_2^-	Potentiometric	1.5–30 μg (1.5%)	Variable time	[223]
Catechol Violet + Chloramine-B	A ($\lambda = 440$ nm)	1.3 ng/ml	Poor reproducibility	[224]
Bromopyrogallol Red + Chloramine-B	A ($\lambda = 530$ nm)	1.3 ng/ml	< 40% and < 20%, respectively	[224]
Malachite Green + Chloramine-R	Potentiometric		Induction period measurements	[225]
Malachite Green + IO_4^-		10–200 ng/ml	EDTA as masking agent	[54]
Hydrazine + IO_4^-	Potentiometric	$> 5 \times 10^{-5}M$	Iodide-selective electrode, Landolt reaction	[226]

Table 2.12— Kinetic catalytic methods for various anions

Anion	Indicator reaction	Detection technique	Detection limit or range (rsd)	Observations	References
Bromide	Decomposition of BrO_3^-	A ($\lambda = 510$ nm)	3 ng/ml	Autocatalysis	[227]
	$I^- + MnO_4^-$	A ($\lambda = 460$ nm)	> 0.04 μg/ml	Extraction of released I_2 and reaction with $Fe^{2+} + SCN^-$	[228]
	o-Phenylenediamine + H_2O_2	A ($\lambda = 440$ nm)	0.1–1 μg/ml ($< 8\%$)	Reaction-rate measurements	[229]
	Rhodamine 6Zh + Chloramine-B	Potentiometric	1.8×10^{-4} μg/ml	Induction period measurements	[230]
Cyanide	Reduction of *o*-dinitrobenzene	A	0.1–1.0 μg/ml		[231]
	PPOH + O_2	F ($\lambda = 350$, 420 nm)	3–180 ng/ml ($< 1\%$)	Interference only by Hg(II)	[232]
Oxalate	Azo dye + chromic acid	A ($\lambda = 510$ nm)	5–15 μg/ml	Data on mechanism	[233]
	$KI + K_2Cr_2O_7$	A ($\lambda = 582$ nm)	10^{-6}–$10^{-5}M$	In the presence of starch	[234]
Silicate	$I^- + MoO_4^{2-}$	A ($\lambda = 530$ nm)	0.2–1.0μM		[235]
Phosphate	$I^- + MoO_4^{2-}$		0.03 μg/ml		[236]
Germanate	$I^- + MoO_4^{2-}$		0.1 μg/ml		[236]
Chloride	Fe(II) + ClO_3^-	A	0.1 μg/ml (7.3%)	Fixed time	[237]
Thio-anions	Decomposition of BrO_3^-	A	5–10 ng/ml		[238]

PPOH: Pyridoxal-5-phosphate oxalyldihydrazone

$$3Br_2 + 3MO \rightarrow 3MOBr + 3Br^- + 3H^+ \qquad \text{(XVII)}$$

where MO denotes Methyl Orange and MOBr its colourless brominated derivative. The recycling of bromide between (XVII) and (XVI) in the presence of excess of bromate results in amplification of the effect of the former. Under these conditions, the decoloration of Methyl Orange rapidly reaches completion.

The catalytic action of oxalic acid on the oxidation of aromatic azo compounds with chromic acid is plausibly explained in a paper by Subba Rao *et al.* [233]. The catalytic effect of the acid is ascribed to the formation of a complex between it and chromium(VI) that increases the redox potential of the Cr(VI)/Cr(V) system. The suggested mechanism is as follows:

$$HCrO_4^- + (COOH)_2 \xrightarrow{K_e} \underset{(X)}{HO_2CCO_2CrO_3^-} + H_2O \text{ (fast)}$$

$$X + H^+ \underset{\text{equil.}}{\overset{\text{fast}}{\rightleftharpoons}} HX^+$$

$$X + \text{–N=N–} \underset{\text{slow}}{\overset{k}{\rightarrow}} (COOH)_2 + \text{Products}$$

$$HX^+ + \text{–N=N–} \underset{\text{slow}}{\overset{k'}{\rightarrow}} (COOH)_2 + \text{Products}$$

consistent with the following rate equation:

$$-\frac{d[\text{azo dye}]}{dt} = K_e[HCrO_4^-][(COOH)_2][\text{azo dye}]\ (k + k'[H^+])$$

Several interesting conclusions can be drawn from the material presented in Tables 2.11 and 2.12, *viz*. (a) fixed-time and initial-rate methods are the commonest kinetic methods used to monitor those reactions; (b) the photometric technique, which offers the best performance, is also the most frequently used; (c) the selectivity and detection limits afforded by these reactions are similar in many cases to those obtained with metal ions and practically independent of the type of indicator reaction (whether organic or inorganic) used.

2.6.1.3 Determination of organic compounds

Determinations of organic compounds based on catalysed reactions usually involve modified catalytic effects (see Chapters 3 and 4). In fact, as shown in the review on this topic by Milovanović [239], primary catalytic effects are rarely the basis for the catalytic determination of these species, most of which can act as bearers of inorganic catalysts.

Thus, the primary catalytic effect of these organic compounds involves two aspects.

Non-intrinsic catalysis. This is promoted by catalytically active inorganic components of the molecule, as is the case with the determination of organic iodine compounds by the Sandell and Kolthoff reaction [240,241] or that of sulphur-containing substances such as cysteine [240], vitamin B_1 [242], thiourea derivatives [243], etc., by use of the reaction between iodine and sodium azide. Cysteine can also be determined in the range 10^{-12}–10^{-11} g/l. through its catalytic action on the reaction between Ag(I) and Fe(II) [244]. Likewise, by an 'absorptiostat' method, Pantel [245] determined thiourea, *N*-methylthiourea, *N*-allylthiourea and thioacetamide in the range 0.1–1m*M* on the basis of their catalytic effect on the oxidation of Bromopyrogallol Red by hydrogen peroxide.

Intrinsic catalysis. This is the name given to the primary catalytic effect exerted by the organic compound as a whole, a representative example of which is the determination of phosphorus- [246] and nitrogen-containing compounds [247] with the *o*-dianisidine/H_2O_2 system. The oxidizing action of hydrogen peroxide on pyrecatechol is used to determine acetonitrile in the range 1.5×10^{-3}–$2.28 \times 10^{-2}M$, with a relative standard deviation of 6% [248]. Another typical indicator reaction in this context is that of Mo(VI) and hydrogen peroxide, which operates through formation of a catalytically active complex of the analyte with molybdate. This allows the determination of biologically active substances such as L-dopa, methyldopa, carbidopa, etc., in the range 0.4–17.1 μg/ml [249], as well as that of D(–)arabinose (46–135 μg/ml) by the tangent method, with photometric detection [250].

Some organic compounds are determined indirectly. Thus, Alekseeva *et al.* [251] determined primary and secondary amines by adding an excess of copper(II) and carbon disulphide to the sample and extracting the resulting Cu(II)bis(dithiocarbamate) complex into chloroform. The copper remaining in the aqueous phase was determined with the Pyrocatechol Violet/H_2O_2 system. This procedure allows primary and secondary amines to be distinguished from tertiary ones.

2.6.2 Chemiluminescence reactions

The analytical uses of chemiluminescence reactions in solution are of great interest in the determination of metal ions at very low concentrations. An interesting survey of chemiluminescence methods in analytical chemistry has been published [252,252a]. According to Mottola [253], determinations in this field are kinetic in nature and those of metal ions are sometimes classed as catalytic. This is a clear mistake, insofar as the 'catalyst' undergoes irreversible changes leading to inactive products in high oxidation states, and the catalytic cycle (if there is one) can hardly be visualized.

Below are described some of the most relevant applications involving these erroneously denominated catalytic reactions reported in the last few years, the best known of which is no doubt the oxidation of luminol (5-amino-2,3-dihydrophthalazine-1,4-dione). According to Guilbault [254], the mechanism by which the luminescence is generated is as follows:

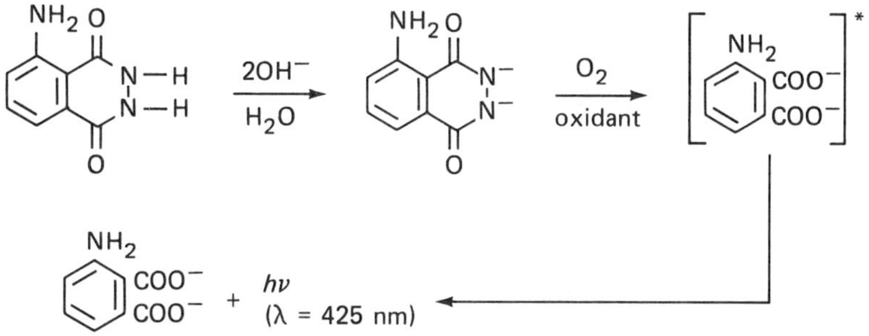

i.e. luminol is converted into a doubly charged anion which is subsequently oxidized to an excited singlet state that emits radiation on decomposing to form the aminophthalate ion.

This reaction has been applied to the determination of metal ions such as Cu(II), Ni(II), Cr(III), Mn(II), Fe(III), Co(II), etc., at the ng/ml or even lower levels, with relatively straightforward instrumentation. The lowest detection limit is achieved in the determination of Co(II) ($<10^{-11}M$).

Hydrogen peroxide is the oxidant most frequently used in this reaction. In the presence of excess of luminol it permits the determination of Cr(III) in the range 0.01–20 μg/ml [255], Fe(III) and Co(II), both individually and in mixtures (by masking iron with a suitable complexone) [256], Pt(IV) in the presence of Pt(II) and Pd(II) [257], as well as other ions, with detection limits ranging from 10^{-10} to $10^{-8}M$ [252,258]. Hydrogen peroxide itself can be determined by use of Cu(II) as a catalyst [259].

There are a great many literature references to this reaction. Thus, it has been used as an indicator in chromatographic processes aimed at determining the influence of various solvents and acids present in the separation process [260]. It has also been adapted to flow systems by Burguera *et al.* [261] who determined zinc and cadmium through their inhibitory effect on the chemiluminescence reaction between Co(II), luminol and H_2O_2. The problem posed by the injection of hydrogen peroxide into the reaction zone in a flow system is readily overcome by electrogenerating the oxidant, as sugested by Haapakka and Kankare [262]. The procedure has been satisfactorily applied to the determination of copper(II) in the range 10^{-7}–$10^{-5}M$, in the presence of glycine.

When the reaction is performed in the absence of hydrogen peroxide it is atmospheric oxygen itself which acts as the oxidant, as in the determination of copper (0.2–600 ng/ml) in the presence of cyanide as activator of the luminescence appearing at 450 nm [263]. In an interesting paper, Lukovskaya *et al.* [264] report the determination of tin with the luminol/oxygen system, on the basis of the reaction of the organic compound with superoxide radicals formed between Sn(II) and oxygen. The detection limit is of the order of 1 ng/ml and the determination is interfered with by the chloro-complexes of Ir(IV), V(V) and Au(III).

Other oxidants used include diperoxyadipic acid for the determination of nickel [265], peroxodisulphate for silver in the presence of amines [266] and periodate for ruthenium [267], iridium [268] and rhodium [269], with the possibility of determining

Ir(IV) in the presence of Rh(III) in the last case.

This special 'indicator' reaction can also be applied to the determination of anions: thus, chlorine dioxide and hypochlorite can be determined in aqueous solutions in the micro or submicromolar range by the pulse technique or in a continuous flow system [270]. An interesting application of this reaction is the determination of phosphorus, arsenic and silicon by reaction of their corresponding heteropoly acids with luminol [271,272]. Phosphorus and arsenic form ternary heteropoly acids with molybdovanadic acid, whereas silicon yields molybdosilicic acid. As a rule, these heteropoly acids are selectively extracted into suitable organic solvents (so that they can be determined in mixtures) and a portion of the extract is reacted with luminol. Detection limits of the order of a few ng/ml are readily achieved.

Luminol, undoubtedly the commonest reagent in chemiluminescence reactions, is followed in importance by lucigenin (*N,N'*-dimethylacridinium nitrate), which is usually employed in the presence of an oxidant such as hydrogen peroxide in alkaline medium, wherein it yields a greenish blue luminescence ($\lambda_{max} = 479$ nm).

Despite the divergences in the various interpretations of the mechanism by which the luminescence of this reaction is generated, Montano and Ingle [273] have compiled 16 widely accepted points on different aspects of this reaction.

As in the reaction between luminol and hydrogen peroxide, Co(II) is the metal ion affecting the luminescence generation to the greatest extent. A logarithmic calibration graph is usually used, and is linear from 20 pg/ml to 100 μg/ml. This procedure has been used for the determination of cobalt in real samples after its extraction to avoid the interference of iron and manganese [274].

This reaction has been applied to the determination of a number of metal ions, with detection limits of the order of a few ng/ml, with photometric detection in many cases [252,273].

The chemiluminescence reaction involving the oxidation of lofine (2,4,5-triphenylimidazole) was described by Radziszewski as far back as 1877. This reaction yields a green luminescence under conditions similar to those described above. The mechanism involved in the generation of the luminescence consists of three essential stages: (a) attack by hydrogen peroxide to form a hydroperoxide, (b) conversion into an intermediate dioxethane and (c) cleavage of the peroxide bond, with emission of radiation.

By varying the reactant concentrations in order to increase the selectivity, MacDonald *et al.* [275] have developed procedures for the determination of Co(II), Cr(III), Cu(II), Fe(II), etc., as well as of anions such as hypochlorite, with detection limits of about $10^{-6}M$.

In their search for chemiluminescence reactions to improve the selectivity afforded by luminol, Stieg and Nieman have used the oxidation of gallic acid with hydrogen peroxide in an alkaline medium [276]. This system emits radiation in two spectral regions, giving rise to an intense band at 643 nm and a weaker one at 478 nm. The emission is effected by a small number of inorganic species, a great asset compared to other luminescent systems previously described, though the sensitivity is somewhat decreased. Of all ions tested, it is again cobalt which provides the highest sensitivity, with detection limits of 0.4 ng/ml, whilst those of Mn(II), Ag(I), Cd(II) and Pb(II) range between 0.4 and 1.0 μg/ml [276].

Of the four systems described, that exploiting the luminol reaction offers the lowest detection limits, cobalt(II) being the metal ion exerting the most marked 'catalytic' effect [252]. Recent studies attribute the effect to the formation of a complex between the metal ion and H_2O_2, which is then responsible for the oxidation of the organic substrate to form a luminescent compound in a rate-determining step.

2.6.3 Ligand-exchange and complex-formation reactions

Catalytic methods (and kinetic methods in general) based on ligand-exchange reactions are quite recent and have thus been studied less. They have a promising future as they allow the determination of non-transition metals such as alkaline-earth metals (individually or in mixtures) as well as of lanthanides, ammonia and some other species.

2.6.3.1 Ligand-exchange reactions

Three general types of ligand-exchange reaction can be considered in this context, as follows.

Exchange of a ligand between two metal ions. Consider the system:

$$NiY^{2-} + Zn^{2+} \underset{k_{Ni}^{ZnY}}{\overset{k_{Zn}^{NiY}}{\rightleftharpoons}} ZnY^{2-} + Ni^{2+} \qquad \text{(XVIII)}$$

described by Margerum *et al.* [277]. This reaction moves slowly to the left and is monitored photometrically at 380 nm by measuring the disappearance of the NiY^{2-} complex. The addition of a small amount of Cu(II) to the system results in two further reactions

$$NiY^{2-} + Cu^{2+} \xrightarrow{k_{Cu}^{NiY}} CuY^{2-} + Ni^{2+}$$

$$CuY^{2-} + Zn^{2+} \xrightarrow{\text{fast}} ZnY^{2-} + Cu^{2+}$$

arising from the fact that the displacement brought about by this ion is 6000 times faster than that caused by Zn(II). The sum of both reactions is equivalent to (XVIII), so Cu(II) obviously acts as a catalyst. In fact, the rate of disappearance of the NiY^{2-} is proportional to the amount of copper added. The plot of the initial rate as a function of the copper concentration is a straight line which can be used as a calibration graph.

Exchange of a metal between two ligands. Consider the reaction between Cu(II) and the ligands ethyleneglycol(2-aminoethyl ether)tetra-acetic acid (EGTA) and 4-(2-pyridylazo)resorcinol (PAR),

$$\text{Cu–EGTA} + \text{PAR} \xrightarrow{Ca^{2+}} \text{Cu–PAR} + \text{EGTA}$$

where charges have been omitted for simplicity. This reaction is catalysed by traces of Ca(II); Mg(II) causes no interference, as its EGTA complex is much less stable than that formed with Ca(II). Calcium can thus be determined at concentrations between 0.4 and 40 μg/ml by monitoring (at 515 nm) the rate of appearance of the Cu–PAR complex [278].

This type of exchange reaction can also be catalysed by a unidentate ligand instead of a metal ion. Such is the case with the system:

$$\text{Hg–CPC} + \text{DCTA} \xrightarrow{NH_3} \text{Hg(II)–DCTA} + \text{CPC}$$

where CPC stands for 3,3-bis-*N*,*N*-di(carboxymethyl)aminoethyl-*o*-cresolphthalein and DCTA denotes diaminocyclohexanetetra-acetic acid. This reaction is catalysed by ammonia, which acts as a ligand-transfer agent, thereby accelerating the substitution reaction according to:

$$\text{Hg–CPC} + NH_3 \rightarrow \text{Hg–}NH_3 + \text{CPC}$$

$$\text{Hg–}NH_3 + \text{DCTA} \rightarrow \text{Hg–DCTA} + NH_3$$

The process is monitored photometrically at 580 nm by following the disappearance of the Hg–CPC complex and permits ammonia to be determined at concentrations above 1 μg/ml [279].

Dual exchange reactions. Olson and Margerum [280] have reported a reaction involving ligand and metal exchanges between the complexes Ni–triethylenetetramine and Cu–EDTA:

$$NiT^{2+} + CuY^{2-} \xrightarrow{\text{slow}} CuT^{2+} + NiY^{2-}$$

This reaction develops very slowly, though it can be accelerated by simply adding an excess of one of the ligands. Thus, a small excess of EDTA (Y^{4-}) attacks the Ni–trien complex to yield Ni–EDTA and free trien (T), which in turn reacts with the Cu–EDTA complex, from which it displaces the ligand:

$$Y^{4-} + NiT^{2+} \rightarrow NiY^{2-} + T$$

$$T + CuY^{2-} \rightarrow CuT^{2+} + Y^{4-}$$

The chain is propagated cyclically and a steady state is rapidly reached. The reaction rate is monitored photometrically by means of the absorbance at 550 nm, which corresponds exclusively to the Cu–trien complex and is proportional to the EDTA (or trien) concentration added. This allows the determination of small concentrations of either ligand, or even indirect determination of metal ions.

Other kinetic catalytic methods for determination of both metal ions and anions are listed in Table 2.13. They normally use photometric detection, and their indicator reactions involve cyano and aminocyano complexes of iron, as well as

Table 2.13 — Kinetic catalytic methods for determination of inorganic species based on ligand-exchange reactions

Species	Indicator reaction	Detection technique	Detection limit or range (rsd)	Observations	References
Mercury	$Fe(CN_6^{4-}$ + bipyridyl/1,10-phenanthroline	A (λ = 522 nm)	$10^{-5}M$	Interference by Ag(I) and Pd(II)	[281]
	$Fe(CN)_6^{4-}$ + *p*-nitrosodiphenylamine	A (λ = 640 nm)	$2 \times 10^{-8}M$	Fixed time (15 min)	[282]
	$[Fe(CN)_5NH_3]^{3-}$ + bipyridyl/1,10-phenanthroline	A	0–0.3 μg/ml (3%)	Interference by Ag(I) and Zn(II)	[283]
	$[Fe(CN)_5NH_3]^{3-}$ + ferrozine	A (λ = 562 nm)	0.01–0.4 μg/ml (2.7%)	Data on mechanism	[284]
Silver	$Fe(CN)_6^{4-}$ + bipyridyl	A (λ = 520 nm)	1.67μ*M*–0.3m*M*	Fixed time	[285,286]
	$[Fe(CN)_5NH_3]^{3-}$ + ferrozine	A (λ = 562 nm)	0.02–0.5 μg/ml (4.7%)		[284]
Gold	$Fe(CN)_6^{4-}$ + bipyridyl	A (λ = 520 nm)	16.5μ*M*–0.15m*M*	Fixed time	[286]
	$[Fe(CN)_5NH_3]^{3-}$ + ferrozine	A (λ = 562 nm)	0.1–3.0 μg/ml (4.3%)		[284]
Iodide	Hg–PAR + DCTA	A (λ = 500 nm)	$0–10^{-7}M$	No interference by Cl^- or Br^-	[287,288]

PAR: 4-(2-pyridylazo)resorcinol; DCTA: 1,2-diaminocyclohexanetetra-acetic acid.

ferroin-like ligands. Mercury(II) is by far the most frequently used catalyst, both in these ligand-exchange reactions and in direct complex-formation reactions (Table 2.14). This is in contrast with the catalytic effect of this ion on redox reactions (as shown in Fig. 2.8, which illustrates the different methods proposed for the determination of metal ions on the basis of the reactions described above).

2.6.3.2 Complex-formation reactions

Complex-formation reactions can be regarded as the exchange of a metal ion between two ligands inasmuch as in solution the metal ion is not in a free state, but is present as complexes with anions in the medium (e.g. chloride, acetate) or as hydrated species. However, the influence of these matrix 'ligands' will not be taken into account in dealing with this type of reaction.

Table 2.14 lists some of the kinetic catalytic methods available for the determination of inorganic species by means of complex-formation reactions.

The complex-formation reaction between Mn(II) and porphyrine derivatives has been used for the kinetic determination of mercury and cadmium, in one of the few references to the determination of the latter by a catalytic method. According to Tabata and Tanaka [290], the catalytic effect of Hg(II) on the formation of the complex between Mn(II) and α,β,γ,δ-tetraphenylporphyrinsulphonate (TPPS) can be described as follows:

$$\mathrm{Hg(II)} + \mathrm{TPPS} \overset{\text{fast}}{\rightleftharpoons} \mathrm{Hg(II)\text{–}TPPS}$$

$$\mathrm{Hg(II)\text{–}TPPS} + \mathrm{Mn(II)} \overset{\text{slow}}{\rightleftharpoons} \mathrm{Mn\text{–}TPPS} + \mathrm{Hg(II)}$$

The catalytic effect of anionic species on this type of reaction is closely related to the formation of the metal–catalyst complex from sub-stoichiometric amounts of the anion. The aquo-complexes of Cr(III) are reported to be kinetically very inert, especially the mixed complexes with chloride, etc., hence the slowness of the reactions involving this metal ion and a ligand, to the point of requiring some heating, as is the case with the Cr(III)–EDTA complex. The catalytic action of carbonate and/or bicarbonate on the formation of the complex between Cr(III) and Xylenol Orange or EDTA is probably related to the prior decomposition of the aquo-complexes. Other anions such as benzoate or *o*-chlorobenzoate also have a catalytic effect on the formation of the Cr(III)–EDTA complex [299].

The situation is very similar in the fluoride-catalysed formation of complexes between Zr(IV) and Xylenol Orange or Methylene Blue. In solution, zirconium is usually found in polymeric forms and this significantly reduces its complex-formation capability. The presence of fluoride in the medium gives rise to species such as ZrF^{3+}, ZrF_2^{2+}, etc., which allows the zirconium ion to react with either indicator [296,297]. Strictly speaking the fluoride ion catalyses the depolymerization of zirconyl ions.

Table 2.14 — Kinetic catalytic methods for determination of inorganic species, based on complex-formation reactions

Species	Indicator reaction	Detection technique	Detection limit or range (rsd)	Observations	References
Mercury	$Fe(CN)_6^{4-} + H_2O \rightarrow [Fe(CN)_5.H_2O]^{3-}$	*A*	8–800 ng/ml	Automated procedure	[289]
	Mn(II) + TPPS	*A* (λ = 413 nm)	1×10^{-8}–$12 \times 10^{-8} M$	Data on the separation of Hg by distillation	[290]
	Mn(II) + R	*A* (λ = 411 nm)	10n*M*	60°C; pH 6.9	[291]
Iron, copper	2,2′-Diquinoxalyl + Sn(II) or Ti(III)	*A* (λ = 685 nm)	0.04–0.35 μg/ml	Subsequent oxidation with H_2O_2	[292]
Cadmium	Mn(II) + TPPS	*A* (λ = 413 nm)	$10^{-7} M$	Pb is separated by co-precipitation	[293]
Carbonate or bicarbonate	Cr(III) + Xylenol Orange	*A* (λ = 550 nm)	1–10 μg/ml (18%)		[294]
	Cr(III) + EDTA	*A* (λ = 550 nm)	100–800 μg/ml	Variable time	[295]
Fluoride	Zr(IV) + Xylenol Orange	*A* (λ = 550 nm)	5–50 ng/ml	Other catalysts: PO_4^{3-}, AsO_4^{3-}, SO_4^{2-}, $C_2O_4^{2-}$ and citrate	[296]
	Zr(IV) + Methylthymol Blue	*A* (λ = 586 nm)	0.01–0.095 μg/ml	Interference by PO_4^{3-}, AsO_4^{3-} and SO_4^{2-}	[297]
Sulphate	Zr(IV) + Methylthymol Blue	*A* (λ = 586 nm)	0.1–2.4 μg/ml	Cationic interferences are removed by ion-exchange	[298]

TPPS: α,β,γ,δ-tetraphenylporphyrinsulphonate; R: tetrakis(4-trimethylaminophenyl)porphyrine iodide.

2.6.4 Other catalytic reactions

In this group are included a number of recently reported reactions not described in Bontchev's earlier review [6].

Gijsbers and Kloosterboer have developed a procedure for the determination of borate on the basis of its catalytic effect on the hydrolysis of *N*-nitrosohydroxylamine-*N*-sulphonate, which is monitored photometrically at 258 nm and allows boron to be determined at concentrations as low as 12 μg/ml [300]. The alkaline hydrolysis of 2-hydroxybenzaldehyde azine is catalysed by copper(II). This effect has allowed Valcárcel *et al.* [301] to determine this ion fluorimetrically ($\lambda_{ex} = 355$ nm, $\lambda_{em} = 465$ nm) in the range 100–500 ng/ml by monitoring the hydrazone formed, with very few interferences.

Ditzler *et al.* [302] have used the decomposition of benzil into benzaldehyde and methyl benzoate to develop a method for the determination of cyanide based on its catalytic effect on this reaction. The reaction is followed by injecting 1 μl of the reacting mixture into a gas chromatograph and monitoring the methyl benzoate formed. Cyanide can thus be determined in the range 0.05–3.0 μg/ml.

The hydration reaction of potassium ferrocyanide has allowed the determination of cobalt, based on its catalytic effect on this in the presence of bipyridyl as activator [303]. Finally it is interesting to note the catalytic effect of pyrophosphate on the precipitation of Ca(II) with $P_3O_{10}^{5-}$ [304]. When an NH_4Cl/NH_3 buffer is used, the calibration graph obtained is linear up to 15 μg/ml calcium.

REFERENCES

[1] A. Guyard, *Chem. Zentralbl.*, 1876, 10.
[2] G. Witz and F. Osmond, *Bull. Soc. Chim. France*, 1885, **45**, 309.
[3] E. B. Sandell and I. M. Kolthoff, *J. Am. Chem. Soc.*, 1934, **56**, 1426.
[4] E. B. Sandell and I. M. Kolthoff, *Mikrochim. Acta*, 1937, 9.
[5] H. A. Mottola, *Crit. Rev. Anal. Chem.*, 1975, **4**, 229
[6] P. R. Bontchev, *Talanta*, 1970, **17**, 499.
[7] H. Landolt, *Ber. Deut. Chem. Ges.*, 1886, **19**, 1317.
[8] E. K. Nikitin, *Izv. Sektora Fiz-Khim. Analiza*, 1940, **13**, 75.
[9] G. Svehla, *Analyst*, 1969, **94**, 513.
[10] H. Weisz and S. Pantel, *Anal. Chim. Acta*, 1975, **76**, 487.
[11] A. M. Zhabotniskii, *Zh. Analit. Khim.*, 1972, **27**, 437.
[12] K. B. Yatsimirskii, L. P. Tikhonova and L. N. Sakrevskaya, *Teor. Eksp. Khim.*, 1973, **9**, 805.
[13] J. D. Ingle, Jr. and S. R. Crouch, *Anal. Chem.*, 1971, **43**, 697.
[14] L. J. Papa, J. H. Patterson, H. B. Mark, Jr. and C. N. Reilley, *Anal. Chem.*, 1963, **35**, 1889.
[15] G. Svehla, *Sel. Ann. Rev. Anal. Sci.*, 1971, **1**, 235.
[16] G. G. Guilbault, in *Treatise on Analytical Chemistry*, I. M. Kolthoff and P. Elving (eds.), 2nd Ed., Part I, Vol. 1, Chapter 11, Wiley–Interscience, New York, 1978.
[17] H. Müller, *Crit. Rev. Anal. Chem.*, 1982, **13**, 313.
[18] M. Valcárcel and F. Grases, *Talanta*, 1983, **30**, 139.
[19] G. López-Cueto and J. A. Casado-Riobó, *Talanta*, 1979, **26**, 127.
[20] H. Weisz, S. Pantel and W. Meeners, *Anal. Chim. Acta*, 1976, **82**, 145.
[21] S. Pantel and H. Weisz, *Anal. Chim. Acta*, 1977, **89**, 47.
[22] G. López-Cueto and J. A. Casado-Riobó, *Talanta*, 1979, **26**, 151.
[23] D. P. Nikolelis and T. P. Hadjiioannou, *Mikrochim. Acta*, 1977 **I**, 124.
[24] V. I. Vershinin and G. L. Bukhbinder, *Chem. Abstr.*, 1980, **92**, 51398t.
[25] P. C. Nigam and V. N. Prasad, *Indian J. Chem.*, 1979, **18A**, 191.
[26] I. F. Dolmanova, O. I. Mel'nikova and T. N. Shekhovtsova, *Zh. Analit. Khim.*, 1978, **33**, 2096.
[27] I. F. Dolmanova and O. I. Mel'nikova, *Vestn. Mosk. Univ. Khim.*, 1978, **19**, 367.
[28] V. I. Martsokha, S. U. Kreingol'd and L. L. Soina, *Metody Anal. Lyumin. Mater. Veshchestv Osoboi Chist.*, 1975, **32**, 29.

[29] P. I. Kuznetsov, B. D. Luft, V. V. Shemet, V. N. Antonov, S. U. Kreingol'd, A. A. Panteleimonova and M. S. Chupakhin, *Zavodsk. Lab.*, 1976, **42**, 657.
[30] S. Gantcheva and P. R. Bontchev, *Talanta*, 1980, **27**, 893.
[31] G. A. Milovanović, L. Trifković and T. J. Janjić, *Bull. Soc. Chim. Beograd*, 1981, **46**, 285.
[32] A. A. Alexiev, P. R. Bontchev and S. Gantcheva, *Mikrochim. Acta*, 1976 **II**, 487.
[33] R. P. Igov, *Chem. Abstr.*, 1979, **91**, 150628d.
[34] D. Costache and G. Costache, *Chem. Abstr.*, 1982, **96**, 228123.
[35] R. P. Igov, M. D. Jaredic and T. G. Pecev, *Talanta*, 1980, **27**, 361.
[36] J. L. Ferrer-Herranz and D. Pérez-Bendito, *Anal. Chim. Acta*, 1981, **132**, 157.
[37] S. Nakano, K. Kuramoto and T. Kawashima, *Chem. Lett.*, 1980, 849.
[38] S. Nakano, H. Enoki and T. Kawashima, *Chem. Lett.*, 1980, 1173.
[39] S. Nakano, M. Tanaka, M. Fushihara and T. Kawashima, *Mikrochim. Acta*, 1983 **I**, 457.
[40] S. Nakano, M. Sakai, M. Tanaka and T. Kawashima, *Chem. Lett.*, 1979, 473.
[41] F. Grases, F. García-Sanchez and M. Valcárcel, *Anal. Chim. Acta*, 1980, **119**, 359.
[42] F. Lázaro, M. D. Luque de Castro and M. Valcárcel, *Analyst*, 1984, **109**, 333.
[43] F. Grases, R. Forteza, J. G. March and V. Cerdá, *Anal. Chim. Acta*, 1984, **158**, 389.
[44] F. Lázaro, M. D. Luque de Castro and M. Valcárcel, *Anal. Chim. Acta*, 1984, **165**, 177.
[45] F. Grases, F. García-Sánchez and M. Valcárcel, *Anal. Chim. Acta*, 1981, **125**, 21.
[46] M. Román Ceba, J. C. Jiménez–Sánchez and T. Galeano Díaz, *Quim. Anal.*, 1983, **2**, 199.
[47] F. Salinas López, J. J. Berzas Nevado and A. Espinosa Mansilla, *Talanta*, 1984, **31**, 325.
[48] S. Pantel, *Anal. Chim. Acta*, 1979, **104**, 205.
[49] R. Baranowski, I. Baranowska and Zb. Gregorowicz, *Microchem. J.*, 1979, **24**, 367.
[50] V. F. Toropova, M. G. Gadbullin, A. R. Garifzyanov and R. A. Cherkasov, *Zh. Analit. Kim.*, 1984, **36**, 267.
[51] R. P. Pantaler, L. D. Alfimova and H. M. Bulgakova, *Zh. Analit. Khim.*, 1975, **30**, 1834.
[52] H. A. Mottola and C. R. Harrison, *Talanta*, 1971, **18**, 683.
[53] H. Weisz and K. Rothmaier, *Anal. Chim. Acta*, 1975, **75**, 119.
[54] Z. Liu and Y. Hu, *Huaxue Shijie*, 1983, **24**, 45.
[55] S. U. Kreingol'd, I. M. Nelen and L. I. Sosenkova, *Zavodsk. Lab.*, 1984, **50**, No. 7, 14.
[56] A. A. Alexiev and K. L. Mutafchiev, *Anal. Lett.*, 1983, **16**, 769.
[57] P. Bartkus, A. Nauekaitis and E. Jasinskiene, *Chem. Abstr.*, 1976, **85**, 28237m.
[58] C. E. Efstathiou and T. P. Hadjiioannou, *Talanta*, 1977, **24**, 270.
[59] C. E. Efstathiou and T. P. Hadjiioannou, *Anal. Chem.*, 1977, **49**, 414.
[60] S. Abe, K. Takahashi and T. Matsuo, *Anal. Chim. Acta*, 1975, **80**, 135.
[61] T. Yamane and T. Fukasawa, *Bunseki Kagaku*, 1977, **26**, 300.
[62] M. Otto, J. Rentsch and G. Werner, *Anal. Chim. Acta*, 1983, **147**, 267.
[63] M. A. Sekheta, G. A. Milovanović and T. J. Janjić, *Mikrochim. Acta*, 1978 **I**, 297.
[64] I. F. Dolmanova, G. A. Zolotova and M. A. Ratina, *Zh. Analit. Khim.*, 1978, **33**, 1356.
[65] A. Ya. Sychev and V. G. Isak, *Chem. Abstr.*, 1980, **92**, 98967a.
[66] A. Moreno, M. Silva, D. Pérez-Bendito and M. Valcárcel, *Talanta*, 1983, **30**, 107.
[67] D. Pérez-Bendito, J. Peinado and F. Toribio, *Analyst*, 1984, **109**, 1297.
[68] S. Rubio, A. Gómez-Hens and M. Valcárcel, *Analyst*, 1984, **109**, 717.
[69] D. Pérez-Bendito, M. Valcárcel, M. Ternero and F. Pino, *Anal. Chim. Acta*, 1977, **94**, 405.
[70] Yu R. Zeng and R. Lin, *Huaxue Xuebao*, 1983, **41**, 960.
[71] E. L. Dickson and G. Svehla, *Microchem. J.*, 1979, **24**, 509.
[72] M. Kataoka, S. Miyagata and T. Kambara, *Nippon Kagaku Kaishi*, 1980, 1520.
[73] T. Yamane and F. Tsutoma, *Bunseki Kagaku*, 1976, **25**, 454.
[74] T. Yamane, T. Suzuki and T. Mukoyama, *Anal. Chim. Acta*, 1974, **70**, 77.
[75] K. Hirayama and N. Unohara, *Bunseki Kagaku*, 1980, **29**, 733.
[76] S. U. Kreingol'd, E. M. Yutal, I. E. Pokrovskaya and Yu. A. Ivanov, *Zavodsk. Lab.*, 1981, **47**, No. 5, 17.
[77] A. Sevillano-Cabeza, J. Medina-Escriche and F. Bosch-Reig, *Analyst*, 1984, **109**, 1559.
[78] M. Hernández–Córdoba, P. Viñas and C. Sánchez-Pedreño, *Analyst*, 1985, **110**, 1343.
[79] S. Nakano, E. Kasahara, M. Tanaka and T. Kawashima, *Chem. Lett.*, 1981, 507.
[80] T. Kawashima, T. Karasumaru, S. Hashimoto and S. Nakano, *Nippon Kagaku Kaishi*, 1981, 175.
[81] K. Hiraki, N. Shimizu, Y. Nishikawa and T. Shigematsu, *Bunseki Kagaku*, 1981, **30**, 780.
[82] M. Otto, J. Stach and R. Kirmse, *Anal. Chim. Acta*, 1983, **147**, 277.
[83] Quiang Weiguo, *Anal. Chem.*, 1983, **55**, 2043.
[84] M. Kataoka, M. Takahashi and T. Kambara, *Bunseki Kagaku*, 1979, **28**, 169.
[85] M. Otto, G. Schöbel and G. Werner, *Anal. Chim. Acta*, 1983, **147**, 287.
[86] F. García-Sánchez, A. Navas, M. Santiago and F. Grases, *Talanta*, 1981, **28**, 833.
[87] R. P. Pantaler, L. D. Alfimova, A. M. Balgakova and I. V. Pulyaeva, *Zh. Analit. Khim.*, 1975, **30**, 946.

[88] H. Hirayama and N. Unohara, *Nippon Kagaku Kaishi*, 1978, 1498.
[89] R. P. Igov, M. D. Jaredic and T. G. Pecev, *Bull. Soc. Chim. Beograd*, 1980, **45**, 365.
[90] R. P. Igov and M. D. Jaredic, *Chem. Abstr.*, 1980, **93**, 87938d.
[91] A. A. Alexiev and M. G. Angelova, *Mikrochim. Acta*, 1980 **II**, 187.
[92] Gh. Ionescu, A. Duca and F. Matei, *Mikrochim. Acta*, 1980 **II**, 329.
[93] A. I. Merkulov and R. I. Ksvortsova, *Zh. Analit. Khim.*, 1981, **36**, 1778.
[94] T. Yamane, *Anal. Chim. Acta*, 1981, **130**, 65.
[95] V. K. Zinchuk, V. S. Besidka, Ya. P. Skorobogatyi and R. F. Markovskaya, *Zh. Analit. Khim.*, 1981, **36**, 701.
[96] P. C. Nigam and R. D. Srivastava, *Indian J. Chem.*, 1980, **19A**, 563.
[97] A. Duca, F. Matei and G. Ionescu, *Talanta*, 1980, **27**, 917.
[98] F. Grases, J. March, R. Forteza and V. Cerdá, *Thermochim. Acta*, 1984, **73**, 181.
[99] F. Matei, G. Ionescu and A. Duca, *Rev. Roum. Chim.*, 1979, **24**, 99.
[100] F. Grases, F. García-Sánchez and M. Valcárcel, *Ann. Quim.*, 1980, **76B**, 402.
[101] M. Kataoka, Y. Yoshizawa and T. Kambara, *Bunseki Kagaku*, 1982, **31**, E171.
[102] K. Okashi, T. Sagawa, E. Goto and Y. Kamamoto, *Anal. Chim. Acta*, 1977, **92**, 209.
[103] C. Papadppoulos, V. Vasilidis and G. S. Vasilikiotis, *Microchem. J.*, 1979, **24**, 23.
[104] H. Schuring and H. Müller, *Acta Hydrochim. Hydrobiol.*, 1979, **7**, 281.
[105] A. P. Gumenyuk and S. P. Mushtakova, *Zh. Analit. Khim.*, 1984, **39**, 1278.
[106] M. A. Ditzlev and W. F. Gutknecht, *Anal. Chem.*, 1980, **52**, 614.
[107] A. Moreno, M. Silva, D. Pérez-Bendito and M. Valcárcel, *Anal. Chim. Acta*, 1984, **157**, 333.
[108] S. Rubio, A. Gómez-Hewns and M. Valcárcel, *Anal. Chem.*, 1984, **56**, 1417.
[109] T. Kawashima, Y. Kozuma and S. Nakano, *Anal. Chim. Acta*, 1979, **106**, 355.
[110] T. Kawashima, N. Hatakeyama, M. Kamada and S. Nakano, *Nippon Kagaku Kaishi*, 1981, 84.
[111] S. Nakano, M. Tanaka and T. Kawashima, *Mikrochim. Acta*, 1983 **I**, 403.
[112] N. V. Trofimov, V. F. Akashina, N. A. Kanev, A. I. Busev and S. V. Zolotova, *Zavodsk. Lab.*, 1975, **41**, 1177.
[113] T. Pérez-Ruiz, C. Martínez-Lozano and V. Tomás, *Analyst*, 1984, **109**, 1401.
[114] B. F. Quin and P. H. Woods, *Analyst*, 1979, **104**, 552.
[115] M. Kataoka, K. Nishimura and T. Kambara, *Talanta*, 1983, **30**, 941.
[116] C. M. Wolf and J. P. Schwing, *Bull. Soc. Chim. France*, 1976, 675.
[117] M. Kataoka, H. Hemmi and T. Kambara, *Bull. Chem. Soc. Japan*, 1984, **57**, 1083.
[118] M. Kataoka and T. Kambara, *Chem. Abstr.*, 1978, **88**, 83145s.
[119] M. Otto and H. Müller, *Talanta*, 1977, **24**, 15.
[120] S. U. Kreingol'd and A. N. Vasnev, *Zavodsk. Lab.*, 1978, **44**, 265.
[121] Yu. K. Shazzo and I. P. Kharlamov, *Zavodsk. Lab.*, 1983, **49**, No. 2, 17.
[122] G. D. Christian and G. J. Patriarche, *Anal. Lett.*, 1979, **12**, 11.
[123] R. N. Voevutskaya, V. K. Pavlova and A. T. Pilipenko, *Zh. Analit. Khim.*, 1979, **34**, 1299.
[124] V. K. Pavlova, A. T. Pilipenko and R. N. Voevutskaya, *Zh. Analit. Khim.*, 1975, **30**, 2190.
[125] W. A. De Oliveira and J. Meditsch, *Rev. Quim. Ind.*, 1976, **45**, 525.
[126] R. L. Wilson and J. D. Ingle, Jr., *Anal. Chem.*, 1977, **49**, 1066.
[127] P. V. Subba Rao, P. S. N. Murty, R. V. S. Murty and B. A. N. Murty, *J. Indian Chem Soc.*, 1978, **55**, 1280.
[128] L. M. Matat, I. B. Mizetskaya, V. K. Pavlova and A. T. Pilipenko, *Zh. Analit. Khim.*, 1982, **37**, 2165.
[129] S. B. Jonnalagadda, *Anal. Chim. Acta*, 1982, **144**, 245.
[130] M. A. Cejas, A. Gómez-Hens and M. Valcárcel, *Mikrochim. Acta*, 1984 **III**, 349.
[131] Z. Zhang, Z. Li, W. Gan and L. Liu, *Fenxi Huaxue*, 1984, **12**, 433.
[132] O. A. Bilenko, N. B. Potekhina and S. P. Mushtakova, *Zh. Analit. Khim.*, 1984, **39**, 804.
[133] Z. Ma, X. Zhu, Y. Tian, H. Ye and X. Han, *Fenxi Huaxue*, 1983, **11**, 759.
[134] F. Grases, F. García-Sánchez and M. Valcárcel, *Anal. Lett.*, 1879, **12**, 803.
[135] F. Grases, J. M. Estela, F. García-Sánchez and M. Valcárcel, *Analusis*, 1981, **9**, 66.
[136] A. Navas and F. Sánchez-Rojas, *Quim. Anal.*, 1983, **2**, 112.
[137] V. V. S. E. Dutt and H. A. Mottola, *Anal. Chem.*, 1976, **48**, 80.
[138] C. Sánchez-Pedreño, M. Hernández–Córdoba and G. Martínez-Tudela, *Anal. Quim.*, 1979, **75**, 536.
[139] C. Sánchez-Pedreño, M. J. Albero-Quinto and M. Hernández- Córdoba, *Afinidad*, 1980, **37**, 313.
[140] C. Sánchez-Pedreño, M. J. Albero-Quinto and M. S. García-García, *Quim. Anal.*, 1985, **4**, 168.
[141] E. G. Khomutova, N. A. Khvorostukhina and I. A. Moskvina, *Zh. Analit. Khim.*, 1983, **38**, 170.
[142] I. N. C. Ling and G. Svehla, *Talanta*, 1984, **61**, 31.
[143] H. Mueller and M. Otto, *Mikrochim. Acta*, 1975 **I**, 519.
[144] N. V. Rao and P. V. Ramana, *Mikrochim. Acta*, 1981 **II**, 269.

[145] A. P. Rysev, L. P. Zhitenko and V. A. Nadezhdina, *Zavodsk. Lab.*, 1981, **47**, No. 6, 20.
[146] R. P. Morozova, L. P. Nischenkova and L. P. Blinova, *Zh. Analit. Khim.*, 1981, **36**, 2356.
[147] Z. Gregorowicz, T. Suwinska and D. Matysek-Majewska, *Chem. Abstr.*, 1980, **92**, 87479j.
[148] N. N. Gusakova and S. P. Mushtakova, *Zh. Analit. Khim.*, 1981, **36**, 317.
[149] R. P. Morozova, M. P. Volynets and M. A. Ginzburg, *Zh. Analit. Khim.*, 1975, **30**, 1836.
[150] L. P. Tikhonova, S. N. Borkovets and L. N. Revenko, *Anal. Abstr.*, 1977, **32**, 205.
[151] A. T. Pilipenko, T. S. Maksimenko and N. M. Lukovskaya, *Zh. Analit. Khim.*, 1979, **34**, 523.
[152] I. I. Alekseeva, G. N. Latysheva, L. E. Romanovskaya and L. P. Tikhonova, *Zavodsk. Lab.*, 1984, **50**, No. 3, 5.
[153] D. P. Nikolelis and T. P. Hadjiioannou, *Mikrochim. Acta*, 1978 **I**, 383.
[154] N. N. Gusakova, S. P. Mashtakova and N. S. Frumina, *Zh. Analit. Khim.*, 1979, **34**, 2213.
[155] F. Grases, J. M. Estela, F. García-Sánchez and M. Valcárcel, *Anal. Lett.*, 1980, **13**, 181.
[156] E. S. Akberdina, L. G. Pavlova and E. F. Speranskaya, *Chem. Abstr.*, 1980, **92**, 121205d.
[157] C. E. Efstathiou and T. P. Hadjiioannou, *Anal. Chim. Acta*, 1977, **89**, 391.
[158] S. Nakano, S. Hinokuma and T. Kawashima, *Chem. Lett.*, 1983, 357.
[159] A. Vanni and P. Amico, *Ann. Chim. (Rome)*, 1977, **67**, 321.
[160] H. Müller, J. Mattusch and G. Werner, *Mikrochim. Acta*, 1980 **II**, 349.
[161] Ya. P. Skoroboyatyi and V. K. Zinchuk, *Zh. Analit. Khim.*, 1978, **33**, 1587.
[162] O. I. Mel'nikova, T. N. Shekhovtsova and I. F. Dolmanova, *Zh. Analit. Khim.*, 1980, **35**, 1960.
[163] M. A. López-Fernández, A. Gómez-Hens and D. Pérez-Bendito, *Anal. Lett.*, 1984, **17**, 507.
[164] R. G. Anderson and B. C. Brown, *Talanta*, 1981, **28**, 365.
[165] S. Rubio, A. Gómez-Hens and M. Valcárcel, *Analyst*, 1984, **109**, 597.
[166] T. E. Gaytán-Placeres, A. Carta-Fuentes and A. Yu. Zhukov, *Chem. Abstr.*, 1981, **94**, 10670y.
[167] V. K. Rudenko, *Chem. Abstr.*, 1983, **98**, 190912t.
[168] R. H. Dinius and J. M. Baker, *Microchem. J.*, 1980, **25**, 209.
[169] L. Lan and Y. Hu, *Fenxi Huaxue*, 1984, **12**, 118.
[170] V. I. Rigin and G. I. Alekseeva, *Zh. Analit. Khim.*, 1975, **30**, 2372.
[171] K. Kuroda, T. Salto, T. Kiuchi and K. Oguma, *Z. Anal. Chem.*, 1975, **29**, 277.
[172] E. P. Diamandis and T. P. Hadjiioannou, *Anal. Chim. Acta*, 1981, **123**, 143.
[173] T. Kawashima, S. Kai and S. Takashima, *Anal. Chim. Acta*, 1977, **89**, 65.
[174] V. K. Rudenko and L. T. Zhukova, *Zh. Analit. Khim.*, 1979, **34**, 605.
[175] M. D. Luque de Castro and M. Valcárcel, *Talanta*, 1980, **27**, 645.
[176] I. F. Dolmanova, I. E. Degtereva, I. A. Il'icheva and L. A. Petrukhina, *Zh. Analit. Khim.*, 1978, **33**, 1779.
[177] I. I. Alekseeva, V. V. Borisova, A. V. Shuginina and L. T. Yuranova, *Zh. Analit. Khim.*, 1981, **36**, 108.
[178] I. I. Alekseeva, K. F. Kharitonovich and L. G. Morozova, *Anal. Abstr.*, 1977, **32**, 5B157.
[179] S. U. Kreingol'd and A. N. Vasnev, *Zavodsk. Lab.*, 1979, **45**, 481.
[180] G. Knapp and H. Spitzy, *Talanta*, 1969, **16**, 1353.
[181] A. Lein and N. Schwartz, *Anal. Chem.*, 1951, **23**, 1507.
[182] J. Deman, *Mikrochim. Acta*, 1964, 67.
[183] P. A. Rodríguez and H. L. Pardue, *Anal. Chem.*, 1969, **41**, 1369.
[184] G. H. Schenk, *J. Chem. Ed.*, 1964, **41**, 32.
[185] W. Qingzhang and C. Dechang, *Fenxi Huaxue*, 1981, **9**, 686.
[186] W. Jedrzejewski and W. Ciesielski, *Chem. Anal. Warsaw*, 1981, **26**, 437.
[187] J. M. Elvecrog and P. W. Carr, *Anal. Chim. Acta*, 1980, **121**, 135.
[188] H. Weisz, K. Rothmaier and H. Ludwig, *Anal. Chim. Acta*, 1974, **73**, 224.
[189] S. Pantel and H. Weisz, *Anal. Chim. Acta*, 1974, **70**, 391.
[190] A. Appleby and R. E. Spillet, *Rep. U.K. Atom. Energy Auth.*, 1967, **9**, RCC-M 210.
[191] P. A. Rodríguez and H. L. Pardue, *Anal. Chem.*, 1969, **41**, 1376.
[192] V. W. Truesdale and P. J. Smith, *Analyst*, 1975, **100**, 111.
[193] S. D. Jones, C. P. Spencer and V. W. Truesdale, *Analyst*, 1982, **107**, 1417.
[194] G. Palumbo, M. F. Tecce and G. Ambrosio, *Anal. Biochem.*, 1982, **123**, 1983.
[195] G. Knapp and H. Leopold, *Anal. Chem.*, 1974, **46**, 719.
[196] M. Timotheou-Potamia, E. G. Sarantonis, A. C. Calokerinos and T. P. Hadjiioannou, *Anal. Chim. Acta*, 1985, **171**, 363.
[197] A. Sakuragawa, T. Harada, T. Okutani and S. Utsumi, *Bunseki Kagaku*, 1980, **29**, 264.
[198] N. Kiba, M. Nishijima and F. Furusawa, *Talanta*, 1980, **27**, 1090.
[199] H. Weisz, W. Meiners and G. Fritz, *Anal. Chim. Acta*, 1979, **107**, 301.
[200] S. Utsumi and T. Okutani, *Nippon Kagaku Kaishi*, 1973, 75.
[201] S. Utsumi, T. Okutani and T. Yamada, *Bunseki Kagaku*, 1975, **24**, 799.
[202] S. R. Crouch and H. V. Malmstadt, *Anal. Chem.*, 1967, **39**, 1084.

[203] R. Szczesny, *Chem. Anal. Warsaw*, 1970, **15**, 157.
[204] E. E. Kriss, V. K. Rudenko and G. T. Kurbatova, *Zh. Analit. Khim.*, 1971, **26**, 1000.
[205] E. E. Kriss, V. K. Rudenko, K. B. Yatsimirskii and V. I. Vershinin, *Zh. Analit. Khim.*, 1970, **25**, 1603.
[206] S. R. Crouch and H. V. Malmstadt, *Anal. Chem.*, 1967, **39**, 1090.
[207] E. M. Cordos, S. R. Crouch and H. V. Malmstadt, *Anal. Chem.*, 1968, **40**, 1812.
[208] S. R. Crouch, *Anal. Chem.*, 1969, **41**, 880.
[209] D. Klockow, G. F. Graf and J. Auffarth, *Talanta*, 1979, **26**, 733.
[210] I. Iwasaki, S. Utsumi and T. Ozawa, *Bull. Chem. Soc. Japan*, 1953, **26**, 108.
[211] J. M. Ottaway, C. W. Fuller and W. B. Rowston, *Anal. Chim. Acta*, 1969, **45**, 541.
[212] J. Bognár and S. Sárosi, *Mikrochim. Acta.*, 1965, 1004.
[213] V. F. Toropova and L. M. Tamarchenko, *Zh. Analit. Khim.*, 1967, **22**, 234.
[214] H. Weisz and K. Rothmaier, *Anal. Chim. Acta*, 1974, **68**, 93.
[215] V. F. Toropova and L. M. Tamarchenko, *Zh. Analit. Khim.*, 1967, **22**, 576.
[216] O. Koichiro, K. Taeko, M. Hideo and K. Yahei, *Bunseki Kagaku*, 1968, **17**, 805.
[217] J. Bognár and L. Nagy, *Mikrochim. Acta*, 1969, 108.
[218] E. I. Yasinskene and O. P. Umbrazhyunaite, *Zh. Analit. Khim.*, 1973, **28**, 2025.
[219] E. I. Yasinskene and O. P. Umbrazhyunaite, *Zh. Analit. Khim.*, 1975, **30**, 1590.
[220] S. U. Kreingol'd, L. I. Sosenkova, A. A. Panteleimonova and L. V. Lavrelashvili, *Zh. Analit. Khim.*, 1978, **33**, 2168.
[221] R. P. Igov, M. D. Jaredic and T. G. Pecev, *Mikrochim. Acta*, 1979 **II**, 171.
[222] M. A. Kouparis and T. P. Hadjiioannou, *Mikrochim. Acta*, 1978 **II**, 267.
[223] M. A. Kouparis and T. P. Hadjiioannou, *Anal. Chim. Acta*, 1978, **96**, 31.
[224] E. I. Yasinskene and O. P. Umbrazhyunaite, *Zh. Analit. Khim.*, 1975, **30**, 962.
[225] J. Barkauskas and R. Ramanauskas, *Chem. Abstr.*, 1982, **97**, 103617p.
[226] R. A. Hasty, *Mikrochim. Acta*, 1973, 925.
[227] R. A. Hasty, F. J. Lima and J. M. Ottaway, *Analyst*, 1981, **106**, 76.
[228] K. Takahashi, M. Yoshida, T. Ozawa and I. Iwasaki, *Bull. Chem. Soc. Japan*, 1970, **43**, 3159.
[229] S. U. Kreingol'd, L. V. Labvrelashvili, I. M. Nelen and E. M. Yutal, *Zh. Analit. Khim.*, 1981, **36**, 303.
[230] J. Barkauskas, *Chem. Abstr.*, 1981, **94**, 95275w.
[231] V. G. Badding and J. L. Durney, *Plat. Surf. Finish*, 1980, **67**, 49.
[232] S. Rubio, A. Gómez-Hens and M. Valcárcel, *Talanta*, 1984, **31**, 783.
[233] P. V. Subba Rao, K. S. Murty, P. S. N. Murty and R. V. S. Murty, *J. Indian Chem. Soc.*, 1979, **56**, 604.
[234] N. M. Ushakova, *Chem. Abstr.*, 1981, **94**, 149768e.
[235] R. P. Morozova and L. V. Il'enko, *Zh. Analit. Khim.*, 1973, **28**, 1835.
[236] I. I. Alekseeva and I. I. Nemzer, *Zh. Analit. Khim.*, 1970, **25**, 1118.
[237] Z. Guang-Yu and P. Zhomg-Hua, *Ti Ch'iu Hua Hsueh*, 1979, 353.
[238] L. M. Tamarchenko, *Zh. Analit. Khim.*, 1978, **33**, 824.
[239] G. A. Milovanović, *Microchem. J.*, 1983, **28**, 437.
[240] S. Pantel, *Anal. Chim. Acta*, 1982, **141**, 353.
[241] S. Pantel and H. Weisz, *Anal. Chim. Acta*, 1977, **89**, 47.
[242] T. Pérez-Ruiz, C. Martínez-Lozano and M. Hernández-Córdoba, *Ann. Quim.*, 1982, **78B**, 241.
[243] S. Pantel, *Anal. Chim. Acta*, 1983, **152**, 215.
[244] A. K. Babko, L. V. Markova and T. S. Maksiemenko, *Zh. Analit. Khim.*, 1968, **33**, 1268.
[245] S. Pantel, *Anal. Chim. Acta*, 1984, **158**, 85.
[246] G. Shapenova, *Chem. Abstr.*, 1981, **95**, 138268m.
[247] I. F. Dolmanova, I. M. Popova, T. N. Sheknovtsova and N. N. Ugarova, *Zh. Analit. Khim.*, 1981, **36**, 673.
[248] G. A. Milovanović and N. A. Bozilović, *Microchem. J.*, 1982, **27**, 345.
[249] G. A. Milovanović, M. A. Sekheta and T. J. Janjić, *Mikrochim. Acta*, 1981 **I**, 241.
[250] G. A. Milovanović, M. A. Sekheta and I. M. Petrović, *Microchem. J.*, 1982, **27**, 135.
[251] I. I. Alekseeva, L. P. Ruzinov, E. G. Khachaturyan and L. P. Chernyshova, *Chem. Abstr.*, 1980, **92**, 65329h.
[252] J. L. Bernal and V. Cerdá, *Quim. Anal.*, 1985, **4**, 205.
[252a] L. J. Kricka and G. H. G. Thorpe, *Analyst*, 1983, **108**, 1274.
[253] H. A. Mottola and H. B. Mark, Jr., *Anal. Chem.*, 1982, **54**, 62R.
[254] G. G. Guilbault, *Practical Fluorescence: Theory, Methods and Techniques*, Dekker, New York, 1973, Chapter 9.
[255] V. I. Rigin and A. S. Bakhmurov, *Zh. Analit. Khim.*, 1976, **31**, 93.
[256] V. I. Rigin and A. I. Blonkhin, *Zh. Analit. Khim.*, 1977, **3**, 312.
[257] V. I. Rigin, A. S. Bakhmurov and A. I. Blonkin, *Zh. Analit. Khim.*, 1975, **30**, 2413.

[258] A. T. Pilipenko, E. V. Mitropolitska and N. M. Lukovskaya, *Ukr. Khim. Zh.*, 1975, **41**, 1196.
[259] G. Kok, T. Holler, M. López, H. Nashtrieb and M. Yuan, *Environ. Sci. Technol.*, 1978, **12**, 1072.
[260] R. Delumyea and A. V. Hartkopf, *Anal. Chem.*, 1976, **48**, 1402.
[261] J. L. Burguera, M. Burguera and A. Townshend, *Anal. Chim. Acta*, 1981, **127**, 199.
[262] K. E. Haapakka and J. J. Kankare, *Anal. Chim. Acta*, 1980, **118**, 133.
[263] W. Dong and Z. Zhong, *Fenxi Huaxue*, 1984, **12**, 186.
[264] N. M. Lukovskaya and N. F. Kuschevskaya, *Ukr. Khim. Zh.*, 1985, **51**, 511.
[265] V. K. Zinchuk and R. N. Gal'chun, *Zh. Analit. Khim.*, 1984, **39**, 56.
[266] A. T. Pilipenko and A. V. Terletskaya, T. A. Bogoslovskaya and N. M. Lukovskaya, *Zh. Analit. Khim.*, 1983, **38**, 807.
[267] N. M. Lukovskaya and A. V. Terletskaya, *Zh. Analit. Khim.*, 1976, **31**, 751.
[268] N. M. Lukovskaya, L. V. Markov and N. F. Evtushenko, *Zh. Analit. Khim.*, 1974, **29**, 767.
[269] N. M. Lukovskaya and N. F. Kushchevskaya, *Ukr. Khim. Zh.*, 1976, **42**, 87.
[270] U. Isacsson and G. Wettermark, *Anal. Chim. Acta*, 1976, **83**, 227.
[271] N. M. Lukovskaya and V. A. Bilochenko, *Ukr. Khim. Zh.*, 1977, **43**, 756.
[272] N. M. Lukovskaya and V. A. Bilochenko, *Zh. Analit. Khim.*, 1977, **32**, 2177.
[273] L. A. Montano and J. D. Ingle, Jr., *Anal. Chem.*, 1979, **51**, 919.
[274] L. A. Montano and J. D. Ingle, Jr., *Anal. Chem.*, 1979, **51**, 927.
[275] A. MacDonald, K. W. Chan and T. A. Nieman, *Anal. Chem.*, 1979, **51**, 2077.
[276] S. Stieg and T. A. Nieman, *Anal. Chem.*, 1977, **49**, 1322.
[277] D. W. Margerum and T. J. Bydalek, *Inorg. Chem.*, 1962, **1**, 852.
[278] S. Funashi, S. Yamada and M. Tanaka, *Anal. Chim. Acta*, 1971, **56**, 371.
[279] M. Tabata, S. Funashi and M. Tanaka, *Anal. Chim. Acta*, 1972, **62**, 289.
[280] D. C. Olsen and D. M. Margerum, *J. Am. Chem. Soc.*, 1963, **85**, 297.
[281] S. Raman, *Indian J. Chem.*, 1975, **13**, 1229.
[282] M. Phull, H. C. Bajaj and P. C. Nigam, *Talanta*, 1981, **28**, 610.
[283] N. D. Lis and N. E. Katz, *Anal. Asoc. Quim. Argentina*, 1981, **69**, 1.
[284] M. K. Gadia and M. C. Mehra, *Microchem. J.*, 1978, **23**, 278.
[285] J. Das and K. Datta, *J. Indian Chem. Soc.*, 1974, **51**, 553.
[286] B. R. Reddy and S. Raman, *Indian J. Chem.*, 1984, **23A**, 48.
[287] S. Funashashi, M. Tabata and M. Tanaka, *Anal. Chim. Acta*, 1971, **57**, 311.
[288] S. Funashashi, M. Tabata and M. Tanaka, *Bull. Chem. Soc. Japan*, 1971, **44**, 1586.
[289] H. Schuring and H. Mueller, *Z. Chemie*, 1975, **15**, 286.
[290] M. Tabata and M. Tanaka, *Anal. Lett.*, 1980, **13**, 427.
[291] N. Xie, M. Xu, Z. Pan and J. Miao, *Fenxi Huaxue*, 1984, **12**, 281.
[292] R. Baronowski, I. Baranowski and Zb. Gregorowicz, *Microchem. J.*, 1979, **24**, 367.
[293] M. Tabata and M. Tanaka, *Mikrochim. Acta*, 1982 **II**, 149.
[294] R. P. Pantaler and I. V. Pulyaeva, *Zh. Analit. Khim.*, 1977, **32**, 394.
[295] J. Meditach and E. C. Barros, *Rev. Quim. Ind.* (*Rio de Janeiro*), 1978, **47**, 7.
[296] M. L. Cabello-Tomás and T. S. West, *Talanta*, 1969, **16**, 781.
[297] R. V. Hems, G. F. Kirkbright and T. S. West, *Talanta*, 1970, **17**, 433.
[298] R. V. Hems, G. F. Kirkbright and T. S. West, *Talanta*, 1969, **16**, 789.
[299] W. E. van der Linden and W. J. J. Ozinga, *Mikrochim. Acta*, 1980 **I**, 107.
[300] J. C. Gijsbers and J. G. Kloosterboer, *Anal. Chem.*, 1978, **50**, 455.
[301] R. Alarcón, M. Silva and M. Valcárcel, *Anal. Lett.*, 1982, **15**, 891.
[302] M. A. Ditzler, F. L. Keohan and W. F. Gutknecht, *Anal. Chim. Acta*, 1982, **135**, 69.
[303] S. Wu and S. Chen, *Fenxi Huaxue*, 1984. **12**, 989.
[304] A. A. Levshina, P. M. Zaitsev and E. N. Savchenko, *Zavodsk. Lab.*, 1981, **47**, No. 4, 5..

3

Activation and inhibition

3.1 INTRODUCTION

Effects which modify the rate of reaction are of great theoretical and practical significance. They are usually associated with homogenous catalysed reactions (their chief field of application in analytical chemistry), though they can also be applied to uncatalysed or even photochemical reactions.

This aspect of kinetic analysis has been dealt with extensively by several authors, as shown in reviews [1, 2], although only two papers have been exclusively devoted to this topic, one by Mottola [3] in 1974 and another by Nikolelis and Hadjiioannou [4], in 1979.

The rate of a catalysed reaction, v_c, can be either increased or decreased (i.e. the reaction can be 'activated' or 'inhibited') by addition of certain substances. These *activation* and *inhibition* phenomena are illustrated in Fig. 3.1, which is a plot of the

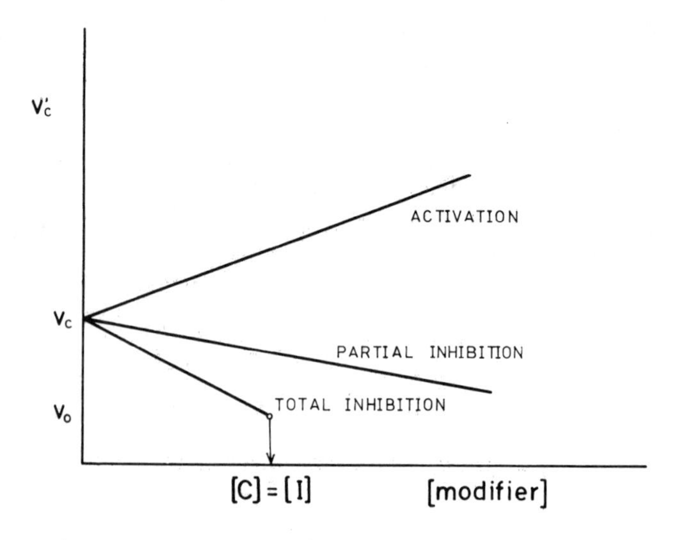

Fig. 3.1 — Variation of the rate of a catalysed reaction as a function of the modifier concentration.

variation of the rate of a modified reaction, v_c', corresponding to a fixed catalyst concentration, C, as a function of the concentration of the modifying substance. As can be seen, the reaction is completely inhibited if the equilibrium constant of the inhibition reaction is sufficiently large, so that the rate equals that of the uncatalysed reaction, v_c, for a stoichiometric ratio of the inhibitor, I, to the catalyst. If the inhibition equilibrium constant is not so large or the reaction product (either a complex or a precipitate) is also active to some extent, then the reaction of interest is only partly inhibited. The plots in the figure are usable as calibration graphs.

Catalytic methods are widely used in trace analysis on account of their high sensitivity and low detection limits — most of them in the range 10^{-9}–$10^{-7}M$, though some methods afford limits as low as 10^{-12}–$10^{-10}M$. Nevertheless, practical needs call for even lower limits. These are only achievable by exploiting the phenomenon of activation with the added advantage of the possibility of determining the activator, which more often than not has no catalytic properties itself.

Analogously, the phenomenon of inhibition allows the determination of some substances acting as inhibitors of catalysed reactions. The modifier can be determined provided that the rate of the modified reaction is proportional to its concentration. One of the major applications of this phenomenon is the determination of the inhibitor by catalytic titration (Chapter 4). Figure 3.2 sums up the

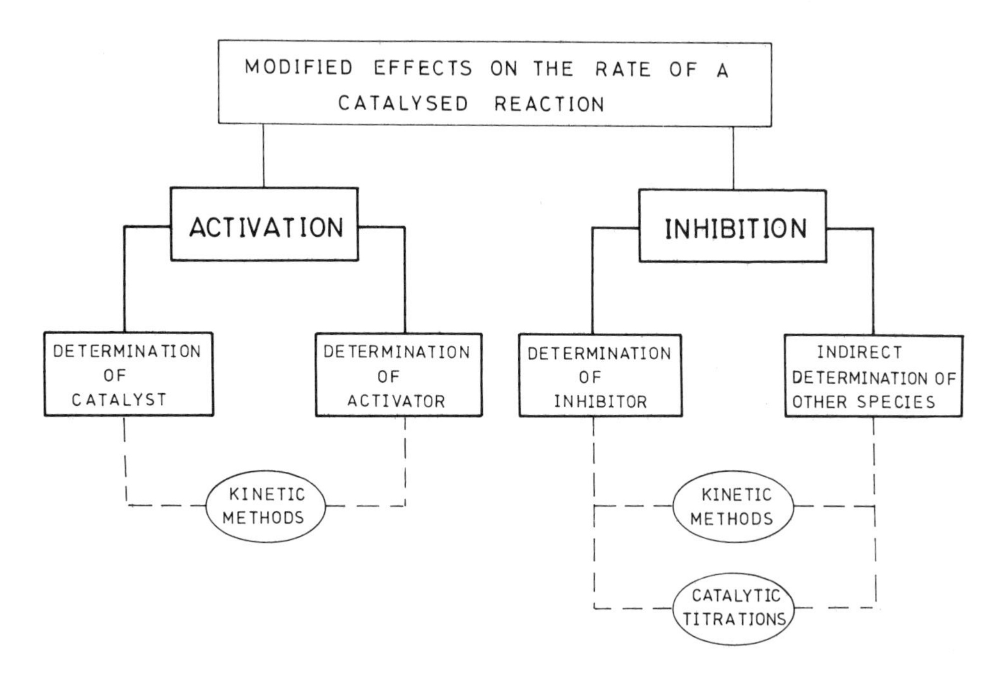

Fig. 3.2 — Uses of effects modifying the rate of a catalysed reaction.

different aspects of the use of effects modifying the rate of a catalysed reaction.

Both inhibitory and activating effects have been applied to enzymatic and non-enzymatic processes. In enzymatic reactions, activation or inhibition arises from the presence of metal ions [5], which can thus be determined on the basis of their effect on the catalytic properties of the enzyme.

The rate of non-enzymatic reactions is usually modified by hydroxylated, polyaminocarboxyl and chromogenic ligands, complexones being preferentially used as inhibitors and the other ligands acting either as inhibitors or activators. Inorganic anions are generally inhibitors of metal-catalysed reactions, while metal ions are used both in inhibition processes involving anion-catalysed reactions (less frequently) and in the activation of metal-catalysed reactions.

A given ligand may play a dual activating or inhibitory role; this allows it to be used in differential kinetic analysis. One such ligand is citric acid, which activates the catalytic action of Mo(VI) and completely inhibits that of W(VI) on the same indicator reaction, namely, the oxidation of iodide by hydrogen peroxide. This permits the determination of both metals in mixtures, as the catalytic effect of both species is additive in the absence of the modifier [6].

Ordinary organic solvents such as ethanol, acetone, dimethylformamide, etc., can also act as modifers and have indeed been used to develop methods of determination [7].

The overall rate equation of a catalytic reaction in the presence of a modifier is the general equation describing the behaviour of catalysed reactions, with an additional term accounting for the reaction rate in the presence of the activator or inhibitor:

$$v = \frac{d[P]}{dt} = k'[C]_0 + k_1' \pm k''[M]$$

where k' and k'' are the pseudo first-order rate constants with respect to the catalyst and the modifier; k_1' includes the rate of the uncatalysed reaction and M denotes the activator or inhibitor according to whether the sign of the last term is positive or negative.

In practice, the kinetic equation is derived by determining the partial order for each reactant, as described in Chapter 1, including that of the activator or inhibitor, as appropriate. Such orders are rarely the same for the catalysed and the modified (activated or inhibited) reaction for a common concentration range.

Activators and inhibitors can be determined kinetically with good sensitivity and satisfactory detection limits (between 10^{-6} and $10^{-5}M$), though these are not so good as those afforded by the determination of the catalyst. The precision, however, is rather poor in many cases.

As with kinetic catalytic analysis, spectrophotometry is by far the most common detection technique. Fluorescence and electrochemical techniques are rarely used (the latter generally involve selective electrodes). Nor are thermometric techniques (so common in catalytic titrations) widely used, owing to the low rate at which accurate enthalpimetric measurements (based on thermal equilibrium) can be obtained in the course of the reaction. Initial-rate and fixed-time methods are the most frequently employed kinetic modes.

The determination of complexing agents at low concentrations has gained significance over the last few years on account of their widespread use in biological, medical and industrial processes. These species can be determined at trace levels on the basis of their activating or inhibitory effect on reactions catalysed by metal ions [8, 9]. They are also of relevance to the determination of histamine, anthihistamine and bacteriostatic agents, vitamins, fungicides and other compounds of pharmacological interest [10].

3.2 ACTIVATION

3.2.1 Definition of activation and promotion

As stated above, activation is understood from a catalytic point of view as the increase in the rate of a catalysed reaction resulting from the action of a chemical species (the activator) that takes part in a step for which the activation energy is lower than that involving the catalyst alone. This beneficial effect not only results in a significant improvement in determination of the catalyst, but also makes possible that of the activator.

The analytical uses of activation are apparently limited to the modification of the rate of catalysed reactions by complexation of metal ions in catalytic processes involving electron transfer (particularly transitions between metal ions and chelating agents). Thus, any change in co-ordination results in a concomitant change in redox properties.

In other cases, these substances accelerate the first reaction step only. In the catalytic cycle, the 'new' catalytic species loses its integrity as a result of taking part in competitive reactions that diminish its 'activating' power. These competitive reactions may eventually destroy the ligand or its complexes with other species present in the system, thus depriving it of all activity. Once the modifier is destroyed or inactivated, the overall rate tends to that of the indicator reaction in its absence. This particular effect is known as *promotion* [11] to distinguish it from true metal-complex catalysis, in which the activator is regenerated and acts as a catalyst of the 'effective' catalyst. Thus, the determination of sodium thiobarbitone is based on its promoting effect on the copper(II)-catalysed reaction between Pyrocatechol Violet and hydrogen peroxide [12].

Induced reactions [13,14] can be considered as being due to a promoting effect. However, not all promoting effects seem to meet one of the chief practical requirements for an induced reaction to take place: "If a substance is slowly oxidized or reduced by a given reagent, the reaction rate is increased by the occurrence of a fast reaction between the reagent and another substance (inducer)" [14]. This is strictly fulfilled in the determination of V(IV) and As(III) based on their inducing effect on the oxidation of ferroin to ferriin by Cr(VI), as the vanadyl or arsenious ion is rapidly oxidized by Cr(VI). However, oxalic and citric acid seem to have a promoting rather than an inducing effect on this reaction since they are oxidized quite slowly by Cr(VI). This reactions allow the determination of the species involved by means of initial-rate measurements [15]. This is what could be termed a primary promoting effect insofar as these species act essentially as 'catalysts'. Promotion is thus a more general term than activation.

Activating and promoting effects can be distinguished by analysing the kinetic curves obtained in the presence and absence of the promoter. Thus, Fig. 3.3 shows the variation of the analytical signal as a function of time for different concentrations of promoter or activator.

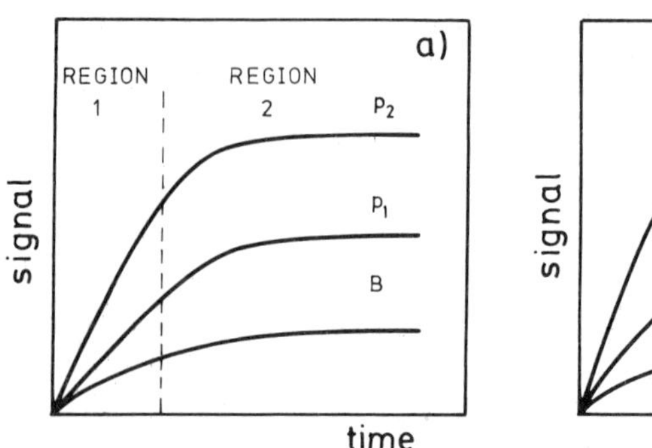

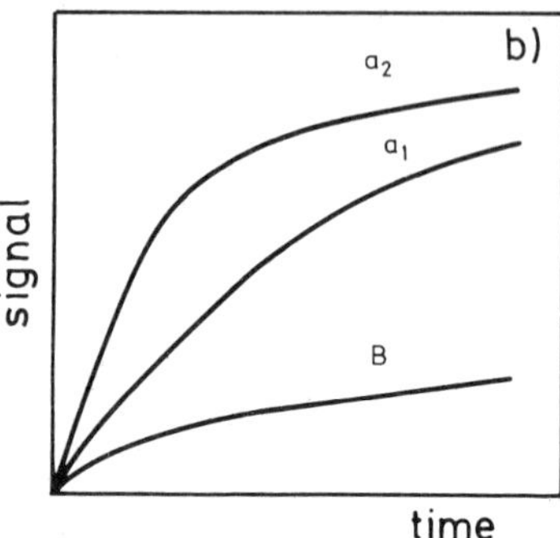

Fig. 3.3 — Variation of the analytical signal as a function of time for different concentrations of (a) promoter and (b) activator. Curve B corresponds to the catalysed reaction in the absence of a promoter or an activator.

The curves are different in shape, and that illustrating promotion (Fig. 3.3a) has two distinct regions. In the first of these, the increase in the rate with respect to that of the indicator reaction is a result of the presence of the promoter; the higher its concentration the greater the effect it has. The effect is not proportional to concentration, as the partial order with respect to the promoter is not usually unity, in contrast to the characteristic pseudo first-order dependence of the activator. The promoter loses some of its activity in the course of the reaction through interaction with a reactant or product, so that the overall reaction rate (region 2, Fig. 3.3a) coincides with that of the indicator reaction. The equilibrium signal (Fig. 3.3b) permits distinction between promotion and activation since its shape varies with the promoter concentration, while it is independent of that of the activator, which only determines the time elapsed until equilibrium is reached (Fig. 3.3b).

3.2.2 Reaction mechanisms

The selection of a suitable activator requires knowledge of the mechanism by which its effect on the catalytic process used as indicator reaction is exerted. Unfortunately, activators are normally chosen empirically, their activating effect being accounted for at a later stage owing to the difficulty involved in elucidating the mechanism.

The participation of the activator in a catalytic process depends essentially on the role played by the catalyst in the reaction, on the nature of the catalyst–activator interaction and on the particular reaction step affected by the presence of the activator [16]. The mechanism by which the activator takes part in the process can not be elucidated if nothing is known about the mode of participation of the catalyst in the overall reaction.

Bontchev [16] classifies activators into three groups according to whether they: (a) act upon the catalyst–substrate interaction; (b) are involved in the catalyst regeneration; (c) participate indirectly in the catalytic process.

Regarding the first group, it is well known that many catalysed reations take place by the formation of a complex between the metal catalyst and one of the reactants of the indicator reaction (substrate), as illustrated for a redox reaction in Fig. 3.4. The

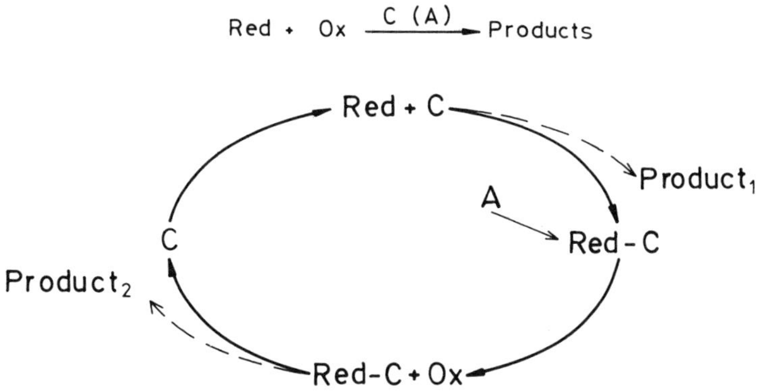

$$\text{Red} + \text{Ox} \xrightarrow{\text{C (A)}} \text{Products}$$

Fig. 3.4 — Influence of the activator on the catalyst–substrate interaction within the catalytic cycle.

activator, A, may act upon the formation of the Red–C complex and/or on its subsequent reactions by the formation of a C–A complex through which it can influence the reactivity of the catalyst in the substrate reaction. Thus, if A in C–A can form hydrogen bonds with Red, then formation of the intermediate Red–C will be facilitated if C and Red adopt a suitable spatial arrangement. This mechanism accounts for the activating effect of glycerol in vanadium(V)-catalysed reactions [17].

This activating effect can also arise from the formation of ternary complexes between the catalyst, the substrate and the activator, such as those detected in the decarboxylation of dibasic β-keto acids (e.g. oxalacetic acid). This is catalysed by Cu(II), Ni(II), Co(II), Mn(II), Al(III) and Fe(III), the effect increasing with increasing ionic charge. Ligands able to form metal–ligand π-bonds tend to diminish the catalyst's effective charge and hence its activity. One such ligand is pyridine, which activates the catalytic effect of Cu(II) on this type of reaction through this mechanism [18].

However, few explicit applications based on this type of interaction have been reported so far. In fact, applications exploiting activating effects belong in most cases to the second group of Bontchev's classification, i.e. catalysed reactions in which the catalyst undergoes a series of alternating oxidations and reductions, acting as an electron transferrer between Red and Ox and even bearing unusual valences (activated form). The activator can either favour the conversion of the catalyst into its activated form or take part in its regeneration, depending on the nature of the rate-determining step (Fig. 3.5). The catalyst is then a transition metal ion that can occur in several oxidation states.

If the rate-determining step is I, any substance capable of complexing the catalyst in its lowest oxidation state will foreseeably shift the reaction to the right, thereby increasing its rate. Thus, chloride activates the Cu(II)-catalysed decomposition of hydrogen peroxide according to:

$$Cu^{2+} + HO_2^- \rightarrow Cu^+ + HO_2^{\cdot}$$

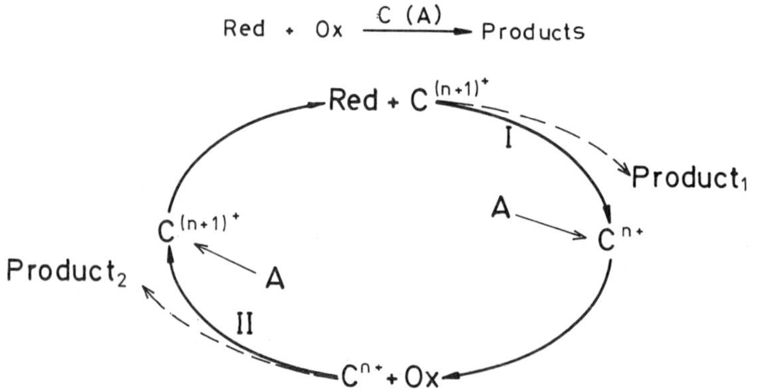

$$\mathrm{Red} + \mathrm{Ox} \xrightarrow{\mathrm{C\ (A)}} \mathrm{Products}$$

Fig. 3.5 — Influence of the activator on the catalyst regeneration within the catalytic cycle.

since Cl^- forms more stable compounds with Cu(I) than it does with Cu(II) and therefore speeds up the reaction.

If the rate-determining step (slower) is II, i.e. the catalyst regeneration, then any multidentate activator capable of complexing C^{n+} and forming hydrogen bonds with Ox will lower the activation energy of the catalyst regeneration and hence increase the reaction rate. Those species forming more stable complexes with $C^{(n+1)1}$ than they do with C^{n+} will also make good activators.

A representative example of this type of activation is the Ag(I)-catalysed oxidation of sulphanilic acid by peroxodisulphate [19]. The rate-determining step of the overall process is the oxidation of Ag(I) to Ag(II). The addition of ligands forming more stable complexes with Ag(II) than with Ag(I) lowers the redox potential of the Ag(II)/Ag(I) pair, thereby accelerating the conversion of Ag(I) into Ag(II). Among such ligands are those containing electron-releasing nitrogen atoms, their effect increasing with the co-ordination number, in the order 1,10-phenanthroline > 2,2′-bipyridyl > ethylenediamine > pyridine > 2-methylpyridine = 4-methylpyridine > 4-aminopyridine > ammonia.

In this group of reactions, the determining step in which is the formation of the activated species of the catalyst, can be included the oxidation of diamine derivatives by hydrogen peroxide, catalysed by Cr(III) and Cr(VI). There is experimental evidence that compounds such as *p*-aminobenzoic acid, which accelerate the reduction of Cr(VI) to Cr(III) (the effective catalyst), are potential activators of the catalytic effect of Cr(VI). Conversely, activators such as γ-picoline, which promote the deactivation of the aquo-complex of the chromic ion $[Cr(H_2O)_6^{3+}]$ and the interaction of its complex with the oxidants, are potential activators [20].

γ-Picoline is a better activator than pyridine, quinoline, 2,2′-bipyridyl and 1,10-phenanthroline. In fact, as these heterocyclic amines are protophilic and the electron density on the nitrogen atom increases from phenanthroline (through 2,2′-bipyridyl, quinoline and pyridine) to γ-picoline, the proton affinity increases in the same order and so does the activating effect on the protolytic dissociation of the aquo-complex of Cr(III) according to:

$$[(H_2O)_xCr(OH_2)]^{3+} + L = [(H_2O)_xCr(OH)\ldots HL]^{3+} + H_2O$$

After successive substitutions of water molecules by L and in the presence of hydrogen peroxide the following process takes place:

$$CrL_6^{3+} + H_2O_2 \rightarrow Cr(V)$$

On the other hand, *p*-aminobenzoic acid, A, forms a complex with Cr(VI) that undergoes an internal electron transfer according to:

$$Cr(VI) + A = [Cr(VI)A]$$
$$[Cr(VI)A] = Cr(V) + A\cdot$$

which takes place with the formation of free radicals. The presence of these is confirmed by demonstrating the kinetic dependence of the *o*-dianisidine/Cr(VI)/*p*-aminobenzoic acid/H_2O_2 system on the concentration of methyl methacrylate, a typical inhibitor of processes involving free radicals. Under these conditions, the reaction rate decreases with the methacrylate concentration. It should be pointed out that the formation of Cr(V) also takes place in the presence of hydrogen peroxide.

In the last group of Bontchev's classification can be included those activators for which it is rather difficult to elucidate the mechanism of action. The activator participates indirectly in the process, in one or several of the numerous steps involved in a catalysed reaction. In this case, the activator can even be involved in side-processes that may affect the main reaction. Thus citric acid, an activator of the vanadium(V)-catalysed oxidation of *p*-phenetidine by halates, acts upon two reaction steps: it accelerates the substrate–catalyst interaction by formation of free radicals from the activator [21] and increases the rate of generation of the catalyst, thus serving as a matrix for the interaction between V(IV) and XO_3^- [22].

The activating effect of non-multivalent metal ions on metal-catalysed reactions is, unfortunately, rather more difficult to account for.

3.2.3 Analytical uses of activating effects

The chief analytical application of activators is in the kinetic determination of the catalyst and also of the activator itself, as described in Fig. 3.2.

3.2.3.1 Kinetic determination of the catalyst

No application has been reported so far, to the determination of anions (as catalysts) in the presence of an activator, though this situation is perfectly possible. Among the rest (i.e. metal-catalysed reactions), redox processes are by far the most commonly employed on account of the wealth of information available.

The use of an activator in catalysed reactions is usually intended to increase the sensitivity and hence (a) lower the detection limit for the catalyst and (b) improve the selectivity and precision of its determination. In augmenting the catalytic activity of the ion to be determined, the action of other possible interfering catalysts is hindered, with simultaneous benefit to the sensitivity, selectivity and detectability.

The use of activating and masking effects to regulate the selectivity of catalysed reactions was first proposed by Bontchev [23] and applied to the reaction between *p*-

phenetidine and chlorate, catalysed by V(V), Fe(II) and Cu(II), with citric acid as activator. The sensitivity of this reaction is thus increased by a factor of 15 for V(V) and the possible interference from copper and iron is completely overcome. This is a result of the aforesaid formation of free radicals from the activator, so that citric acid does not hinder the activity of vanadium(V), even in the presence of a large excess of iron and copper, which are masked by the activator and thus prevented from interacting with the substrate.

The above-mentioned Ag(I)-catalysed oxidation of sulphanilic acid by peroxodisulphate affords the determination of 5 ng of silver in the presence of 2,2′-bipyridyl (i.e. the reaction is 5000 times as sensitive as in the absence of the activator). In addition, the masking of copper and iron by the activator results in a lessening of their interfering effect by factors of 2 and 10, respectively.

Activating effects [24] and selectivity [25] can be further enhanced by suitably combining several activators.

While preserving the high sensitivity and low detection limits of kinetic methods, activation increases their precision. Thus, the determination of manganese based on its catalytic effect on the reaction between phosphinate and periodate has a relative standard deviation of 2.2%, which is roughly halved in the presence of an activator such as NTA (nitrilotriacetic acid) [26].

Pyridine and ascorbic acid are the two activators most commonly used in Cu(II)-catalysed reactions. The beneficial action of pyridine [27] results from the formation of a 1:1 complex with copper, responsible for its catalytic activity [28]. Ascorbic acid enhances the catalytic power of copper, e.g. in the oxidation of sodium 2-thiosemicarbazone-1,2-naphthoquinone-4-sulphonate by hydrogen peroxide [29], lowering the detection limit of copper from 50 to 0.25 ng/ml.

On the other hand, NTA is probably the most common activator of manganese(II)-catalysed reactions.

3.2.3.2 Kinetic determination of the activator

There are relatively few methods available for the determination of activators (see Table 3.1). According to Mottola, the term 'indicator reaction' has a wider meaning in this field as it includes the catalyst. The activator most commonly determined is probably NTA, which acts upon the Mn(II)-catalysed reactions of periodate with Malachite Green [8], phosphinate [26] and Sb(III), and is detectable at concentrations of the order of $10^{-6}M$ by the variable-time method.

Ascorbic acid can also be determined at concentrations above 0.5 μg/ml by the V(V)-catalysed reaction between I^- and ClO_3^- [31]. Citric acid, which has so far been used more often as a promoter or inducer than as an activator, has so far not been determined as such. However, it has been determined on the basis of its promoting effect on the reaction between ferroin and Cr(VI) [15].

With the exception of the copper(II)-catalysed determination of adrenaline, tyrosine and other related compounds with the Pyrocatechol Violet/H_2O_2 system [32], no substances of biological interest have been investigated in this field to date.

Metal ions have also rarely been determined on the basis of their activating effects. Among the few references available on this subject, those worth mentioning are to the determination of Zn(II) in the range 50–400 ng/ml by the initial-rate method [33], and of In(III) [34], Ga(III) [35], selenium and tellurium [36], and

Table 3.1 — Kinetic determination of activators

Activator	System	Determination range	References
NTA	Malachite Green/IO_4^-/Mn(II)	10^{-7}–$10^{-6}M$	[8]
NTA	Phophinate/IO_4^-/Mn(II)	10^{-6}–$10^{-5}M$	[26]
NTA	Sb(III)/IO_4^-/Mn(II)	10^{-7}–$10^{-6}M$	[30]
Ascorbic acid	I^-/ClO_3^-/V(V)	0.5 μg/ml	[31]
Citric acid[a]	Ferroin/Cr(VI)	0.9–4.5 μg/ml	[15]
Adrenaline, tyrosine, etc.	Pyrocatechol Violet/H_2O_2/Cu(II)	1–10%	[32]
Thiobarbitone[b]	Pyrocatechol Violet/H_2O_2/Cu(II)		[12]
Zn(II)	R*/H_2O_2/Mn(II)	50–400 ng/ml	[33]
In(III)	R**/H_2O_2/Cu(II)	0.2–1.0 μg/ml	[34]
Ga(III)	R**/H_2O_2/Cu(II)	0.5–1.5 μg/ml	[35]
Se(IV)	Co(III)/EDTA/Sn(II)/Au(III)	2.5–80 ng/ml	[36]
Te(IV)	Co(III)/EDTA/Sn(II)/Au(III)	4.0–100 ng/ml	[36]
Hg(II)	R***/NaH_2PO_2/Au(III)	1.0–30 ng/ml	[37]

R*: 2-hydroxybenzaldehyde thiosemicarbazone.
R**: 4,4′-dihydroxybenzophenone thiosemicarbazone.
R***: molybdophosphoric acid.
[a] Primary promoter. [b] Secondary promoter.

Hg(II) [37], with different systems.

Although the combination of two activators has indeed been studied in order to clarify their operational mechanism [24], there is only a single precedent for the kinetic analysis of a mixture, that of Ga(III) and In(III) in a 1:12.5 ratio, on the basis of their activating effect on the Cu(II)-catalysed oxidation of 4,4′-dihydroxybenzophenone thiosemicarbazone by hydrogen peroxide [35].

3.3 Inhibition

Any species interacting with the catalyst to diminish (partial inhibition) or cancel (total inhibition) its catalytic activity can be determined on the basis of its inhibitory effect provided that this is proportional to its concentration.

As a rule, the detection limits achieved by kinetic methods are quite good, though the inhibitor can usually be determined only over a short concentration range — hence the growing use of catalytic titration techniques for this purpose.

Unlike activation (by definition a catalytic effect on the catalyst), inhibition can also be applied to uncatalysed reactions, which are dealt with in Chapter 5. Thus chloride and bromide are determined on the basis of their inhibitory effect on the oxidation of 4,4′-dihydroxybiphenyl by Tl(III) [38], and bromide on that of Methyl Orange with bromate [39]. Nevertheless, only catalysed reactions are dealt with in this section.

Most of the kinetic methods for the determination of inhibitors involve a reaction with the catalyst (generally complex-formation) which hinders the catalytic cycle. However, the inhibitor may interact with the substrate and diminish the reaction rate. Such is the case with the determination of magnesium, which interacts with 1,4-

dihydroxyphthalimidodithiosemicarbazone in the course of its Mn(II)-catalysed aerial oxidation [40], and with cyanide, which interacts with hydrogen peroxide [rather than with Cu(II)] in the decomposition of the oxidant by this metal ion [41]. It is even possible to determine inhibitors on the basis of their interaction with an activator present in the catalysed reaction, e.g. fluoride, which inhibits the activating effect of Al(III) on the thiosulphate used as a catalyst for the reaction between Indigo Carmine and iodide [42].

The interaction between the inhibitor and the catalyst is usually a complex-formation and very rarely a redox or ligand-displacement reaction.

3.3.1 Influence of the formation constant of the catalyst–inhibitor complex

If it is assumed that the catalyst–inhibitor complex is catalytically inactive, the decrease in the catalytic activity must be a function of the inhibitor concentration and the formation constant of such a complex, K_{IC}. The comparison of the decrease in the catalytic activity brought about by several inhibitors will thus provide a measure of their respective K_{IC} values and allow for comparisons to be made between the relative stabilities of the different complexes.

The inhibitor can be satisfactorily determined from changes of the order of 10–20% in the catalyst concentration, provided that such changes result in appreciable variations in the measured parameter, which in turn depends on the particular measuring technique applied. For a typical catalyst concentration, $[C]_0$, of $4 \times 10^{-7} M$, and assuming formation of a 1:1 complex with the inhibitor, the latter will only be detectable if its initial concentration, $[I]_0$, is greater than $4 \times 10^{-8} M$.

On the other hand, assuming that 99% of the inhibitor is complexed by the catalyst, the conditional formation constant, K'_{IC}, should be of the order of 3×10^8 for the inhibitor to be detectable. Thus, assuming that the inhibitor forms a 1:1 complex with the catalyst and that both are present at equal concentrations in the reaction medium, the catalysed reaction would not take place (total inhibition) for $K'_{IC} > 10^4/[C]_0$. In such a case, $K'_{IC} \approx 10^{11}$ for $[C]_0 \approx 10^{-7} M$.

The considerations above can be summed up as follows.

(a) If K'_{IC} is $> 10^{11}$ inhibition will be complete for $[C]_0 = [I]_0$. Under these conditions, the calibration curve should be run for inhibitor concentrations equal to or lower than that of the catalyst.

(b) If K'_{IC} is $< 10^{11}$ inhibition will not be complete and, as a rule, $[I]_0 > [C]_0$. This is the commoner situation since the determination of the inhibitor is normally less sensitive than that of the catalyst. In practice, the sensitivity is raised by increasing the catalyst concentration, i.e. by decreasing K'_{IC}.

According to the statements above, the detection limits achieved in the determination of EDTA and related ligands will be lower than those obtained for ligands forming complexes with larger co-ordination number. This is a result of the sharp increase in K'_{IC}, which amounts to about 10^{50} in the formation of fluoride complexes such as ZrF_6^{2-} in Zr(IV)-catalysed reactions.

3.3.2 Construction of calibration curves

The calibration graphs used in the kinetic determination of inhibitors are usually run by plotting percentage inhibition as a function of the inhibitor concentration. The former quantity, $I_{\%}$ is calculated from:

$$I_{\%} = \frac{100(\Delta \upsilon)_{\text{sample}}}{(\Delta \upsilon)_{\text{blank}}}$$

where $(\Delta \upsilon)_{\text{sample}}$ and $(\Delta \upsilon)_{\text{blank}}$ denote the measured variable (initial rate, absorbance, etc.) for the catalysed reaction with respect to the inhibited and uncatalysed reactions, respectively.

If the relationship between the reaction rate and the inhibitor concentration is non-linear, or the linear range of the calibration curve is to be broadened, an evaluation method originally proposed for Landolt reactions [43] and based on an empirical calibration equation of the form $t_i \approx \exp(-\text{RM})$ (t_i = induction period; RM = molar inhibitor-to-catalyst ratio) can be used. The RM range is obtained by plotting $(1/t_0) - (1/t_i)$ *vs*. RM, (t_0 coincides with the start of the Landolt reaction and t_i with the appearance of the product). This principle is equally applicable to other types of reactions [44] by simply plotting the property measured at a time t, as a function of exp(− RM), the RM range being determined in this case by plotting the same parameter as a function of RM. In both cases, the RM range is obtained from the portion of the curve corresponding to the greatest inhibitor effect (Fig. 3.6).

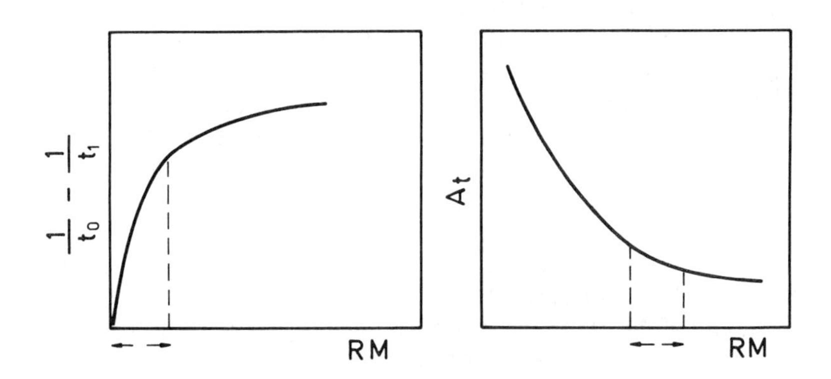

Fig. 3.6 — Selection of the RM range for determination of the inhibitor: (a) for a Landolt reaction; (b) for a different type of reaction.

An alternative method of obtaining linear calibration graphs entails plotting the reciprocal of the initial rate [45] or of the measured parameter corresponding to a preselected time [46], as a function of the inhibitor concentration.

3.3.3 Analytical uses of the inhibitory effect

In dealing with the analytical applications of the inhibitory effect, this section has been divided according to the nature of the inhibitor (a metal ion or a non-metal).

3.3.3.1 Kinetic determination of metal ions

The number of applications involving the kinetic determination of metal ions is as small as that of reactions catalysed by non-metals (Table 3.2). Such applications

Table 3.2 — Determination of metals as inhibitors of catalysed reactions

Inhibitor	System	Determination range	References
Hg(II)	$As(III)/Ce(IV)/I^-$	1–60 ng	[47]
Au(III)	$As(III)/Ce(IV)/I^-$	0.28–0.38 μg/ml	[48]
Ag(I)	$As(III)/Ce(IV)/I^-$	0.2–0.8 μg/ml	[49]
Cu(II)	Sodium azide/I_2/thioammeline	0.06–1.0 μg/ml	[50]
Au(III)	Sodium azide/I_2/2-mercaptopurine	1–50 μg	[51]
Ni(II)	Sodium azide/I_2/DDTC	28–280 ng/ml	[52]
Sn(II)	R*/H_2O_2/Fe(III)	10–80 ng/ml	[53]
Mg(II)	R**/O_2 (air)/Mn(II)	$3.3–5.4 \times 10^{-5} M$	[40]
Ag(I)	R***/H_2O_2/Fe(III)	50–400 ng/ml	[54]

DDTC: diethyldithiocarbamate.
R*: *N,N'*-diethyl-*o*-phenylenediamine sulphate.
R**: 1,4-dihydroxyphthalimide dithiosemicarbazone.
R***: 2-hydroxybenzaldehyde thiosemicarbazone.

usually involve the 'advantageous' determination of metals that cannot be determined on the basis of primary catalytic effects.

Two of the reactions most frequently used in the indirect determination of metals are that described by Kolthoff and Sandell [i.e. the iodide-catalysed oxidation of As(III) by Ce(IV)], and the oxidation of sodium azide by iodine, which is catalysed by sulphur-containing substances (mostly ligands). In both types of process, the metal inhibits the reaction through interaction with the iodide or the sulphur compound, respectively.

The first reaction has been used for the determination of Hg(II) in the range 1–60 ng [47] by use of the fixed-time method, with absorbance measurements made at 275 nm [absorption wavelength of Ce(IV)]. It has also been applied to the determination of Au(III) [48], and of Ag(I) in silicate rocks in the range 0.2–0.8 μg/g [49].

The second reaction has been successfully applied to the determination of copper in zinc salts, based on its inhibitory complexing effect upon thioammeline, a catalyst of the reaction between sodium azide and iodine [50]. This reaction is also catalysed by (among other compounds) 2-mercaptopurine and diethyldithiocarbamate, the action of which is inhibited by Au(III) [51] and Ni(II) [52], respectively, and these can be determined (the latter in margarine samples) on the basis of this effect.

Other applications of interest in this context are the determination of Sn(II) based on its inhibitory effect on the Fe(III)-catalysed reaction between *o*-phenylenediamine and H_2O_2 [53], that of magnesium(II) (which interacts with the substrate rather than with the catalyst) alluded to above [40] and that of Ag(I) involving spectrofluorimetric detection [54].

3.3.3.2 Kinetic determination of non-metals

The scarcity of reactions catalysed by non-metals compels the use of indirect determinative methods. On the other hand, metal-catalysed reactions are plentiful: this, added to the ability of metal species to form stable complexes with anions and organic ligands, facilitates the determination of these non-metallic species.

Determination of multidentate organic ligands. There is a host of systems available for the determination of EDTA and other similar complexones (Table 3.3a). That proposed by Mottola is based on the reaction between Malachite Green,

Table 3.3a — Kinetic determination of aminopolycarboxylic acids as inhibitors of catalysed reactions

Inhibitor	System	Determination range	References
EDTA	Malachite Green/IO_4^-/Mn(II)	$2–14 \times 10^{-6}M$	[9]
DCTA	Phosphinate/IO_4^-/Mn(II)	$10^{-7}–10^{-6}M$	[30]
EDTA, DTPA	Sb(III)/IO_4^-/Mn(II)	$10^{-7}–10^{-6}M$	[30]
EDTA	Ascorbic acid/O_2(air)/Cu(II)	$2–8 \times 10^{-6}M$	[56]
EDTA	Acetylacetone/IO_4^-/Mn(II)	$10^{-7}–10^{-6}M$	[55]
EGTA, NTA	Thiosulphate/IO_4^-/Cu(II)	$10^{-7}–10^{-6}M$	[57]
EDTA	*p*-Phenetidine/H_2O_2/Fe(III)		[58]
EDTA, EGTA, DTPA	Perborate/I^-/Fe(II) (inducer)		[59]
EDTA, EGTA	R*/H_2O_2/Cu(II)	$1–6 \times 10^{-6}M$	[60]

R*: 4,4′-dihydroxybenzophenone thiosemicarbazone.

periodate and manganese(II) and allows the determination of EDTA in the range 0.74–5.0 μg/ml, with a relative error of 5–8%. This method features good selectivity resulting from the formation of a complex between protonated EDTA and Mn(II) at a relatively low pH. It makes use of the first-order rate-constant (i.e. the slope of the log of absorbance *vs*. time plot), which is plotted against the EDTA concentration, as well as of the fixed-time method [9]. Other systems used for the determination of EDTA in a similar concentration range include acetylacetone/periodate/Mn(II) [55], Sb(III)/periodate/Mn(II) (also used for DTPA) [30], ascorbic acid/O_2(air)/Cu(II) [56], *p*-phenetidine/H_2O_2/Fe(III) in the presence of 1,10-phenanthroline as activator [58], perborate/iodide/Fe(III) (the inducer) in the presence of the same heterocyclic compound, with use of an iodide-selective electrode [59]. The last system also allows the determination of EGTA and DTPA.

The phosphinate/periodate/Mn(II) system has been applied to the determination of DCTA [30] in the range $10^{-7}–10^{-6}M$ by the fixed-time method. Over this range EGTA and NTA are detectable with the thiosulphate/periodate/Cu(II) system [57]. Even phosphorus-containing complexones can be determined on the basis of their inhibitory effect on the *o*-dianisidine/H_2O_2/Cr(III) system [61].

Other organic substances determined by this methodology (Table 3.3b) include 1,10-phenanthroline, which inhibits the catalytic action of copper on the decomposition of hydrogen peroxide [62], and of oxine and dimethylglyoxime [63], which inhibit the catalytic effect of Co(II) on the reaction between Alizarin S and H_2O_2. This technique has also been applied to the determination of sulphur-containing ligands such as thiourea, dithiocarbamates, mercaptans, pesticides and carbon

Table 3.3b — Kinetic determination of other organic substances as inhibitors of catalysed reactions

Inhibitor	System	Determination range	References
1,10-Phenanthroline	H_2O_2/Cu(II)	0.25–2.5 μg/ml	[62]
8-Hydroxyquinoline	Alizarin S/H_2O_2/Co(III)	0.1–1.4 μg/ml	[63]
Dimethylglyoxime	Alizarin S/H_2O_2/Co(III)	0.02–5.0 μg/ml	[63]
Gentisic acid	Azorubin/H_2O_2/Mo(VI)	6.3–49.3 μg/ml	[64]
Chromotropic acid	Azorubin/H_2O_2/Mo(VI)	13.2–105.5 μg/ml	[64]
Antibiotics (tetracyclines)	Azorubin/H_2O_2/Mo(VI)	18–160 μg/ml	[65]
Cysteine	*p*-NDA*/CN^-/Hg(II)	$1.0–2.5 \times 10^{-6}M$	[44]
Thioglycollic acid	*p*-NDA*/CN^-/Hg(II)	$1.0–4.6 \times 10^{-7}M$	[44]
Amino-acids	Pyrocatechol Violet/H_2O_2/Cu(II)	$10^{-6}M$	[9]
Nitrogen-containing drugs	Quinol/H_2O_2/Cu(II)	2.0 μg/ml	[67]
Citric acid	2,4-Diaminophenol/H_2O_2/Fe(III)	0.04–0.30 μg/ml	[45]

**p*-NDA: *p*-nitrosodiphenylamine.

disulphide, all of which hinder the catalytic activity of the *p*-phenetidine/periodate system and can be determined at concentrations of the order of 10 ng/ml [66].

The only precedent for the use of a ligand-exchange reaction instead of a redox reaction as indicator is the basis for the determination of micro amounts of cysteine and thioglycollic acid (in addition to thiosulphate), which inhibit the catalytic action of Hg(II) on the substitution of *p*-nitrosophenylamine (*p*-NDA) by cyanide [44].

Some substances of biological interest have also been determined on the basis of their inhibitory effect, e.g. amino-acids such as glycine, serine, phenylalanine, glutamic acid or arginine, which form 1:1 complexes with Cu(II), a catalyst for the oxidation of Pyrocatechol Violet by hydrogen peroxide. These amino-acids can be determined in this manner at the micromolar level, although the determination range is somewhat narrow and the errors involved are rather large (between 5 and 9%, depending on the particular amino-acid) [9]. Some tetracyclines [65] and nitrogen-containing drugs [47] have also been determined in this fashion.

The use of inhibitors allows not only their own determination, but also that of non-catalytic metals. Such indirect determinations are based on the cancellation of the catalyst activity by complexation with EDTA or another inhibitor and its gradual restoration by the metal to be determined, which reacts with the ligand to release the catalyst. The determination does not necessarily entail a complete substitution. In fact, it is equally feasible to measure the reaction rate corresponding to a given amount of inhibitor surpassing that of the metal to be determined. In this case, the inhibitor–catalyst complex may be more stable than that formed between the inhibitor and the metal analyte.

An illustrative example of the first alternative is the determination of mercury-(II), lead and cadmium by means of the *p*-phenetidine/periodate/Fe(II) system in the presence of 2,2′-bipyridyl as activator and thiourea as inhibitory ligand [68], and that of mercury(II) and EDTA in waste waters by the use of the diphenylcarbazone/H_2O_2/Co(II) and *o*-dianisidine/periodate/Mn(II) systems, respectively. The detection limit achieved is quite low (2–5 ng/ml) and the relative error is of the order of 3–10%, though the method is free from interference by the heavy metals commonly found in waste waters [69].

Nickel and zinc have also been determined by using EDTA and EGTA as inhibitory ligands and the indicator system made up of 4,4′-dihydroxybenzophenone tiosemicarbazone, hydrogen peroxide and copper(II), with no apparent signs indicating the occurrence of a displacement reaction [60].

All things considered, the catalytic titration technique is clearly superior in the indirect determination of metals (see Chapter 4).

Determination of inorganic anions. The few applications devoted to inorganic anions in this field deal almost exclusively with cyanide and fluoride (Table 3.4).

Table 3.4 — Determination of anions as inhibitors of catalysed reactions

Inhibitor	System	Determination range	References
CN^-	$H_2O_2/Cu(II)$	0.3–3.0 μg/ml	[41]
	$R/H_2O_2/Cu(II)$	0.7–2.4 μg/ml	[70]
	Acid Chrome Black ET or H-acid/$H_2O_2/Fe(III)$	2–9 μg	[46]
F^-	Perborate/$I^-/Zr(IV)$	0.3–4.0 ng/ml	[71, 72]
	$DAP/H_2O_2/Fe(III)$	0.4–4.2 μg/ml	[45]
	Indigo Carmine/$I^-/S_2O_3^{2-}$ (Al^{3+} as activator)		[42]
	$BrO_3^-/Br^-/Zr(IV)$ (Ascorbic acid; Landolt reaction)	12.5–31 μg	[73]
	$Ce(IV)/As(III)/I^-$	10–100 μg/ml	[74]
S^{2-}	Sulphanilic acid/$S_2O_8^{2-}/Ag(I)$	0.12–4 μg/ml	[75]
$C_2O_4^{2-}$	$DAP/H_2O_2/Fe(III)$	0.01–0.17 μg/ml	[45]

R: 4,4′-dihydroxybenzophenone thiosemicarbazone.
DAP: 2,4-diaminophenol.

Cyanide is determined through its inhibitory effect on Cu(II)-catalysed reactions. Thus, it hinders the copper-catalysed decomposition of hydrogen peroxide in the presence of ammonia. The cyanide is oxidized by the peroxide according to:

$$CN^- + H_2O_2 \xrightarrow{Cu^{2+}} CNO^- + H_2O$$

$$2H_2O_2 \xrightarrow{Cu^{2+}} 2H_2O + O_2$$

This oxidation is also catalysed by copper. By thermometric monitoring of the decomposition of H_2O_2 it is possible to determine cyanide (with a fixed amount of copper) from the delay time. The start of the decomposition (once cyanide has been completely oxidized) is marked by an increase in temperature (Fig. 3.7). Depending on the amount of copper present, cyanide can be determined in three different concentration ranges: 0.3–3.0 μg/ml, 0.6–6.0 μg/ml and 0.8–8.0 μg/ml [41]. In this case, the cyanide ion, despite its affinity for copper, does not interact with the catalyst, but with the substrate. This also seems to be the case with the oxidation of 4,4′-dihydroxybenzophenone thiosemicarbazone by hydrogen peroxide, which permits its determination over various narrow ranges between 0.7 and 2.4 μg/ml [70].

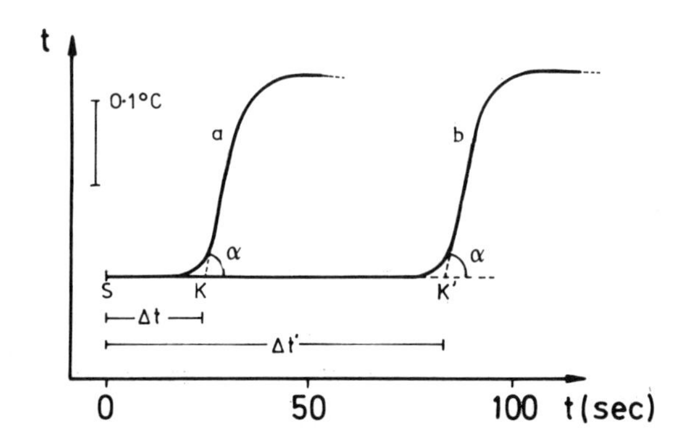

Fig. 3.7 — Variation of temperature as a function of time for (a) 35 μg and (b) 48 μg of cyanide, with 40 μg of Cu(II) in both cases. (Reproduced from [41] with the permission of the copyright holders, Elsevier Science Publishers).

Fluoride, another anion of analytical interest, is determined by kinetic methods based on its inhibitory effect on Zr(IV) (a catalyst of various reactions), with which it forms stable complexes. The procedure for the determination of this halide, based on the oxidation of iodide by perborate, utilizes potentiostatic measurements [71, 72]. The electrochemical potential of the solution is given by the $[I_2]/[I^-]^2$ ratio, which is kept constant by the automatic injection of iodide as it is depleted (continuous steady-state method). Under these conditions, the reaction rate remains constant and the speed of injection is proportional to the concentration of fluoride. This method is applicable to the determination of fluoride in air, water and biological media, in the range 0.3–4.0 ng/ml.

Despite the small number of anions determined by this type of procedure, it is a very promising method because of the number of metal-catalysed reactions available and the low detection limits achievable in the determination of the inhibitory anion.

REFERENCES

[1] H. A. Mottola, *CRC Crit. Rev. Anal. Chem.*, 1975, **4,** 229.
[2] H. Müller, *CRC Crit. Rev. Anal. Chem.*, 1982, **13,** 313.
[3] H. A. Mottola, *Anal. Chim. Acta*, 1974, **71,** 443.
[4] D. P. Nikolelis and T. P. Hadjiioannou, *Rev. Anal. Chem.*, 1979, 81.
[5] G. G. Guilbault, in *MTP Internat. Rev. Sci.: Phys. Chem.*, Ser. 1, Vol. 12, *Analytical Chemistry*, Part 1, T. S. West (ed.), Butterworths, Oxford, 1973, Chapter 5.
[6] I. I. Alekseeva, L. P. Ruzinov, E. G. Khachaturyan and L. M. Chernyshova, *Zh. Analit. Khim.*, 1980, **35,** 60.

[7] I. F. Dolmanova, O. I. Mel'nikova, G. I. Tsizin and T. N. Shekhovtsova, *Zh. Analit. Khim.*, 1980, **35,** 728.
[8] H. A. Mottola and G. L. Heat, *Anal. Chem.*, 1972, **44,** 2322.
[9] H. A. Mottola and H. Freiser, *Anal. Chem.*, 1967, **39,** 1294.
[10] T. J. Janjić and G. A. Milovanović, *Anal. Chem.*, 1973, **45,** 390.
[11] A. E. Martell, *Pure Appl. Chem.*, 1968, **17,** 129.
[12] G. A. Milovanović and M. J. Protolipać, *Bull. Soc. Chim. Beograd*, 1981, **46,** 695.
[13] H. A. Laitinen and W. E. Harris, *Chemical Analysis*, 2nd Ed., McGraw-Hill, New York, 1975, p. 297.
[14] I. M. Kolthoff and V. A. Stenger, *Volumetric Analysis*, Vol. I, Interscience, New York, 1942, p. 174.
[15] V. V. S. E. Dutt and H. A. Mottola, *Anal. Chem.*, 1974, **46,** 1090.
[16] P. R. Bontchev, *Talanta*, 1972, **19,** 675.
[17] P. R. Bontchev, G. Nikolov and B. Lilova, *Ann. Univ. Sophia (Khimia)*, 1964–5, **59,** 87.
[18] P. R. Bontchev and V. Michaylova, *Proc. Symp. Coord. Chem.*, 3rd, 1970, **1,** 403. B. Mihaly (ed.), Akadémiai Kiadó, Budapest.
[19] P. R. Bontchev and A. A. Alexiev, *Teor. Eksp. Khim.*, 1973, **9,** 191.
[20] I. F. Dolmanova and T. N. Shekhovtsova, *Zh. Analit. Khim.*, 1977, **32,** 1154.
[21] P. R. Bontchev and K. B. Yatsimirskii, *Zh. Fiz. Khim.*, 1965, **39,** 1995.
[22] P. R. Bontchev and A. A. Aleksiev, *Ukr. Khim. Zh.*, 1966, **32,** 1044.
[23] P. R. Bontchev, *Mikrochim. Acta*, 1964, 79.
[24] P. R. Bontchev and B. Evtimova, *Mikrochim. Acta*, 1968, 492.
[25] I. E. Kalinichenko, *Ukr. Khim. Zh.*, 1969, **35,** 755.
[26] D. P. Nikolelis and T. P. Hadjiioannou, *Analyst*, 1977, **102,** 591.
[27] A. Aleksiev, P. R. Bontchev and S. Bancheva, *Mikrochim. Acta*, 1976 **II,** 487.
[28] M. Otto, J. Lerchner, T. Pap, H. Zwanziger, E. Hoyer, J. Inczédy and G. Werner, *J. Inorg. Nucl. Chem.*, 1981, **43,** 1101.
[29] R. P. Igov, M. D. Jaredić and T. G. Pecev, *Talanta*, 1980, **27,** 361.
[30] D. P. Nikolelis and T. P. Hadjiioannou, *Anal. Chem.*, 1978, **50,** 205.
[31] K. B. Yatsimirskii, E. E. Kriss and G. T. Kurbatova, *Zh. Analit. Khim.*, 1976, **31,** 598.
[32] G. A. Milovanović, T. J. Janjić, S. Petrović and G. Kuzmanović, *Microchem. J.*, 1980, **25,** 380.
[33] A. Moreno, M. Silva, D. Pérez-Bendito and M. Valcárcel, *Analyst*, 1983, **108,** 85.
[34] J. L. Ferrer and D. Pérez-Bendito, *Quim. Anal.*, 1984, **3,** 40.
[35] A. Marín, M. Silva and D. Pérez-Bendito, *Anal. Chim. Acta*, 1987, **197,** 77.
[36] S. P. Klochkovskii and G. D. Klochkovskaya, *Zh. Analit. Khim.*, 1977, **32,** 736.
[37] S. P. Klochkovskii and G. D. Klochkovskaya, *Zh. Analit. Khim.*, 1978, **33,** 1749.
[38] E. Metasti and E. Pellizzetti, *Anal. Chim. Acta*, 1975, **78,** 227.
[39] R. A. Hasty, E. J. Lima and J. M. Ottaway, *Analyst*, 1977, **102,** 313.
[40] M. Ternero, F. Pino, D. Pérez-Bendito and M. Valcárcel, *Microchem. J.*, 1980, **25,** 102.
[41] H. Weisz, S. Pantel and W. Meiners, *Anal. Chim. Acta*, 1976, **82,** 145.
[42] H. Weisz, S. Pantel and G. Marquardt, *Anal. Chim. Acta*, 1982, **143,** 177.
[43] D. Klockow, J. Auffarth and C. Kopp, *Anal. Chim. Acta*, 1977, **89,** 37.
[44] P. Phull and P. C. Nigam, *Talanta*, 1983, **30,** 401.
[45] G. S. Vasilikiotis, C. Papadopoulos, D. G. Themelis and M. C. Sofoniou, *Microchem. J.*, 1983, **28,** 431.
[46] Z. Gregorowicz, T. Suwinska and P. Gorka, *Chem. Anal. (Warsaw)*, 1974, **19,** 447.
[47] P. J. Ke and R. J. Thibert, *Mikrochim. Acta*, 1972, 768.
[48] T. I. Fedorova, K. B. Yatsimirskii and T. G. Ermolaeva, *Zh. Analit. Khim.*, 1975, **30,** 59.
[49] Y. I. Grosse and A. D. Miller, *Metody Anal. Redkometal Miner. Rud. Gorn. Porod.*, 1971, **2,** 52.
[50] V. Kurzawa and Z. Kurzawa and Z. Swit, *Chem. Anal. (Warsaw)*, 1976, **21,** 791.
[51] Z. Kurzawa, H. Matusiewicz and K. Matusiewicz, *Chem. Anal. (Warsaw)*, 1976, **21,** 797.
[52] Z. Kurzawa and E. Kubaszewski, *Chem. Anal. (Warsaw)*, 1976, **21,** 565.
[53] R. P. Igov and G. Z. Miletic, *Microchim. Acta*, 1981 **I,** 355.
[54] A. Moreno, M. Silva and D. Pérez-Bendito, *Anal. Lett.*, 1983, **16,** 747.
[55] D. P. Nikolelis and T. P. Hadjiioannou, *Anal. Chim. Acta*, 1978, **97,** 111.
[56] H. A. Mottola, M. S. Haro and H. Freiser, *Anal. Chem.*, 1968, **40,** 1263.
[57] D. P. Nikolelis and T. P. Hadjiioannou, *Mikrochim. Acta*, 1977 **I,** 125.
[58] M. Müller, H. Schuring and G. Werner, *Talanta*, 1979, **26,** 785.
[59] L. A. Lazarou and T. P. Hadjiioannou, *Anal. Chem.*, 1979, **51,** 790.
[60] T. Raya Saro and D. Pérez-Bendito, *Mikrochim. Acta*, 1984 **I,** 467.
[61] M. A. Ratina, G. A. Zolotova and I. F. Dolmanova, *Zh. Analit. Khim.*, 1980, **35,** 1366.
[62] S. Pantel and H. Weisz, *Anal. Chim. Acta*, 1977, **89,** 47.
[63] V. N. Antonov and S. U. Kreingol'd, *Zh. Analit. Khim.*, 1976, **31,** 193.

[64] M. A. F. Sekheta and G. A. Milovanović, *Bull. Soc. Chim. Beograd*, 1980, **45**, 41.
[65] M. A. F. Sekheta, G. A. Milovanović and T. J. Janjić, *Bull. Soc. Chim. Beograd*, 1979, **44,** 447.
[66] I. F. Dolmanova, G. A. Zolotova, T. N. Mazko, G. D. Dymshakova and P. P. Trunov, *Zh. Analit. Khim.*, 1977, **32,** 807.
[67] I. F. Dolmanova, G. A. Zolotova, I. M. Popova and E. B. Smirnova, *Zh. Analit. Khim.*, 1980, **35,** 1372.
[68] I. F. Dolmanova, G. A. Zolotova and T. N. Mazko, *Zh. Analit. Khim.*, 1977, **32,** 1025.
[69] I. A. Il'icheva, N. L. Antul'skaya, I. F. Dolmanova and L. A. Petrukhina, *Metody Anal. Kontrolya Kach. Pord. Khim. Prom-sti*, 1978, **8,** 43.
[70] J. L. Ferrer, *Doctoral Thesis*, University of Córdoba, 1980.
[71] D. Klockow, H. Ludwig and M. A. Giraudo, *Anal. Chem.*, 1970, **42,** 1268.
[72] J. Auffarth and D. Klockow, *Anal. Chim. Acta*, 1979, **111,** 89.
[73] F. F. Gaál, V. Soros and V. D. Canic, *Mikrochim. Acta*, 1975 **II,** 689.
[74] J. De Oliveria and E. Cunha, *Rev. Quim. Ind.* (*Rio de Janeiro*), 1979, **48,** 561.
[75] H. Ludwig, H. Weisz and T. Lenz, *Anal. Chim. Acta*, 1974, **70,** 359.

4

Catalytic titrations

4.1 INTRODUCTION

A major application of the inhibitory effect in catalysed reactions is titrimetric end-point detection. This technique is a combination of kinetic and equilibrium aspects, resulting in a new end-point detection procedure.

The possibility of using catalysed reactions for end-point detection in titrations was originally examined by Yatsimirskii and Fedorova [1] in 1962. These authors proposed the name 'catalytic' or 'catalymetric' titrations and applied it to the determination of silver with iodide by a 'simulated' titration procedure involving the iodide-catalysed reaction between Ce(IV) and As(III). In 1966, Weisz and Muschelknautz [2] formally introduced catalytic titrations in applying the novel methodology to various systems such as that of Tiron and perborate, used in the titration of EDTA with Co(II) (the catalyst of the indicator reaction), as well as to a number of back-titrations involving metal ions. As these authors point out, their catalytic detection of the end-point involves an open system, whereas Yatsimirskii's technique is based on a closed system.

To date, the chief contributions in this field have originated from the research teams headed by Weisz, Hadjiioannou and Gaál, based in Germany, Greece and Yugoslavia, respectively. The principles and applications of this technique have been thoroughly described in two reviews [3, 4] and a paper by Weisz [5]. Another review was exclusively devoted to thermometric catalytic titrations [6].

The instrumentation available today allows the use of automated, or more often, semi-automated, procedures that result in increased rapidity and reproducibility as well as in greater accuracy.

4.2 DEFINITION OF CATALYTIC TITRATION

Catalytic titrations are based on the inhibitory effect of a substance (whether an anion, ligand or metal) on a metal- or anion-catalysed reaction and are applied to the determination not only of the inhibitor, but also of the catalyst or other species (generally metal ions) with no catalytic or inhibitory properties.

The application of this methodology involves two consecutive reactions that can be schematically expressed as follows:

(1) *Titration reaction*

$$\text{titrant (catalyst)} + \text{analyte (inhibitor)} \rightarrow \text{products}$$

(2) *Indicator reaction*

$$A + B \xrightarrow{\text{excess of titrant (catalyst)}} \text{products}$$

In the titration reaction the catalyst reacts with the unknown species. This requires the complex or insoluble product to have a large formation constant or a small solubility product as the case may be. In addition, the reaction must meet the typical requirements for a conventional titration reaction (i.e. it should be fast, stoichiometric and go to completion). The indicator reaction, the ingredients of which are added to the unknown solution, allows the instrumental monitoring of the titration by means of one of its reactants or products.

The process develops as follows: the catalyst added from the burette starts the titration reaction. The indicator reaction is only initiated by excess of titrant (catalyst) after the end-point, and can be monitored photometrically, fluorimetrically, potentiometrically, conductimetrically, thermometrically or even visually. Thus, the catalyst acts first as a reagent for the analyte and then as a catalyst for the indicator reaction, of which the analyte is a transient inhibitor. The indicator reaction is aimed at end-point detection and thus plays the role of the typical chemical or physicochemical indicators employed in classical titrimetry.

On the other hand, it is essential that the optimum conditions for development of the titration reaction (pH, solvent, ionic strength) should be the same as those for the indicator reaction.

A distinction should be made at this point between catalytic titrations proper and catalysed titrations. In the latter, the titration reaction is accelerated by addition of a catalyst which does not take part in the indicator reaction. They should also be distinguished from 'kinetic titrations' and those using kinetic end-point detection (whether catalytic or non-catalytic).

Kinetic titrations involve passage of titrant and analyte streams, through a Y-junction for mixing, into a flow-through cell where an analytical signal can be measured. The flow-rate of titrant is changed, while that of the analyte is kept constant, until the analytical signal in the cell reaches a value corresponding to the titration end-point. The analyte concentration is then a linear function of the titrant flow-rate, and a standard solution of analyte can be used for calibration. A typical example of this is the automatic potentiometric titration proposed by Blaedel and Laessig [7]. The set-up used is straightforward and needs no burette (Fig. 4.1). The

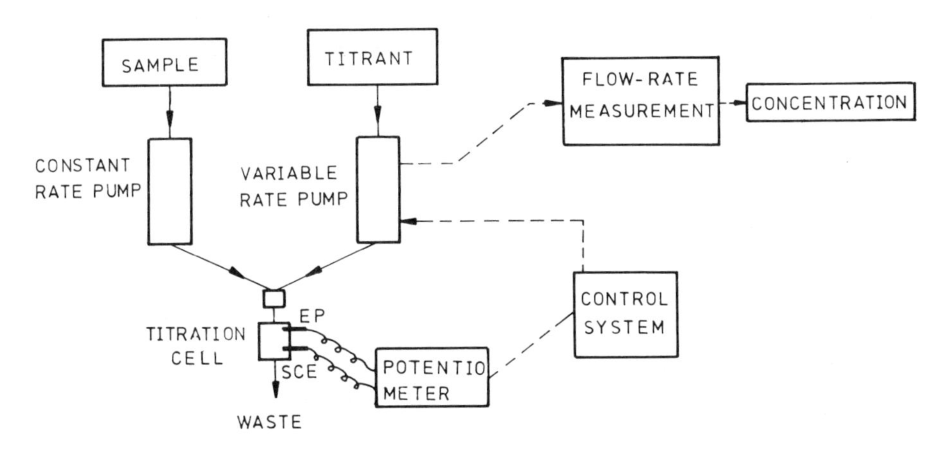

Fig. 4.1 — Scheme of a kinetic titration.

indicator system (potentiometer) sends the signal (potential) to the central unit, which compares it with that corresponding to the end-point. This difference starts a servo mechanism controlling the titrant flow-rate, which is thus conveniently measured. This type of titration is similar to 'stat' methods in open systems, which are described in detail in Chapter 7.

A representative titration involving kinetic end-point detection is that of polyhydroxy compounds with periodate (a slow stoichiometric reaction) with a microprocessor-controlled set-up [8]. The procedure involves keeping constant the monitored signal obtained upon addition of the sample in an initial volume of titrant. The gradual addition of increasingly smaller volumes of titrant (controlled by the computer) results in decreases of the potentiometric signal and hence in increasingly lower analyte concentrations and reaction rates. Once the signal has stabilized, the addition of a volume of titrant restores its initial level. The end of the titration, indicated by the microprocessor, is marked by reduction of the signal intensity to one thousandth of the original (Fig. 4.2). After introduction of suitable volume corrections, the volume of titrant required to reach this situation is directly proportional to the analyte concentration. Pardue and Fields [9] have established clear-cut differences between both titration techniques.

Catalytic titrations can be implemented in direct, reverse or indirect modes. In direct titrations, the components of the indicator reaction and the analyte are placed in the titration vessel and titrated with a standardized catalyst solution. Reverse titrations, applied to the determination of the catalyst, involve titrating a known volume of standard inhibitor solution with the analyte (catalyst). Indirect or back-titration is the most interesting of the three modes, as it allows the determination of metals with no catalytic or inhibitory properties. It involves adding a known and excessive amount of inhibitor to the metal to be determined, and titration of the excess with standard catalyst solution. It is therefore essential that the metal and the inhibitor form a compound of greater stability than that formed between the inhibitor and the catalyst, in order to avoid an unwanted displacement reaction.

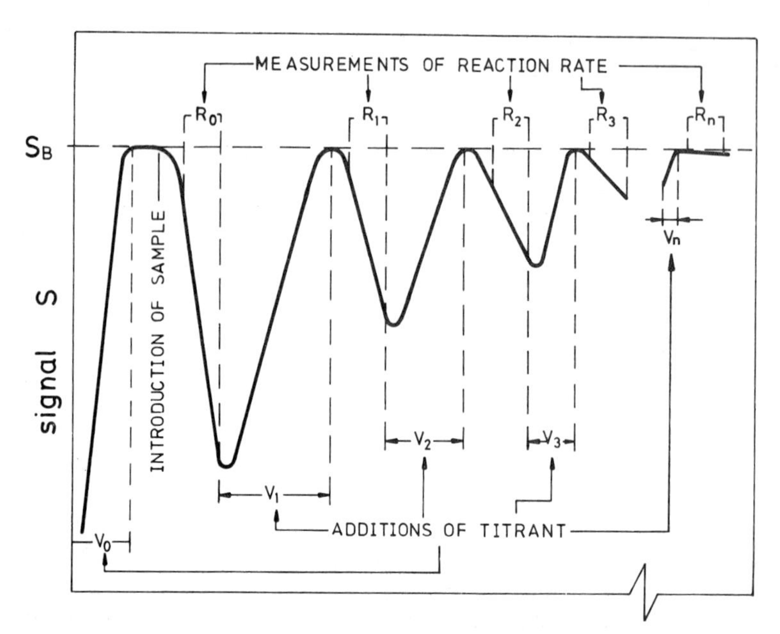

Fig. 4.2 — Graphical presentation of the sequence of events taking place during a titration with kinetic end-point detection. The scale of the time axis is not uniform throughout. V_0, V_1, etc. stand for the time periods during which reaction rate measurements are made. (Reproduced by permission, from C. E. Efsthathiou and T. P. Hadjiioannou, *Talanta*, 1983, **30**, 145. Copyright 1983, Pergamon Press).

4.3 TITRATION CURVES

Titration curves are normally run by plotting the analytical signal (relative measurements) as a function of the volume of titrant added. If the relationship between both parameters is linear, then the titration curve has two linear segments which can be extrapolated to intersect at the end-point. The curve (Fig. 4.3) is rising or falling

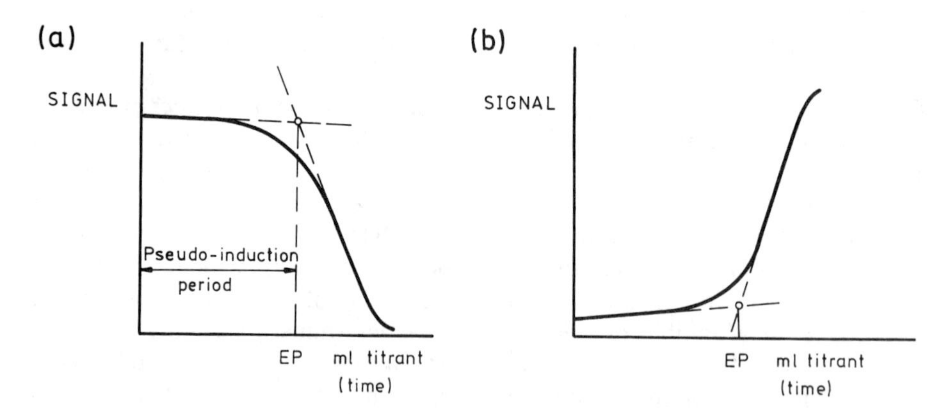

Fig. 4.3 — General types of titration curves corresponding to the monitoring of (a) a reactant or (b) a product of the indicator reaction.

according to whether the monitored species is a reactant or a product of the indicator reaction. The horizontal segment corresponds to the addition of the catalyst, intended to react with the analyte in the presence of the indicator reaction, which in practice tends to develop to some extent in the absence of the catalyst, a shortcoming which can be circumvented by joint addition of the titrant and one of the components of the indicator reaction. The other segment of the titration curve corresponds to the development of the catalysed reaction and its slope depends on the rate of the catalysed reaction. As stated above, the intersection of the two segments corresponds to the titration end-point. Sometimes the titration reaction does not develop to completion, so the excess of catalyst present starts the indicator reaction before the equivalence point. This results in a curved segment in the vicinity of the end-point, which thus has to be determined by extrapolation of the two linear segments.

By use of a semi-automatic device capable of delivering the titrant at constant speed and continuously monitoring the signal yielded, the x-axis of the titration plot can be scaled in time units. In this case, the time elapsed between the start of the titration and the end-point is known as the *pseudo-induction period* and is proportional to the inhibitor concentration.

The plot of tan α (the reaction rate) as a function of titrant volume is of historical interest, as it was the first to be described for this type of titration [1]. It was originally used in the titration of Ag(I) with I^- by the classical Sandell–Kolthoff reaction involving the Ce(IV)/As(III) system. In practice, to a series of standard flasks containing equal volumes of the Ag(I) solution to be analysed, and also the components of the indicator reaction, increasing amounts of iodide are added, and the solutions are diluted to the same volume. The indicator reaction is monitored individually in each flask and a kinetic curve is obtained for each. The slope of the curve remains constant as long as there is excess of Ag(I) present. When it is iodide which is in excess, the slope increases in proportion to its concentration. The plot of tan α as a function of the amount of iodide added (Fig. 4.4) consists of two straight segments intersecting at the equivalence point, at which $[\text{Ag(I)}] = [\text{I}^-] = \sqrt{K_{sp}} = 10^{-8}M$.

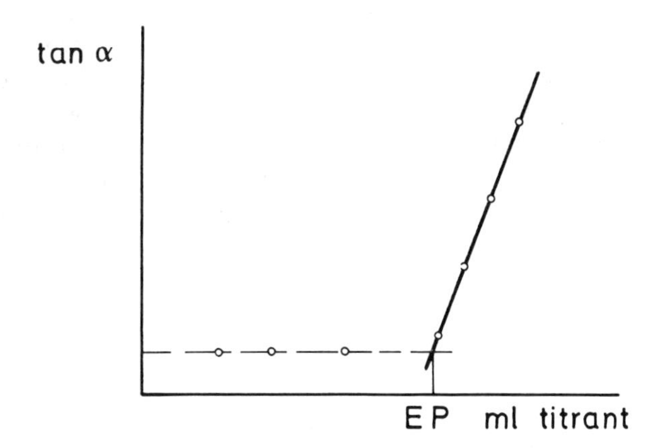

Fig. 4.4 — Plot of the reaction rate as a function of the volume of titrant added.

Mottola and Freiser [10] have devised another procedure to plot the evolution of a catalytic titration and to determine its end-point, based on kinetic considerations of the indicator reaction. The procedure has been applied to the determination of cyanide with copper(II). These ions act as inhibitor and catalyst, respectively, of the indicator reaction (viz. the autoxidation of ascorbic acid, H_2A), which is monitored photometrically.

The rate equation of this reaction can be expressed as:

$$-\mathrm{d}[H_2A]/\mathrm{d}t = k_1[H_2A] + k_2[H_2A][Cu^{2+}]$$

where k_1 and k_2 are the rate constants of the overall uncatalysed and catalysed reactions, respectively. If the titrant is added at constant speed, then $\mathrm{d}[Cu^{2+}]/\mathrm{d}t = Q$ = const. and the equation can be written as:

$$-\mathrm{d}[H_2A]/\mathrm{d}[Cu^{2+}] \propto -\mathrm{d}[H_2A]/\mathrm{d}t = k_1[H_2A] + k_2[H_2A][Cu^{2+}]$$

whence

$$-\mathrm{d}(\ln[H_2A]) = k_1' + k_2'\mathrm{d}[Cu^{2+}]^2 \qquad (4.1)$$

with $k_1' = k_1/Q$ and $k_2' = k_2/2Q$.

Integration of Eq. (4.1) after neglecting the first term on the right-hand side (the rate of the catalysed reaction is far higher than that of the uncatalysed reaction) yields:

$$-\ln\,[H_2A] = k_2''\,[Cu^{2+}]^2$$

There is therefore a direct relationship between the logarithm of the concentration of the monitored species and the square of the catalyst concentration. Insofar as there is also a direct relationship between $[H_2A]$ and the absorbance measured, the plot of $\log A = f[Cu^{2+}]^2$ (Fig. 4.5) will be a set of straight lines of different slopes, intersecting the horizontal segment corresponding to the time elapsed in the titration of CN^- with Cu(II), throughout which the concentration of H_2A remains constant. The slopes differ because the titration is done at a speed of addition suited to the gradually increasing concentrations of CN^- used and this speed is one of the factors included in Eq. (4.1). Logically, this segment is not obtained when no inhibitor has been added. At the intersection, the $[CN^-]/[Cu^{2+}]$ ratio is 2 and corresponds to the reaction

$$Cu^+ + CN^- \rightarrow CuCN \xrightarrow{CN^-} Cu(CN)_2^-$$

which is preceded by the reduction of Cu(II) to Cu(I) by ascorbic acid.

The titration curve in Fig. 4.3 (the most common) allows the analyte to be

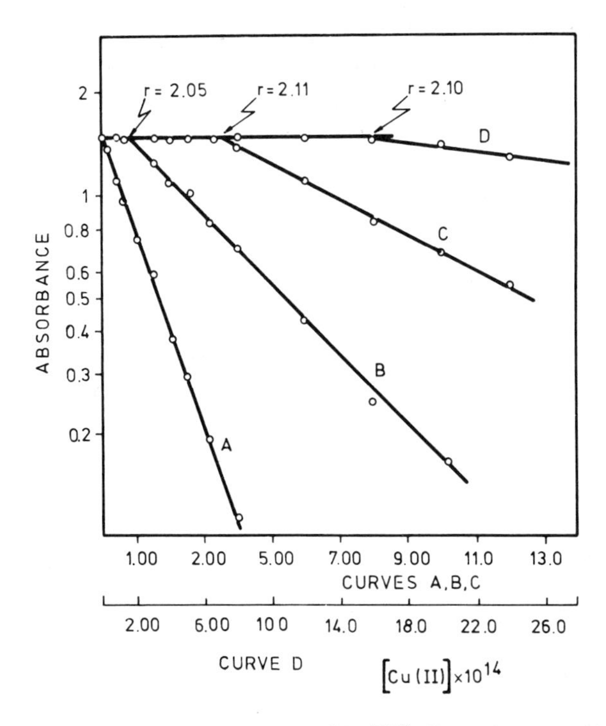

Fig. 4.5 — Typical plot of log absorbance *vs*. $[Cu(II)]^2$. Curve A, no cyanide added; curve B, $2 \times 10^{-7}M$ cyanide; curve D, $8 \times 10^{-7}M$ cyanide. $r = [CN^-]/[Cu(II)]$. (Reproduced with permission, from H. A. Mottola and H. Freiser, *Anal. Chem.*, 1968, **40**, 1266. Copyright 1968, American Chemical Society).

determined by the proportional or the working-curve method. The former, very simple method, requires knowledge of the stoichiometry of the titration reaction as it is based on the relationship between the volume of titrant added (pseudo-induction period) and the analyte concentration. The latter method involves the prior construction of a calibration curve for the volumes used for a series of samples containing known amounts of the inhibitory analyte.

Insofar as the excess of catalyst reacts catalytically rather than stoichiometrically after the end-point, no quantitative kinetic evaluation of the results is required; a qualitative interpretation of the changes in the measured parameter is usually sufficient.

4.3.1 Theoretical aspects

As with other analytical procedures, experimental and methodological work has preceded theoretical studies in the field of catalytic titrations. The mathematical treatment of the titration curves has the added interest of revealing the kinetics of the indicator reaction and of clarifying the influence of factors affecting their shape.

Endeavours aimed at mathematical definition of catalytic titration curves have mainly been based on complexometric titrations meeting a number of requirements, namely (a) the rate constant of the indicator reaction should be rather large; (b) the titrant should be added continuously; (c) the temperature should have no effect on the thermodynamics or kinetics of the reactions involved; (d) both the rate of the uncatalysed reaction and any changes caused in the measured parameter by the titration reaction should be negligible; (e) the indicator reaction should have no induction period; (f) stirring should be optimum.

The earliest mathematical treatment of the different parts of a catalytic titration curve was made by Mottola in 1970 on the basis of the determination of aminopolycarboxylic acids with a flow system [11].

If the concentration of all the species involved in the process except the catalyst and the monitored product is assumed to remain constant throughout the titration, the analytical signal will be proportional to the concentration of the monitored product, which disappears in the course of the reaction. If this is monitored photometrically, then:

$$-\mathrm{d}D/\mathrm{d}t = -\varepsilon l \mathrm{d}[\mathrm{A}]/\mathrm{d}t$$

where D is the absorbance and A is the monitored species. Figure 4.6 shows the

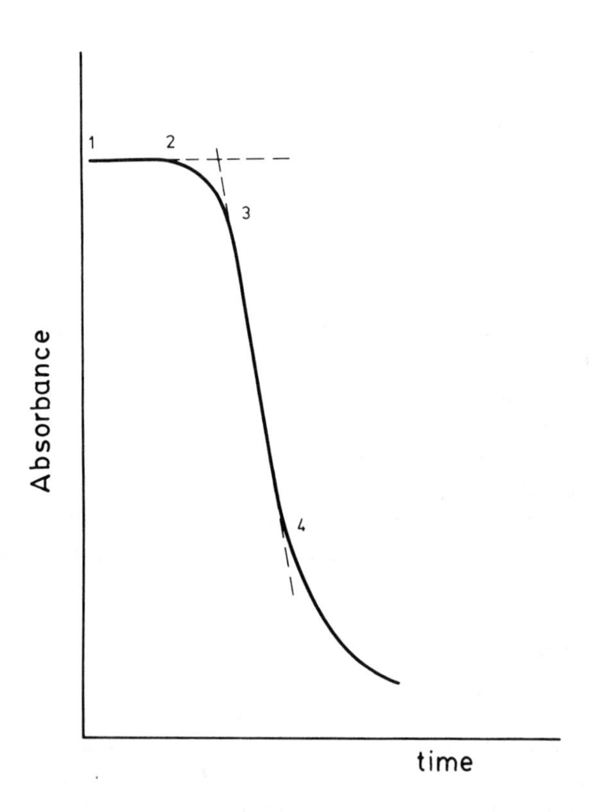

Fig. 4.6 — Plot of absorbance *vs*. time for a constant speed of titrant addition. (Adapted with permission, from H. A. Mottola, *Anal. Chem.*, 1970, **42**, 630. Copyright 1970, American Chemical Society).

typical titration curve obtained. If v is the volume of titrant added, then:

$$\mathrm{d}v = q\mathrm{d}t \tag{4.2}$$

and

$$[\mathrm{C}] = mv \tag{4.3}$$

where [C] is the overall titrant concentration and q and m are proportionality constants.

If the only catalytically active species is the free metal ion, then:

$$-\mathrm{d}D/\mathrm{d}t = k_0'D + k_1'[\mathrm{C}]D \tag{4.4}$$

This equation conforms to the shape of the titration curve after substitution of the catalyst concentration at equilibrium.

The catalyst concentration corresponding to the initial segment (1–2 in Fig. 4.6) will be virtually zero provided that the formation constant of the catalyst–inhibitor complex is sufficiently large and that some inhibitor is present in the reaction medium. Under these conditions, substitution of Eqs. (4.2) and (4.3) into (4.4) yields:

$$-\mathrm{d}D/D = k_0'q'\mathrm{d}v \tag{4.5}$$

where $q' = 1/q$.

Integration of Eq. (4.5) between D_0 and D with $v_o = 0$ yields:

$$\ln (D_0/D) = k_0'q'v$$

or, in exponential form

$$D = D_0 \exp(-k_0'q'v) \tag{4.6}$$

Further, if the uncatalysed reaction is assumed to make no contribution, k_0' will tend to zero and hence $D_0 = D$ (initial straight segment).

The equilibrium catalyst concentration, [C], corresponding to the final segment (3–4 in Fig. 4.6) can be expressed as:

$$[\mathrm{C}] = mv - mv_e = m(v - v_e)$$

where v_e is the volume of titrant corresponding to the equivalence point. By substituting this expression into Eq. (4.4) and again assuming no contribution from the uncatalysed reaction, we obtain

$$-\,\mathrm{d}D/\mathrm{d}t \;=\; [k_1'm(v-v_e)]D$$

The integration of this equation between D_e and D (corresponding to v_e and v, respectively) leads to:

$$\ln\,(D_e/D) \;=\; k_1'mq'(v-v_e)^2/2$$

or in exponential form,

$$D \;=\; D_e\,\exp[-\,k_1'mq'(v-v_e)^2/2]$$

According to this equation, D varies markedly with v, so the exponential curve approximates to a straight line. The catalyst concentration corresponding to this segment is sufficiently large to result in the maximum rate being independent of its concentration.

The shape of the titration curves changes somewhat after point 4 (Fig. 4.6) as a result of the decrease in the concentration of the monitored species.

A more rigorous treatment has been reported by Gaál and Abramović [12]. These authors take into account the catalyst concentration from the start of the titration, as well as the dilution factor. The mathematical treatment involves deriving an equation for the catalyst concentration throughout the titration. Such a concentration is substituted into the rate equation for the indicator reaction, integration of which over the interval considered provides a mathematical equation decribing in detail the titration curve.

The catalyst concentration, [C], is calculated from the formation constant of the catalyst–inhibitor complex. If this is assumed to be a 1:1 complex:

$$K_f \;=\; [\mathrm{CI}]/[\mathrm{C}][\mathrm{I}] \tag{4.7}$$

where [I] and [CI] are the equilibrium concentrations of inhibitor and complex, respectively. On the other hand,

$$[\mathrm{I}] \;=\; \frac{V_0[\mathrm{I}]_0}{V_0+rt} - [\mathrm{CI}]$$

$$[\mathrm{CI}] \;=\; \frac{vt[\mathrm{C}]_0}{V_0+rt} - [\mathrm{C}]$$

where V_0 is the initial volume of the solution to be titrated (expressed in litres), $[\mathrm{I}]_0$ and $[\mathrm{C}]_0$ are the initial concentrations of the inhibitor and catalyst (titrant), respectively, r is the rate of titrant addition (l./sec) and t is the time (sec). Substitution of these expressions into Eq. (4.7) yields a convenient equation for calculation of the equilibrium concentration of catalyst, [C].

On the other hand, the rate equation of an indicator reaction $A + B \rightarrow P$, assumed to be of first order in reactant A, can be expressed as:

$$-\mathrm{d}[A]/\mathrm{d}t = (k_0' + k_1'[C])[A] + \frac{[A]r}{V_0 + rt} \tag{4.8}$$

The last term in this equation describes the change in the rate of the indicator reaction ($\Delta A/\Delta t$) resulting from the increase in volume of the reacting mixture in the course of the titration.

Substituting [C] into Eq. (4.8) and integrating from $[A] = [A]_0$ for $t = 0$ and $[A] = [A]_t$ for time t, yields an expression allowing the calculation of [A] at a given time t:

$$[A]_t = [A]_0 \frac{V_0}{V_0 + rt} \exp(-\phi) \tag{4.9}$$

ϕ being a complex function [12] comprising all the parameters associated with the titration.

Since the product is generally the species monitored and its concentration is given by

$$[P]_t = [A]_0 \frac{V_0}{V_0 + rt} - [A]_t$$

Eq. (4.9) can be rewritten as:

$$[P]_t = [A]_0 \frac{V_0}{V_0 + rt} [1 - \exp(-\phi)] \tag{4.10}$$

i.e. the overall equation of the titration curve can be reduced to simpler equations equally efficient in predicting the shape of the calibration curve.

One such simplification involves assuming that the concentration of catalyst arising from the complex dissociation will be zero before the end-point (t_e) is reached. Under such conditions, Eq. (4.10) becomes:

$$[P]_t = [A]_0 \frac{V_0}{V_0 + rt} [1 - \exp(-k_0't)] \tag{4.11}$$

which is similar to Eq. (4.6), if the volume correction is neglected and V is assumed to be equivalent to t in the exponential term.

This equation corresponds to the initial segment of the titration curve insofar as $\exp(-k_0't) \rightarrow 0$ when the contribution of the uncatalysed reaction is practically negligible and hence $[P] \rightarrow 0$.

A second approximation is concerned with titration times after the end-point.

The catalyst concentration is then given by:

$$[\mathrm{C}] = r(t-t_e)[\mathrm{C}]_0/(V_0+rt)$$

Substituting this into Eq. (4.8) and applying the same treatment as above, gives the simplified equation.

$$[\mathrm{P}] = [\mathrm{A}]_0 \frac{V_0}{V_0-rt}[1-\exp([\mathrm{M}])] \tag{4.12}$$

where

$$[\mathrm{M}] = -k_0't - k_1't[\mathrm{C}]_0(t-t_e) + k_1'[\mathrm{C}]_0[(V_0/r)+t_e] \ln \frac{V_0+rt}{V_0+rt_e}$$

Equation (4.12) is representative of the final segment of the titration curve.

These simplified equations have been checked experimentally with the aid of a computer [12]. The study allowed several interesting conclusions to be drawn, as follows.

(a) The complex formation constant should be of the order of 10^8 or larger if the catalyst concentration at equilibrium is to be negligible so that Eq. (4.11) can be applied.
(b) The reactant concentrations have no effect on the shape of the titration curve. Equations (4.11) and (4.12) reveal a linear relationship between the measured parameter and such concentrations.
(c) If the complex formation constant is sufficiently small the catalysed reaction may take place before the equivalence point, thus invalidating Eq. (4.11). On the other hand, the end-point can be readily determined for a given set of conditions for k_1' values greater than 10 l.mole^{-1}.sec^{-1} (or lower if $[\mathrm{C}]_0$ or r is sufficiently large).
(d) The effect of the rate constant of the uncatalysed reaction is particularly significant in the period preceding the equivalence point, Eq. (4.11). Thus, the end-point for a pseudo first-order indicator reaction can be satisfactorily determined for $k_0' \leq 10^{-3}$ sec^{-1}.
(e) Likewise, the titrant and analyte concentrations should not be less than 5×10^{-4} and $5 \times 10^{-5}M$, respectively, for reliable results to be obtained. Analyte solutions more dilute than this can be titrated, or less concentrated titrants used, only if the titration reaction has a rather large rate constant and is quantitative at the equivalence point.

A similar theoretical treatment has been applied to the catalytic titration curves corresponding to precipitation or redox [13] and neutralization reactions [14].

4.4 TITRATION MODES

Three titration modes have been described so far. The one given above in describing the catalytic titration technique could be termed the *normal mode*. It involves using the catalyst as titrant and placing the analyte and the components of the indicator reaction in the titration vessel under preselected working conditions.

The so-called *substitution titrations* [15,16] allow the components of the indicator reaction, the catalyst and the analyte-inhibitor (which also masks the catalyst) to be added to the titration vessel and the last-named to be titrated with a suitable standard solution. As soon as the end-point is reached, the catalyst is demasked, thereby starting the indicator reaction. In this manner Zn(II) (inhibitor of the Mn(II)-catalysed reaction between resorcinol and H_2O_2) has been determined with EDTA. All the reagents involved are placed in the vessel, with the exception of the Zn(II), which is added from a burette [15].

There is also a less frequently encountered method called the *brake mode* [17,18]. This uses the inhibitor as titrant, to which some of the components of the indicator reaction are added. The titration consists of two basic stages: in the first the titration and indicator reaction take place simultaneously; in the second, only the uncatalysed reaction occurs once the stoichiometric equivalence between the catalyst and the inhibitor (usually EDTA, EGTA or NTA) has been reached. The measured signal varies throughout the first stage as a result of the development of the catalysed reaction; however, the inhibitor added consumes catalyst from the sample, thus 'braking' the effect on the signal (hence the name). The signal finally becomes practically constant in the second stage once the active form of the catalyst has disappeared, since the rate of the uncatalysed reaction is generally very low.

The titration curve therefore consists of a rising linear segment followed by a horizontal. This type of titration is suitable for the direct determination of the catalyst and allows the indirect determination of the inhibitor. A correct application of this mode calls for an appreciable catalytic effect (otherwise negative errors occur), as well as for a fast titration reaction with a high equilibrium constant (in the case of complexation titrations) if delayed end-points are to be avoided and linear first segments are to be obtained. If these conditions are met, the volume of titrant (inhibitor) needed for the end-point to be reached will be directly proportional to the concentration of catalyst in the sample. This mode has been used, among others, in the determination of copper as catalyst for the decomposition of H_2O_2 [17] and the reaction between hydrazine and H_2O_2 [18], both monitored thermometrically.

Finally, a novel mode known as *substrate-inactivation titration* [19] is worth mentioning, in which one of the reactants taking part in the indicator reaction acts as the titrant for the analyte (the inhibitor). The analyte is placed in the titration vessel together with the other reactant and the catalyst. In the first stage, the titrant added is consumed by reaction with the analyte. Once the latter has been fully depleted the indicator reaction takes place, and is accelerated by the catalyst. This mode has been applied to the determination of Hg(II), an inhibitor of the Cu(II)-catalysed oxidation of 4,4′-dihydroxybenzophenone thiosemicarbazone by complexation with Hg(II), and only once this reaction is complete does it act as a reactant in the indicator reaction, which is thus initiated. The titration curve usually obtained has an initial straight, nearly horizontal segment, and a second rising one corresponding to the catalysed reaction. This mode is more limited than the others above, but gives higher selectivity.

A combination of the brake and substrate-inactivation modes has allowed the resolution of binary mixtures of mercury with copper or cadmium by simultaneous titration with the above-mentioned system (see Fig. 4.7, corresponding to the Hg–Cu mixture) [19].

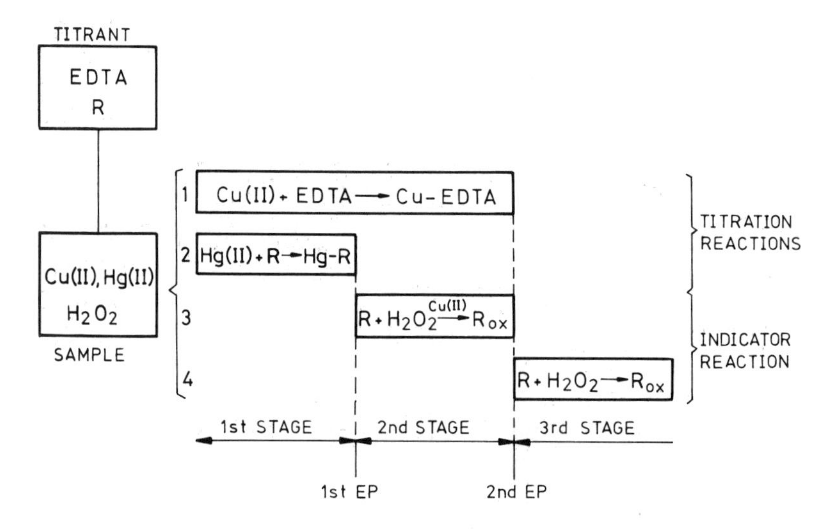

Fig. 4.7 — Foundation of the titration of Hg and Cu by a combination of the brake and substrate-inactivation modes. (Reproduced by permission, from T. Raya Saro and D. Pérez-Bendito, *Anal. Chim. Acta*, 1986, **182**, 163. Copyright 1986, Elsevier Science Publishers).

The titration involves three stages (Fig. 4.8). In the first both titration reactions take place (1 and 2 in Fig. 4.7). In the second stage, which begins with depletion of the Hg(II), reactant R is oxidized by H_2O_2 in a reaction catalysed by excess of Cu(II) (reaction 3 in Fig. 4.7), which continues to react with EDTA. Reaction 4 (third stage, Fig. 4.7) takes place only when the Cu(II) has been fully consumed.

4.5 INDICATOR REACTIONS

Redox reactions are by far the commonest processes used as indicator reactions, followed by polymerization and hydrolysis (and to a lesser extent complexometric) reactions in that order.

The redox reactions commonly used for indicating purposes in catalytic titrations usually involve the Ce(IV)/As(III) system or the oxidation of an organic compound.

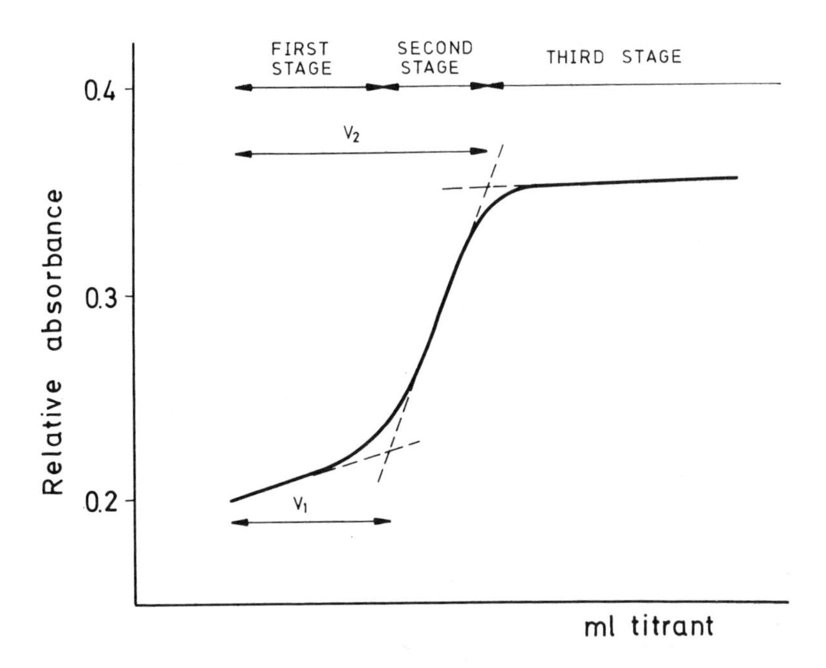

Fig. 4.8 — Simultaneous titration of copper and mercury by a combination of the brake and substrate-inactivation modes.

The catalytic titration technique originated in the *Sandell–Kolthoff reaction* [1, 2], which has since then been the basis for a host of applications in this field. Insofar as it is an iodide-catalysed reaction, it is applicable to the direct titration of metal inhibitors such as Hg(II), Hg(I), Pd(II), Au(III) and Ag(I) (the last being the first that was determined in this manner), with either photometric or thermometric detection of the end-point. This technique can also be applied to the indirect determination of sulphur-containing organic compounds or even the catalyst itself and other halides [6].

The oxidation of an organic compound is more frequently used as the indicator reaction of a catalytic titration because of the increased photometric sensitivity which can be achieved. Ascorbic acid, diphenols, arylamines, Malachite Green, Methyl Orange and azomethine derivatives are the commonest reagents used in this field and hydrogen peroxide, persulphate, iodate, periodate, bromate and dissolved oxygen their most frequent oxidants. These reactions are chiefly catalysed by Mn(II) and, less frequently, by Cu(II), Co(II) or Ni(II), all of which act as titrants. Aminopoly-carboxylic acids (EDTA, DCTA, EGTA, NTA) are the inhibitors frequently titratable. These applications are described at length below.

The polymerization of acetone has been exploited for the catalytic titration of an acid with KOH (as both titrant and catalyst), both dissolved in this organic solvent. An excess of hydroxide ions catalyses the polymerization of acetone to yield 4-hydroxy-4-methyl-2-pentanone according to:

$$2CH_3\text{-}CO\text{-}CH_3 \xrightarrow{OH^-} CH_3\text{-}CO\text{-}CH_2\text{-}\underset{\displaystyle OH}{\underset{|}{C}}\text{-}(CH_3)_2$$

This is an exothermic reaction and can be monitored thermometrically [20].

The hydrolysis of acetic anhydride, catalysed by traces of perchloric acid,

$$(CH_3CO)_2O + H_2O \xrightarrow{HClO_4} 2CH_3COOH$$

can also be followed thermometrically and is used as the indicator reaction in the titration of tertiary amines and organic salts with perchloric acid in glacial acetic acid media (2% water and 8% acetic anhydride) [21].

Despite the attractive end-point detection systems available for ligand-exchange reactions, no applications of these reactions to kinetic catalytic analysis (suggested by Margerum and co-workers in 1963 [22–25]) have been reported so far. Surprisingly, the vast potential of these and enzymatic reactions was envisaged as early as 1969 [3].

In contrast to indicator reactions, complex-formation and precipitation processes are the most frequent titration reactions. Acid–base and redox titration reactions are much less common, so much so that the first catalytic end-point detection in a redox titration was not reported until 1980 [26].

The complexometric reactions used normally involve EDTA and related compounds, cyanide, halides, etc., an organic compound/oxidant indicator system and a metal such as copper, manganese or cobalt as titrant and catalyst. It should be noted that catalytic titrations are more sensitive than conventional titrations involving metallochromic indicators (10^{-5}–$10^{-3}M$ ligand) and afford the determination of catalytically inactive cations such as those of alkaline earth metals, Mg, Cd, Zn, Ni, Al, Sn and In at low concentrations (of the order of a few μg/ml).

The use of precipitation reactions for this purpose is limited to those of Ag(I) and Pd(II) with I^-, involved in the direct titration of these metals by use of the Kolthoff–Sandell reaction.

There are also few catalytic titrations based on acid–base reactions (viz. those of strong acids with KOH in acetone and organic amines with perchloric acid).

Redox titration reactions must meet a twofold requirement: on the one hand, they should be reasonably fast; on the other, the titrant/catalyst, in the same oxidation state, should take part in another redox reaction (the indicator reaction). Hence their security, only one having been reported so far. Moreover, the components of the indicator reaction should not react with the analyte, a serious restriction on account of the difficulty of finding an active and reasonably selective redox reagent. Weisz and Pantel [26] have reported the titration of ascorbic acid with potassium dichromate, which catalyses the oxidation of *o*-dianisidine by hydrogen peroxide. The titration is done in the presence of EDTA (intended to mask the metal impurities that catalyse the aerial oxidation of ascorbic acid) and citric acid [an activator of Cr(VI) that enhances its oxidizing power after the equivalance point].

Vanadium(V), cerium(IV), thallium(III) and some other oxidants can be determined in an indirect manner with this system.

4.6 INSTRUMENTATION AND EXPERIMENTAL TECHIQUES

The vast development of instrumental techniques has allowed the complete or partial automation of catalytic titrations. Though visual detection is feasible when a perceptible colour change takes place at the start of the indicator reaction, semi-automatic instrumental procedures are the commonest detection tools for this type of titration. A typical titration affording visual detection is that of chloride, bromide and iodide with Ag(I), which catalyses the oxidation of Malachite Green to its leuco form by potassium peroxodisulphate in the presence of 2,2′-bipyridyl as activator [27, 28]. The equivalence point is marked by the appearance of a green or green-bluish colour which is readily visible, even in the presence of a large amount of precipitated silver halide. A few other titrations involving visual detection that have been automated to a greater or lesser extent, have been reported recently by Weisz and Schlimpf [29, 30]. Abe *et al.* [31] employed olfactory detection in the inverse titration of Mn(II) with EDTA, using the reaction between *p*-tolidine and hydrogen peroxide as indicator. This is initiated by the catalytic effect of Mn(II) once the end-point has been reached, and yields isonitriles or carbylamines of easily recognizable nauseating odour.

Figure 4.9 shows a semiautomatic catalytic titration system made up of four

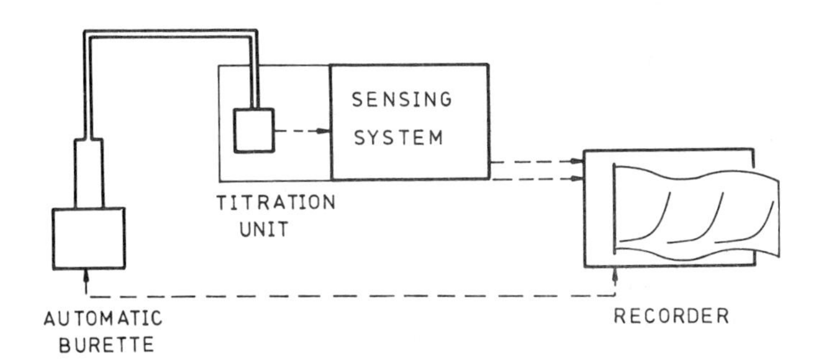

Fig. 4.9 — Scheme of the instrumentation required for semi-automatic catalytic titrations.

elements: titration unit, detection system, recorder and printer (optional).

The *titration unit* is an automatic piston burette acting as titrant dispenser and featuring adjustable delivery speed (0.1–4 ml/min). It delivers the reagent at a constant rate, thereby allowing the pseudo-induction period to be related to the analyte concentration with the aid of a timer. Some models can accommodate interchangeable borosilicate glass burettes of different volumes (5, 10 and 15 ml),with automatic filling. The titration unit includes a titration vessel to which the titrant is added from a Teflon capillary fitted to the burette outlet. The vessel can be a parallelepiped, cylinder or truncated cone made of glass or silica and holds different volumes of liquid (50, 100 ml) according to the particular detection system in use. It normally includes a helical stirrer and occasionally a circulation chamber for temperature control. The titration unit can be furnished with a storage system where the different operational programs and instrumental variables characterizing the titration can be kept.

The *detection system*, often accommodated with the titration unit in a single module (e.g. in thermometric titrations), monitors the signal from the titration vessel throughout the titrant addition. The detector can be photometric, fluorimetric, potentiometric or thermometric .

The *recorder* continuously registers the variation of the analytical signal as a function of time, which is related to the volume of titrant added. The different chart speeds available and the possibility of enlarging the recorded plot permit the titration curve to be run under the optimum conditions in every case.

Some manufacturers supply their titration systems with 20 cpl *printers* with typical data-transfer times of less then 1 sec. They provide a hard copy of the most relevant features of the procedure used (if a storage system has been used) and the results obtained.

As can be seen from Fig. 4.9, the titration unit/detection system module is fitted to the recorder, which in turn is synchronized with the dispensing unit (autoburette). Titration units furnished with memory-storage devices normally govern the operation of the complete system.

The titrant is generally added at a constant speed, though better defined curves can be obtained by use of a servo system allowing a more gradual addition in the vicinity of the end-point.

4.6.1 Detection systems

Optical (particularly photometric), electrical and thermometric detectors are undoubtedly the detection systems most frequently used in catalytic titrations. They normally consist of a series of elements, described in detail below.

4.6.1.1 Photometric detectors

These detectors, also known as 'phototitrators', are generally assembled in the same module as the titration unit and operate in much the same way as in ordinary titrations. As a rule, they include a titration cell with a light-path (2–4 cm) and volume (15–100 ml) that can be chosen according to particular needs. A variable-speed helical stirrer is introduced together with the burette tip into the titration vessel (Fig. 4.10), placed in the photometer in such a way that the light-path traverses it near its base.

The circulation device may consist of a peristaltic pump forcing the solution through the photometer flow-cell. Fibre optic photometric probes, which allow the absorbance to be measured in any type of titration vessel (Fig. 4.10,) have been recently introduced.

4.6.1.2 Fluorimetric detectors

Fluorimetric catalytic titrations, the earliest example of which dates from only 1983 [32], are based on the use of a conventional spectrofluorimeter furnished with a detector and fitted to the titration unit. The influence of temperature on the

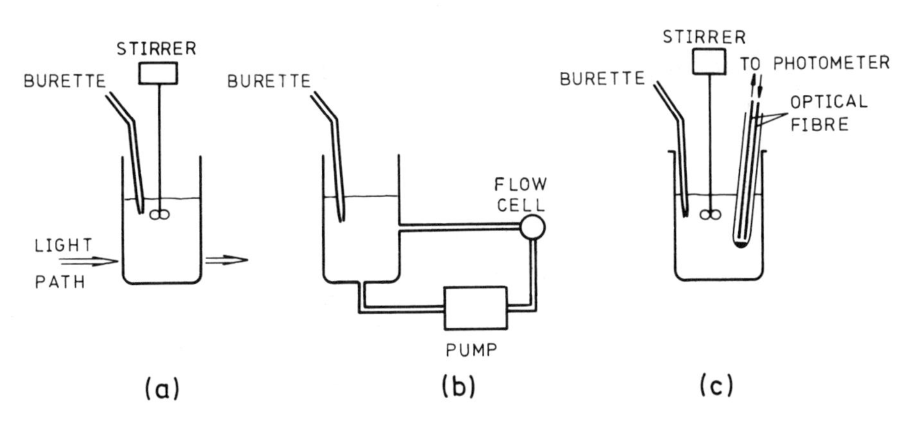

Fig. 4.10 — Generic devices for implementation of catalytic photometric titrations: (a) direct; (b) by circulation; (c) with a photometric probe.

detection technique and the indicator reaction calls for the use of 100-ml silica cells, either cylindrical or parallelepiped in shape, which are placed in the fluorimeter light-path, and kept at constant temperature.

In sensitivity and selectivity, fluorimetry excels other detection techniques used in this field. The marketing of fluorimetric titrators will foster the development of catalytic fluorimetric titrations, which at present are limited to the direct determination of EDTA [32] and DCTA [33], and the indirect determination of various metals such as the alkaline-earth metals, with the 2-hydroxybenzaldehyde thiosemicarbazone/H_2O_2 system and Fe(III) and Mn(II) as the titrant/catalysts.

4.6.1.3 Potentiometric and biamperometric detectors

These two types of detector are commonplace in catalytic titrations. Thus, potentiometric detectors have been employed in the titration of Ag(I) and Hg(II) with I^-, detection being by use of the As(III)/Ce(IV) system and measurement of the rapid decrease in potential between a platinum and a calomel electrode [13, 34]. The same chemical and electrochemical systems have been used in the titration of iodide by the brake technique; I^- and As(III) are placed in the titration vessel and Hg(II) and Ce(IV) are added from the burette. Once the equivalence point has been attained, excess of Ce(IV) gives rise to a sharp increase in the measured potential [13, 28]. Ion-selective electrodes, especially those for iodide, are also rather common in catalytic photometric titrations.

Biamperometric detectors have been employed in catalytic titrations involving hydrogen peroxide, with use of two platinum electrodes. Such is the case with the inverse titration of Mn(II) with EDTA. The metal catalyses the decomposition of H_2O_2, which is first oxidized and then reduced and can thus be detected biamperometrically [27, 28].

4.6.1.4 Thermometric detectors

The number of catalytic titrations involving thermometric detection approaches that of photometric applications. Such is their significance that they have been the subject of a comprehensive review [6]. Instrumentally, thermometric titrators consist of a temperature detector (thermistor) placed in the titration cell (10 ml), which is thermally isolated from its surroundings by means of a Dewar flask. The titration unit and the detector are thus built in a single module (Fig. 4.11).

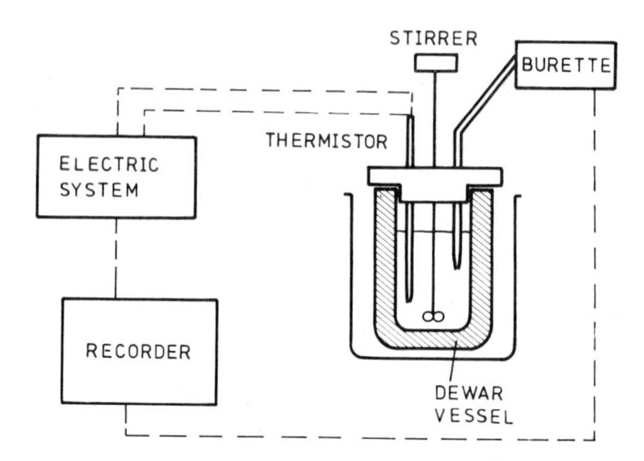

Fig. 4.11 — Generic device used for semi-automatic catalytic thermometric titrations.

Obviously, the indicator reaction used must be markedly endo- or exothermic [35]. The above-mentioned titrations of acids with an alkali in acetone medium and of tertiary amines with perchloric acid in glacial acetic acid, the indicator reactions for which (acetone polymerization and hydrolysis of acetic anhydride) are very exothermic, are characteristic examples. Ag(I), Pd(II) and Hg(II) directly, and Cl^-, Br^-, I^- and SCN^- in an indirect fashion, can also be determined thermometrically with satisfactory results.

The titration curve (temperature as a function of volume of titrant added), shows such a sharp break at the end-point that no graphical extrapolation is needed [36].

Several complexometric titrations have also been implemented with thermometric detection, e.g. the inverse titration of copper with EDTA, based on the catalytic effect of the metal on the oxidation of hydrazine by hydrogen peroxide [37]. The process involves two exothermic reactions, yielding mainly oxygen and nitrogen. The heat released is such that the titration can be done in an ordinary beaker by measuring the temperature with a thermistor of 2.5–100 kΩ resistance.

4.7 APPLICATIONS

The most significant applications of catalytic titrations can be divided into two general categories, namely determination of organic ligands and determination of metals.

4.7.1 Determination of ligands

Multidentate organic ligands such as aminopolycarboxylic acids are the commonest subjects of these determinations. Virtually all complexones (EDTA, EGTA, DCTA, DTPA, HEDTA, NTA, etc.) have been determined in this manner, by use of indicator reactions catalysed by various metals, especially Mn(II). The sensitivity achieved is similar in every case, as can be seen in Table 4.1. The aforesaid ligands can be determined at concentrations of the order of a few μg/ml (10^{-7}–$10^{-5}M$). Photometric and thermometric detection techniques are the commonest in these applications. Hydrogen peroxide, either as oxidant of another species or even in the absence of a substrate (autoxidation), is the reactant most frequently involved in the indicator reaction.

Table 4.2 shows the salient features of the determination of other organic compounds, including that of ascorbic acid by a redox titration [26], the only method so far reported for use of this type of system.

Table 4.3 lists the applications involving the direct determination of unidentate inorganic ligands, and the indirect determination of iodide. It also includes the indirect determinations of some inorganic ligands as proposed by Kiba and Furosawa [38], as well as the thermometric determination of Cl^-, Br^- and SCN^- by the Sandell–Kolthoff reaction [6].

There is only one reference to the analysis of mixtures of aminopolycarboxylic acids (EDTA–EGTA and EGTA–NTA), based on differences in their behaviour towards magnesium [39] (Table 4.1).

4.7.2 Determination of metals

Catalytic determinations of metals by both direct and indirect modes and in their normal and brake variants are plentiful.

Direct determinations by the normal mode use the Ce(IV)/As(III) system and I^- as titrant/catalyst almost exclusively. This facilitaties the determination of metal inhibitors such as Ag, Hg and Pd with thermometric detection, a sensitivity of the order of a few micrograms and good reproducibility. Table 4.4 summarizes the most important direct determinations of metals. The brake mode is much less usual in the direct determination of metals, but the substrate-inactivation mode has a noteworthy representative in the direct titration of mercury.

Indirect titrations or back-titrations by the normal mode are second to direct determinations in importance and are normally based on metal-catalysed reactions inhibited by EDTA or some other complexone. This mode has been the basis for determination of a larger variety of metals, on account of their complex-forming capability with these ligands.

The inverse titration of metals is much less frequent, whether by the normal or by the substitution mode.

In Table 4.5 are gathered the most significant indirect titrations of metals, including inverse titrations (the particular mode, when different from the normal one, is indicated at the bottom of the Table). As can be seen, transition metal determinations are commonplace, though there are also a number of applications involving the alkaline-earth metals, determined with the aid of Cu(II), Mn(II) and Fe(III)-catalysed indicator systems and photometric or thermometric detection. It is worth noting that the indirect titration of V(V), Tl(III) and Ce(IV) is based on a redox rather than a complex-formation process.

4.7.3 Resolution of metal mixtures

In contrast to the individual determination of species by catalytic titration, there is a scarcity of references to the resolution of metal mixtures, and all of these use one of

Table 4.1 — Direct determinations of aminopolycarboxylic acids by catalytic titration

Inhibitor	Indicator system	Detection technique	Determination range and precision	Reference
EDTA	Tiron/perborate/Co(II)	A		[2]
	Alizarin S/H_2O_2/Mn(II)	A		[41]
	Thorin/H_2O_2/Mn(II)	A		[18]
	H_2O_2/Cu(II)	P, Am		[42]
	R/H_2O_2/Fe(III)	Fluorimetric (λ_{ex} 365, λ_{em} 440 nm)	1×10^{-5}–$4 \times 10^{-5} M$	[32]
DCTA	R/H_2O_2/Mn(II)	Fluorimetric (λ_{ex} 365 nm, λ_{em} 440 nm)	3.8×10^{-5}–$2.0 \times 10^{-4} M$	[33]
	Metol/H_2O_2/Co(II) Hydroquinone/H_2O_2/Co(II) H_2O_2/Cu(II)	A, P, Am	3.36–7.70 μg	[43]
EDTA	H_2O_2/Mn(II)	T		[44]
DCTA	Diethylaniline/IO_4^-/Mn(II)	A (semi-automatic)	1×10^{-7}–$8 \times 10^{-4} M$ (1–2%)	[40]
	R*/IO_3^-/Ni(II)	A (510 nm, semi-automatic)	700–6000 μg (1%)	[45]
EDTA	R**/O_2 (air)/Mn(II)	A (600 nm, semi-automatic)	750–4000 μg (0.6%)	[46]
EGTA	R***/H_2O_2/Cu(II)	A (415 nm, semi-automatic)	370–1100 μg 380–1140 μg	[47]
NTA	R***/H_2O_2/Cu(II)	A (415 nm, semi-automatic)	1.25×10^{-5}–$5.8 \times 10^{-5} M$	[39]

EDTA	Resorcinol/H_2O_2/Mn(II)	T(S) (non-aqueous medium)		[16]
DCTA	H_2O_2 + Cu(II) or Pb(II)	T		[48]
NTA	R′/H_2O_2/Cu(II) Metol/H_2O_2/Cu(II) and hydroquinone/H_2O_2/Cu(II)	T A	8.30–25.0 μg 3.97–11.86 μg 4.72–14.25 μg (0.5%)	[49] [50]
EDTA DCTA DTPA HEDTA	Malachite Green/IO_4^-/Mn(II)	A (by circulation through flow-cell)	0.5×10^{-5}–$3.2 \times 10^{-5} M$	[11]
EDTA DCTA EGTA DTPA	$S_2O_3^{2-}$/IO_4^-/Cu(II)	P	0.7–600 μg 7–7000 μg 0.8–800 μg 4–800 μg (0.4–1%)	[51]
EGTA + EDTA EGTA + NTA	R***/H_2O_2/Cu(II)	A (415 nm, semi-automatic)	0.5–3.5 μmole (7:1–1:7) 0.5–3.5 μmole (4:1–1:7)	[39]

A: photometric, P: potentiometric, Am: amperometric, T: thermometric, S: substitution titration, R: 2-hydroxybenzaldehyde thiosemicarbazone, R*: 1,4-dihydroxyphthalimide dioxime, R**: 1,4-dihydroxyphthalimide dithiosemicarbazone, R***: 4,4′-dihydroxybenzophenone thiosemicarbazone, R′: resorcinol, catechol, hydroquinone or dimethyl-*p*-phenylenediamine.

Table 4.2 — Direct determinations of other organic inhibitors by catalytic titration

Inhibitor	Indicator system	Detection technique	Determination range and precision	Reference
Oxine	Hydroquinone/H_2O_2/Cu(II)	A		[52]
Tertiary amines and organic salts	Acetic anhydride/H_2O/$HClO_4$	T	2.5×10^{-2}–$5.7 \times 10^{-2} N$	[21]
Ascorbic acid (redox titration)	*o*-Dianisidine/H_2O_2/Cr(VI)	A (415 nm)	3–30 μmole (in 22.5 ml)	[26]
Dithiocarbamates	Acetone/KOH	T	0.5–20 μmole (0.5%)	[53]
Permanganate	Sodium azide/I_2/S^{2-}	Decoln. time	0.21–2.1 μg (in 0.2 ml)	[29]

A: photometric, T: thermometric.

Table 4.3 — Determinations of inorganic anions by catalytic titration

	Inhibitor	Indicator system	Detection technique and precision	Determination range	Reference
DIRECT	CN^-	Ascorbic acid/O_2(air)/Cu(II)	A (265 nm)	4×10^{-8}–2×10^{-7}*M*	[10]
	Cl^-, Br^- and I^-	$S_2O_8^{2-}$/Ag(I)	A		[2]
	F^- and SiF_6^{2-}	I^-/H_2O_2/Th(IV)	A, P, Am	3–10 mg 4–840 mg	[54]
	S^{2-}	Sodium azide/I_2/S^{2-} Ascorbic acid/I_2/uncat.S^{2-}	Am (Measurement of titration time with and without S^{2-})	32–160 ng	[55]
REVERSE	I^-	Ce(IV)/As(III)/I^-/Hg(II)	A, P	0.1–1200 μg (in 27 ml)	[56]
		Ce(IV)/As(III)/I^-/Hg(II)	A, P	1–20 μg/ml (1%)	[57]
		Ce(IV)/As(III)/I^-/Hg(II) (simulated titration)	A	80 μg/ml (1%)	[58]
		Ce(IV)/As(III)/I^-/Hg or Ag	Visual		[59]
		Ce(IV)/As(III)/I^-/Hg or Ag (with ferroin)	Visual		[60]
		Ce(IV)/As(III)/I^-/Hg or Ag (Landolt reaction)	Visual		[61]
INDIRECT	S^{2-}, $S_2O_3^{2-}$ SCN^-, CN^-	Mn(III)/As(III)/I^-/Hg or Ag			[38]

A: photometric, T: thermometric, Am: amperometric, P: potentiometric.

Table 4.4 — Direct determinations of metals by catalytic titration

Inhibitor	Indicator system	Detection technique	Determination range and precision	Reference
Ag	Ce(IV)/As(III)/I^-	A		[1, 2]
	Ce(IV)/As(III)/I^-	P (semi-automatic)	1–2000 μg	[62]
Hg	Ce(IV)/As(III)/I^-	T		[36, 56]
	Ce(IV)/As(III)/I^-	P		[63]
Ag, Hg	Ce(IV)/As(III)/I^-	P	14–160 mg 59–100 mg	[34]
	Ce(IV)/As(III)/I^-	T	0.1–1000 μg/ml (0.3–1%)	[64]
Pd, Au	Ce(IV)/As(III)/I^-	T	1–700 μg 18–4000 μg (1%)	[65]
Ag, Hg, Pd	Ce(IV)/As(III)/I^-	P	30–300 μg 130–1300 μg (as Cl^-) 150–1500 μg (as NO_3^-)	[66]
		Am	90–900 μg	[66]
	Mn(III)/As(III)/I^-	T	0.5–500 μg 0.5–500 μg 0.2–500 μg (3%)	[38]
Cu (brake titration)	H_2O_2/Cu(II)	T		[17]
	Hydrazine/H_2O_2/Cu(II)	T		[18]
Sb, Ni, Fe (direct, brake displacement)	Sodium azide/I_2/S^{2-}	Visual (decoln. time)	3.2–23.5 μg/0.2 ml 2.3–13.9 μg/0.34 ml 1.6–13.6 μg/0.2 ml	[29]
Hg (direct)	R/H_2O_2/Cu(II)	A (415 nm, semi-automatic)	1.4×10^{-7}–$5.2 \times 10^{-7} M$	[19]
Hg + Cu	R/H_2O_2/Cu(II)	A (415 nm, semi-automatic)	0.45–9.4 μmole (1:4–1:20)	[19]
Hg + Cd	R/H_2O_2/Cu(II)	A (415 nm, semi-automatic)	0.04–1.2 μmole (1:1–1:27)	[19]

A: photometric, T: thermometric, Am: amperometric, P: potentiometric, R: 4,4′-dihydroxybenzophenone thiosemicarbazone.

Table 4.5 — Indirect determinations of metals by catalytic titration

Inhibitor	Indicator system	Detection technique	Determination range and precision	Reference
Various transition metals			5.5–42.3 mg	[2]
	Diethylaniline/IO_4^-/Mn/ EDTA or DCTA	A (470 nm, semi-automatic)	0.06–1500 μg	[40]
Hg + Zn, Hg + Pb, Hg + Cu	Diethylaniline/IO_4^-/Mn/ EDTA or DCTA	A (470 nm, semi-automatic)	6–100 μg (0.1:1–5:1) (1.5–5.2%)	[40]
Co + Ni	Diethylaniline/IO_4^-/Mn/ EDTA or DCTA	A (470 nm, semi-automatic)	1–2.3 μg (1:1) 3.5%	[40]
Alkaline earths (subst. titration)	Resorcinol/H_2O_2/Mn(II)	Visual		[15]
Various transition metals	Oxidation of various dyes by H_2O_2 or BrO_3^-			[67]
	H_2O_2/Cu(II)	T		[44]
	Resorcinol/H_2O_2/Cu(II)	T		[44]
	H_2O_2/Cu(II) or Pb(II)	T		[48]
	Resorcinol/H_2O_2/Cu(II)	T		[49]
	Catechol/H_2O_2/Cu(II) or Fe(III)	T		[49]
In, Ga, Th	DPD/H_2O_2/Cu(II) or Fe(III)			[49]
Cu	$S_2O_3^{2-}$/IO_4^-/Cu(II)/EDTA	P (semi-automatic)	0.02–32 μg (0.72%)	[51]
Co			0.17–29.5 μg (0.2%)	[51]
Mn	R/O_2(air)/Mn(II)/EDTA	A (600 nm)	39–99 μg/ml (inverse) 20–45 μg/50 ml (indirect)	[46]
Ni	R/O_2(air)/Mn(II)/EDTA	A (600 nm)	40–1000 μg/50 ml	[46]

Ca + Mg, Sr, Ba	R/O_2(air)/Mn(II)/EDTA	A (600 nm)	10–150 μg 20–250 μg 20–450 μg (0.2–0.8%)	[68]
Ni (indirect, inverse), Hg, Fe	R*/IO_3^-/Ni(II)/EDTA and DCTA	A (510 nm)	125–1000 μg (indirect) 20–100 μg/50 ml (inverse) 68–345 μg/50 ml 19–95 μg/50 ml (1%)	[45]
Fe, Ca, Mg, Sr, Ba	R**/H_2O_2/Fe(III)/EDTA	Fluorimetric (λ_{ex} 365 nm, λ_{em} 440 nm)	20–70 μg/70 ml 20–100 μg/70 ml 20–100 μg/70 ml 20–140 μg/70 ml 20–100 μg/70 ml (0.5%)	[32]
Cu, Ni, Mn	R***/H_2O_2/Cu(II)/EDTA, EGTA	A (415 nm)	10–103 μg/50 ml 25–118 μg/50 ml 23–114 μg/50 ml (0.5%)	[47]
Fe + Cu, Fe + Ni, Fe + Mn	R***/H_2O_2/Cu(II)/EDTA, EGTA	A (415 nm)	0.2–3.0 μmole (1:14–13:1)	[70]
V(V), Tl(III), Ce(IV)	*o*-Dianisidine/H_2O_2/Cr(VI)/ ascorbic acid	A (415 nm)	5–50 μmole 2–20 μmole 5–50 μmole	[26]
Mn (inverse)	R′/H_2O_2/Mn(II)/EDTA *p*-Tolidine/H_2O_2/Mn(II)/EDTA	Visual Olfactory		[31] [31]

A: photometric, T: thermometric, P: potentiometric, R: 1,4-dihydroxyphthalimide dithiosemicarbazone, R*: 1,4-dihydroxyphthalimide dioxime, R**: 2-hydroxybenzaldehyde thiosemicarbazone, R***: 4,4′-dihydroxybenzophenone thiosemicarbazone, R′: *p*-dimethylaminobenzaldehyde, DPD: dimethyl-*p*-phenylenediamine.

the three indicator systems described below.

One such system is that of sodium azide/I_2, catalysed by S^{2-}. In addition to allowing the direct determination of Sb, Ni and Fe, this system permits the resolution of Hg/Cd and Pb/Cu mixtures with Na_2S [29]. The hydrogen sulphide released (once the metal has been fully precipitated in the acid medium) is swept by a nitrogen stream into another vessel containing the solution of sodium azide, iodide and starch, which is decolorized at once, thus indicating the end-point. The resolution of binary mixtures is based on the sequential precipitation of the metal sulphides at suitable pH, or masking one of the metals with EDTA. The detection limits achieved are quite low, as can be seen in Table 4.4.

The Mn(II)-catalysed diethylaniline/IO_4^- system is used in the presence of EDTA or DCTA as inhibitor in the indirect analysis of Ni/Co mixtures or those of Hg with Zn, Pb, Cu or Cd in suitable ratios [40] (Table 4.5). The titration requires two aliquots: one is used to determine the sum of both metals, while the other, in which Hg(II) is masked by addition of cyanide, or from which nickel is separated by extraction after precipitation with dimethylglyloxime, is employed to determine the metal remaining in solution, the concentration of the other being calculated by difference.

The third system usually employed in the catalytic resolution of mixtures of species is that of 4,4′-dihydroxybenzophenone thiosemicarbazone and hydrogen peroxide, catalysed by Cu(II) in the presence of EDTA or EGTA. It permits the analysis of mixtures of Fe with Cu, Ni or Mn by use of masking agents such as triethanolamine or fluoride, with good results. This system has been satisfactorily applied to the determination of the above-mentioned metals in metallurgical samples [69].

A combination of the brake and the substrate-inactivation modes allows the resolution of Hg–Cu and Hg–Cd mixtures by simultaneous titration with the aforesaid indicator system, thus avoiding the use of two aliquots and the determination by difference of one of the components [19]. The results obtained are shown in Table 4.4.

REFERENCES

[1] K. B. Yatsimirskii and T. I. Fedorova, *Proc. Acad. Sci. USSR*, 1962, **143**, 143.
[2] H. Weisz and U. Muschelknautz, *Z. Anal. Chem.*, 1966, **215**, 17.
[3] H. A. Mottola, *Talanta*, 1969, **16**, 1265.
[4] T. P. Hadjiioannou, *Rev. Anal. Chem.*, 1976, 82.
[5] H. Weisz and S. Pantel, *Anal. Chim. Acta*, 172, **62**, 361.
[6] T. F. A. Kiss, *Talanta*, 1983, **30**, 771.
[7] W. J. Blaedel and R. H. Laessig, *Anal. Chem.*, 1964, **36**, 1617.
[8] C. E. Efstathiou and T. P. Hadjiioannou, *Talanta*, 1983, **30**, 145.
[9] H. L. Pardue and B. Fields, *Anal. Chim. Acta*, 1981, **124**, 39; 1981, **124**, 65.
[10] H. A. Mottola and H. Freiser, *Anal. Chem.*, 1968, **40**, 1266.
[11] H. A. Mottola, *Anal. Chem.*, 1970, **42**, 630.
[12] F. F. Gaál and B. F. Abramović, *Talanta*, 1984, **31**, 987.
[13] B. F. Abramović, F. F. Gaál and D. J. Paunić, *Talanta*, 1985, **32**, 549.
[14] F. F. Gaál and B. F. Abramović, *Talanta*, 1985, **32**, 559.
[15] D. Klockow and L. García-Beltrán, *Z. Anal. Chem.*, 1970, **249**, 304.
[16] T. F. A. Kiss, *Z. Anal. Chem.*, 1970, **252**, 12.
[17] H. Weisz, S. Pantel and H. Ludwig, *Z. Anal. Chem.*, 1972, **262**, 269.
[18] H. Weisz and S. Pantel, *Z. Anal. Chem.*, 1973, **268**, 389.

[19] T. Raya Saro and D. Pérez-Bendito, *Anal. Chim. Acta*, 1986, **182**, 163.
[20] G. A. Vaughan and J. J. Swithenbank, *Analyst*, 1965, **90**, 594.
[21] V. J. Vajgand and F. F. Gaál, *Talanta*, 1967, **14**, 345.
[22] D. C. Olson and D. W. Margerum, *J. Am. Chem. Soc.*, 1963, **85**, 297.
[23] D. W. Margerum and R. K. Steinhaus, *Anal. Chem.*, 1965, **37**, 222.
[24] R. H. Stehl, D. W. Margerum and J. J. Latterell, *Anal. Chem.*, 1967, **39**, 1346.
[25] D. W. Margerum and R. H. Stehl, *Anal. Chem.*, 1967, **39**, 1351.
[26] H. Weisz and S. Pantel, *Anal. Chim. Acta*, 1980, **116**, 421.
[27] H. Weisz and S. Pantel, *Anal. Chim. Acta*, 1972, **62**, 361.
[28] H. Weisz and S. Pantel, *Anal. Chim. Acta*, 1973, **64**, 389.
[29] H. Weisz and J. Schlimpf, *Anal. Chim. Acta*, 1980, **121**, 257.
[30] H. Weisz and J. Schlimpf, *Anal. Chim. Acta*, 1983, **147**, 247.
[31] S. Abe, S. Kon and T. Matsuo, *Anal. Chim. Acta*, 1978, **96**, 429.
[32] A. Moreno, M. Silva, D. Pérez-Bendito and M. Valcárcel, *Analyst*, 1984, **109**, 249.
[33] A. Moreno, M. Silva, M. Valcárcel and D. Pérez-Bendito, *Quim. Anal.*, 1985, **4**, 39.
[34] H. Weisz and D. Klockow, *Z. Anal. Chem.*, 1967, **232**, 321.
[35] E. J. Greenhow, *Chem. Rev.*, 1977, **77**, 835.
[36] H. Weisz, T. Kiss and D. Klockow, *Z. Anal. Chem.*, 1969, **247**, 248.
[37] H. Erlenmeyer, C. Flierl and H. Sigel, *J. Am. Chem. Soc.*, 1969, **91**, 1065.
[38] N. Kiba and M. Furosawa, *Anal. Chim. Acta*, 1978, **98**, 343.
[39] T. Raya Saro and D. Pérez-Bendito, *Quim. Anal.*, 1985, **4**, 259.
[40] E. A. Piperaki and T. P. Hadjiioannou, *Chim. Chronica*, 1977, **6**, 375.
[41] B. E. Reznik, V. I. Chniko and V. I. Vershinin, *Zh. Analit. Khim.*, 1972, **27**, 395.
[42] F. F. Gaál, B. F. Abramović, F. B. Szebenyi and V. D. Canić, *Z. Anal. Chem.*, 1977, **286**, 222.
[43] F. F. Gaál, B. F. Abramović and V. D. Canić, *Z. Radova PMF, Novi Sad B*, 1978, 199.
[44] H. Weisz and T. F. A. Kiss, *Z. Anal. Chem.*, 1970, **249**, 302.
[45] A. Gómez-Hens, M. Ternero, D. Pérez-Bendito and M. Valcárcel, *Mikrochim. Acta*, 1979 **I**, 375.
[46] M. Ternero, F. Pino, D. Pérez-Bendito and M. Valcárcel, *Anal. Chim. Acta*, 1979, **109**, 401.
[47] T. Raya-Saro and D. Pérez-Bendito, *Analyst*, 1983, **108**, 857.
[48] T. F. A. Kiss, *Mikrochim. Acta*, 1972, 420.
[49] T. F. A. Kiss, *Mikrochim. Acta*, 1973, 847.
[50] F. F. Gaál, M. J. Csanyi and B. F. Abramović, *Zb. Radova PMP, Novi Sad*, 1979, **9**, 387.
[51] M. Tomotheu, M. A. Kouparis and T. P. Hadjiioannou, *Mikrochim. Acta*, 1982 **II**, 433.
[52] J. F. Dolmanova and V. M. Peskova, *Zh. Analit. Khim.*, 1984, **39**, 297.
[53] N. Kiba, Y. Sawada and M. Furusawa, *Talanta*, 1982, **29**, 418.
[54] F. F. Gaál, B. F. Abramović and V. D. Canić, *Talanta*, 1978, **25**, 113.
[55] D. Klockow, J. Auffarth and G. F. Graf, *Z. Anal. Chem.*, 1982, **311**, 244.
[56] T. P. Hadjiioannou and E. A. Piperaki, *Anal. Chim. Acta*, 1977, **90**, 329.
[57] M. M. Timotheu and T. P. Hadjiioannou, *Mikrochim. Acta*, 1983 **II**, 59.
[58] T. I. Fedorova, K. B. Yatsimirskii, E. V. Vasileva and V. G. Markichev, *Zh. Analit. Khim.*, 1977, **32**, 1951.
[59] J. Bognár and S. Sárosi, *Mikrochim. Acta*, 1963, 1072.
[60] J. Bognár and S. Sárosi, *Mikrochim. Acta*, 1966, 534.
[61] J. Bognár and S. Sárosi, *Mikrochim. Acta*, 1969, 463.
[62] T. P. Hadjiioannou, E. A. Piperaki and S. S. Papastathopoulos, *Anal. Chim. Acta*, 1974, **68**, 447.
[63] F. F. Gaál and B. F. Abramović, *Mikrochim. Acta*, 1982 **I**, 465.
[64] K. C. Burton and H. M. N. H. Irving, *Anal. Chim. Acta*, 1970, **52**, 441.
[65] T. P. Hadjiioannou and M. M. Timotheu, *Mikrochim. Acta*, 1977 **I**, 61.
[66] F. F. Gaál and B. F. Abramović, *Talanta*, 1980, **27**, 733.
[67] H. Weisz and T. J. Janjić, *Z. Anal. Chem.*, 1967, **227**, 1.
[68] M. Ternero, D. Pérez-Bendito and M. Valcárcel, *Microchem. J.*, 1981, **26**, 61.
[69] T. Raya Saro and D. Pérez-Bendito, *Anal. Chim. Acta*, 1985, **172**, 273.

5

Uncatalysed reactions

5.1 INTRODUCTION

The use of uncatalysed reactions for determination of a single species is to be avoided if an adequate non-kinetic method (equilibrium or static) is available, since little is gained in sensitivity or accuracy over modern methods involving automated instrumentation. In addition, kinetic methods are normally somewhat more laborious than their conventional counterparts.

All things considered, kinetic methods offer some advantages over equilibrium methods, particularly in two special situations.

(a) When the classical method is based on a slow or incomplete process that may involve side-reactions as it approaches equilibrium. Thus, acetylacetone cannot be titrated with hydroxylamine hydrochloride since the reaction is very slow (50 hr) and does not develop to completion [1]. The monitoring of the initial rate [2] by high-frequency conductimetry, photometry or potentiometry allows this substance to be determined in a matter of a few minutes.

(b) When the method of determination is rather laborious or time-consuming. Such is the case with the conventional determination of phenols by bromination, the kinetic counterpart of which takes only a few minutes [3].

This type of reaction is of especial relevance to the analysis of mixtures of closely related compounds, for which differential reaction-rate methods have been developed. These methods have been widely used in organic analysis, as well as in the resolution of metal mixtures on the basis of differences in the rate of substitution of some complexes. They are described in length in Chapter 6.

5.2 METHODS OF DETERMINATION

5.2.1 Determination of a single species

Consider a chemical species A interacting with another reactant, B, to yield product P:

$$A+B \rightarrow P$$

This determination can involve two limiting situations.

(a) If B is in large excess (over 50-fold) over the analyte, A, then the reaction will be pseudo first-order in A.
(b) If the concentration of B is similar to or less than fifty times that of A, the reaction will be second-order.

5.2.1.1 Pseudo first-order reactions

The rate of a reaction of pseudo first-order with respect to the analyte, A, is given by:

$$\upsilon = -\mathrm{d}[A]/\mathrm{d}t = \mathrm{d}[P]/\mathrm{d}t = k[A][B]_0 = k'[A] \qquad (5.1)$$

where $k' = k[B]_0$. This expression is normally the basis for the kinetic determination of A.

The mathematical treatment of the fixed-time and variable-time methods uses the exponential version [4] of Eq. (5.1), namely:

$$\upsilon = -\mathrm{d}[A]/\mathrm{d}t = k[A]_0\exp(-kt) \qquad (5.2)$$

Integration of this equation over an interval of time $t_2 - t_1$ yields

$$-\Delta[A] = [A]_0[\exp(-kt_1) - \exp(-kt_2)]$$

whence, taking into account that $t_2 = \Delta t + t_1$,

$$[A]_0 = -\Delta[A]/\exp(-kt_1)[1 - \exp(k\Delta t)] \qquad (5.3)$$

This equation is of relevance both to differential (initial-rate) and integration methods.

In the case of initial-rate methods (pseudo zero-order kinetics), Eq. (5.1) can be expressed in incremental form as:

$$\upsilon = \Delta[P]/\Delta t = k[A]_0 \qquad (5.4)$$

from which the concentration of A is readily calculated.

Let us assume $[A]_0$ to be linearly related to time. The range over which this assumption is valid can be determined by expanding the exponential function in Eq. (5.3) as a Maclaurin series:

$$\exp(-k\Delta t) = 1 - k\Delta t + [k^2(\Delta t)^2/2!] - [k^3(\Delta t)^3/3!] + \ldots$$

If $k\Delta t$ is small, then only the first few terms in the series will be significant, so Eq. (5.3) will become

$$[A]_0 = -\Delta[A]/k\Delta t \exp(-kt_1) \tag{5.5}$$

The equation can be used to derive the mathematical expressions representative of the fixed-time and variable-time differential methods by simply making Δt or $\Delta[A]$ constant. It becomes Eq. (5.4) if $t_1 = 0$, so that $[A]_0$ and Δt or $\Delta[A]$ are linearly related, whatever the shape of the kinetic curve.

The mathematical treatment of integration methods involves the integration of Eq. (5.2), which yields:

$$\ln([A]_2/[A]_1) = k\Delta t \tag{5.6}$$

This expression is illustrative of the variation of the analyte concentration as a function of time. If $t_1 = 0$ (i.e. $[A]_1 = [A]_0$), Eq. (5.6) can be rewritten as:

$$\ln[A]_t = \ln([P]_\infty - [P]_t) = \ln[A]_0 - kt \tag{5.7}$$

which allows $[A]_0$ to be calculated from the intercept of the curve obtained by plotting $\ln [A]_t$ as a function of time.

As Eq. (5.5) is valid only for small values of Δt, which involves neglecting the factorial terms in the series expansion, the accuracy in the determination of A will decrease with increasing variation in its concentration with time (i.e. the results will be reliable as long as the assumption made above is valid).

Variable-time integration methods involve keeping $\Delta[A]$ constant. Since Δt and t_1 depend on $[A]_0$ according to Eq. (5.3), $[A]_0$ will never be linearly related to $1/\Delta t$ unless a small $\Delta[A]$ value is introduced into Eq. (5.5). As in the previous case, minimum errors will be made when $[A]_1$ virtually coincides with $[A]_0$ and hence $\exp(-kt_1) \rightarrow 1$.

The application of the fixed-time method involves keeping t_1 *and* Δt constant, so that, according to Eq. (5.3), $\Delta[A]$ and $[A]_0$ will always be linearly related, even if the curve is non-linear in the measured interval. In short, Δt should not be too large, in order to ensure the validity of both types of integration methods (fixed- and variable-time).

Insofar as the measurements made in the implementation of integration methods are not made exclusively at the start of the reaction, fixed-time methods will give wider dynamic ranges than will variable-time methods for first- or pseudo first-order reactions.

The rate equation of a reversible reaction of the type

$$A + B \underset{k_{-1}}{\overset{k_1}{\rightleftharpoons}} P$$

is, according to Laidler [5],

$$\ln\frac{[P]_e}{([P]_e - [P])} = (k_1 + k_{-1})t$$

where $[P]_e$ is the equilibrium concentration of the product. This is the starting point for a mathematical treatment which allows the derivation of an expression for reversible reactions that is similar to Eq. (5.3):

$$[A]_0 = \frac{(k_1 + k_{-1})[P]/k_1}{\exp[-(k_1 - k_{-1})t_1]\ \{1 - \exp[-(k_1 + k_{-1})\Delta t]\}}$$

from which the analyte concentration can be readily calculated by any of the methods described above.

5.2.1.2 Second-order reactions

The rate equation of a second-order reaction involving the analyte, A, and another reactant, B, is of the form:

$$\upsilon = d[P]/dt = k_A([A]_0 - [P])[B]_0 \quad (5.8)$$

where k_A is the second-order rate constant for reactant A.

If initial-rate measurements are made, then [P] can be safely neglected with respect to [A] in the expression above, so

$$[A]_0 = (d[P]/dt)/k_A[B]_0$$

which is the basis for the determination of the analyte by the initial-rate method.

The incremental form of this equation

$$[A]_0 = \Delta[P]/k_A[B]_0\Delta t$$

is the basis for the fixed-time and variable-time methods based on initial-rate measurements. Under these conditions, $[A]_0$ and $[B]_0$ are similar.

If the analyte is determined by an integration method and the concentration of B is up to fifty times that of A, then the integration of the rate equation

$$-d[A]/dt = k_A[B]_0[A]$$

over the interval of time from t_1 to t_2 yields

$$[A]_1/[A]_2 = k_A[B]_0(t_2 - t_1)$$

If $t_1 = 0$, then $[A]_1 = [A]_0$ and

$$\ln[A]_t = \ln[A]_0 - k_A[B]_0 t$$

which is analogous to Eq. (5.7) and, like it, allows $[A]_0$ to be calculated from the intercept of the plot of $\ln[A]_t$ as a function of time.

Similarly, Eq. (5.3) becomes

$$[A]_0 = -\Delta[A]/[B]_0 \exp(-kt_1)[1 - \exp(-k\Delta t)]$$

for fixed-time and variable-time methods applied to second-order reactions.

If the concentration of A is up to fifty times that of B, then the integration of Eq. (5.8) yields

$$[A]_0 = \frac{\ln([B]_1/[B]_2)}{k_A(t_2 - t_1)} \tag{5.9}$$

where $[B]_1$ and $[B]_2$ are the concentrations of B measured at times t_1 *and* t_2, respectively. This equation is equally applicable with the fixed- and variable-time methods.

The tangent method can also be applied to uncatalysed reactions by plotting $\Delta(\ln[B])$ as a function of time. According to Eq. (5.9), the slope of the straight line thus obtained will be proportional to the initial concentration of A.

5.2.2 Determination of a single species in a mixture

The determination of a given species in a mixture is feasible as long as it reacts with the operative reagent at a rate sufficiently different from that of the other component(s) of the mixture, i.e. if the analyte alone contributes significantly to the overall reaction rate over a given time interval.

The determination of the analyte with a given accuracy will be possible if the ratios of its rate constant to those of the other components of the mixture are well above or below a certain value, i.e. if the analyte rate constant is much smaller or larger than those of the fastest or slowest reacting components, respectively. In dealing with these two possibilities, the reactions involved will be considered to be pseudo first-order with respect to the analyte.

Consider the determination of A in the presence of B. Their respective pseudo first-order equations will be

$$-\mathrm{d}[A]/\mathrm{d}t = k_A[A] \text{ and } -\mathrm{d}[B]/\mathrm{d}t = k_B[B]$$

or, integrated over time:

$$\ln([A]_0/[A]_t) = k_A(t - t_0)$$

$$\ln([B]_0/[B]_t) = k_B(t - t_0)$$

If the extent of reaction of B over this time interval is negligible, $[A]_0$ can readily be determined merely by measuring $[A]_t$. However, if B reacts to an appreciable extent in this interval, the degree of interference with the analyte reaction can be calculated from the quotient of the two expressions. Thus:

$$\frac{\ln([A]_0/[A]_t)}{\ln([B]_0/[B]_t)} = \frac{k_A}{k_B} \tag{5.10}$$

This equation is time-independent and establishes a relationship between the extents of both reactions on the bases of their rate constants.

The error made in calculating $[A]_0$ from the rate-constant ratio for a given extent of reaction of A will be:

$$\%\ \text{error}_A = \frac{[B]_t}{[A]_t + [B]_t} \times 100 \sim \frac{[B]_t}{[A]_t} \times 100 \tag{5.11}$$

since $[A]_t \gg [B]_t$.

From Eq. (5.10), and taking $[A]_t = \text{const.}$,

$$\ln([B]_0/[B]_t) = (k_B/k_A)\ln([A]_0/[A]_t$$

whence

$$\frac{[B]_0}{[B]_t} = ([A]_0/[A]_t)^{k_B/k_A}$$

and

$$[B]_t = [B]_0([A]_t/[A]_0)^{k_B/k_A}$$

Substitution of $[B]_t$ into Eq. (5.11) yields

$$\%\ \text{error}_A = \frac{[B]_0}{[A]_t}([A]_t/[A]_0)^{k_B/k_A} \times 100$$

or alternatively

$$\%\ \text{error}_A = 100 \times [B]_0[A]_t^{\left[\left(\frac{1}{k_A/k_B}\right) - 1\right]}[A]_0^{\left(\frac{-1}{k_B/k_A}\right)} \tag{5.12}$$

Figure 5.1 is a plot of $\%\ \text{error}_A = f(k_A/k_B)$ for different percentages of A transformed. As can be seen: (a) the error in the determination decreases with increasing rate-constant ratio; (b) for a given ratio, the error increases exponentially with the fraction of A transformed (Fig. 5.2), so for a ratio of 100, % error$_A$ decreases sharply between 70 and 90%; (c) for a constant k_A/k_B ratio and say a 60% fraction of A transformed (Fig. 5.3), the percentage error increases linearly with increase in the initial concentration of B, consistent with Eq. (5.12).

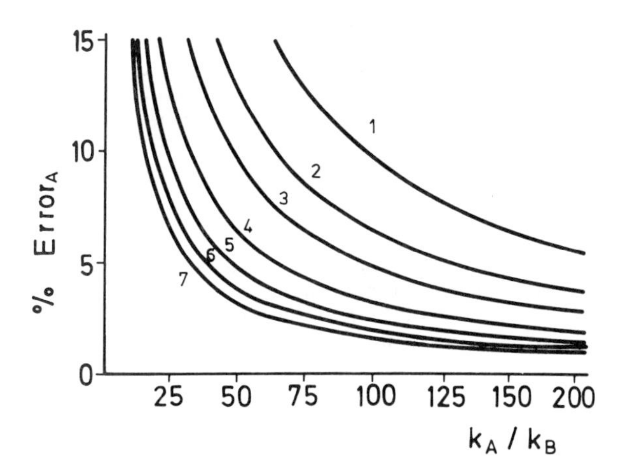

Fig. 5.1 — Variation of the relative error in the determination of reactant A in a mixture with B, as a function of their rate-constant ratio. Curves 1–7 correspond to 90, 85, 80, 70, 60, 50 and 40% of A transformed, respectively. $[A]_0 = [B]_0 = 10^{-3}M$.

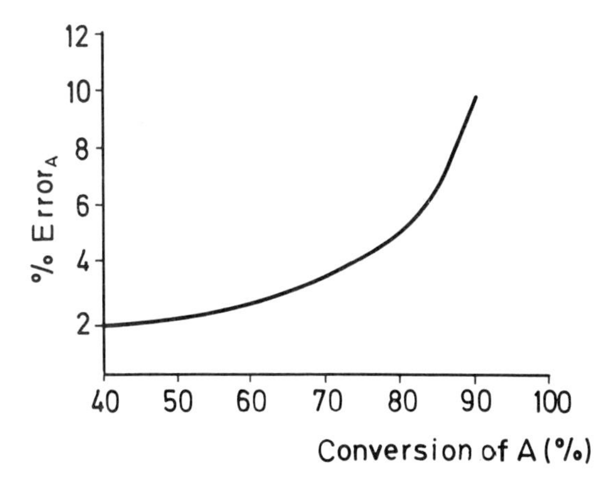

Fig. 5.2 — Variation of the relative error in the determination of A, as a function of the fraction of A transformed: $k_A/k_B = 100$; $[A]_0 = [B]_0 = 10^{-3}M$.

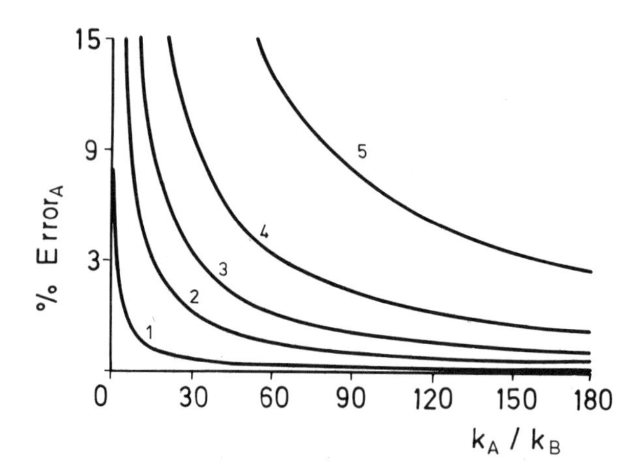

Fig. 5.3 — Variation of the relative error in the determination of A, as a function of the rate-constant ratio for different initial concentrations of B. Curves 1—5 correspond to 10^{-4}, 5×10^{-4}, 10^{-3}, 2×10^{-3} and $5\times10^{-3}M$ B, respectively. $[A]_0=10^{-3}M$; 60% of A transformed.

If a maximum error of 1% is acceptable (i.e. if there is still 1% of A present when less than 1% of B has reacted), the k_A/k_B ratio in Eq. (5.10) should be at least

$$k_A/k_B=\ln 100/\ln 1.01=463$$

In practice, a minimum ratio of 500 is commonly accepted.

On the other hand, A should be measured between $t=0$ and $t=(\ln 100)/k_A$, during which it is the only reacting component. Obviously, B should be measured once this interval has elapsed.

If B is the slower reactant and is the one to be determined, the sum of the concentrations of A and B to yield a product P in time t will be given by:

$$[P]_\infty-[P]_t=[A]_t+[B]_t=[A]_0\exp(-k_At)+[B]_0\exp(-k_Bt) \qquad (5.13)$$

If the reaction rate of A is very high compared to that of B, then the first term in (5.13) will be negligible, as $[A]_t\rightarrow0$ at time t. This expression allows calculation of $[B]_0$:

$$[B]_0=([P]_\infty-[P]_t)/\exp(-k_Bt)$$

provided that k_B is known, since $[P]_\infty$ can be readily determined either by another method or by allowing the reaction to proceed to completion so that the overall concentration $[A]_0+[B]_0$ is measured.

The relative error made in the determination of B can be calculated from:

$$\% \text{ error}_B = \frac{[A]_t}{[A]_t + [B]_t} \times 100 = \frac{[A]_0 \exp(-k_A t)}{[B]_0 \exp(-k_B t)} \tag{5.14}$$

insofar as $[B]_t \gg [A]_t$ at the time of measurement. A rearrangement of Eq. (5.14) yields:

$$\% \text{ error}_B = 100 \times ([A]_0/[B]_0)\exp[-(k_A - k_B)t]$$

$$= 100 \times ([A]_0/[B]_0)\exp\{-[(k_A/k_B) - 1]k_B t\} \tag{5.15}$$

This, unlike Eq. (5.12), is time-dependent, and allows the calculation of the time t required to obtain a preselected error for a given pair of rate-constant and concentration ratios.

Figure 5.4 shows the variation of the relative error as a function of the k_A/k_B ratio for a constant k_B value, $[A]_0 = [B]_0$ and measurement times ranging from 1 to 4 min. As can be seen, the error decreases with increase in the rate-constant ratio.

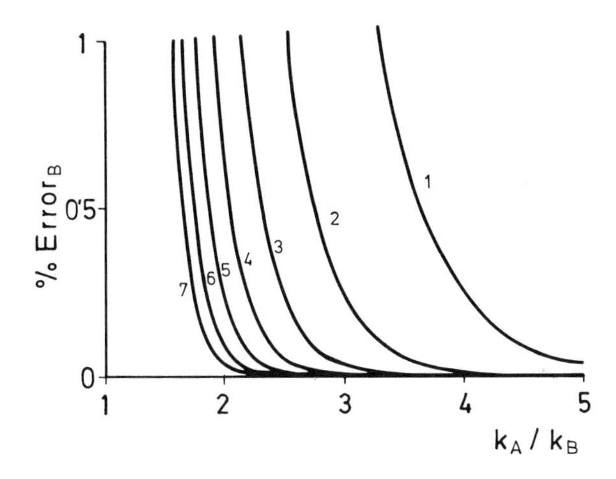

Fig. 5.4 — Variation of the relative error in the determination of B in a mixture with A, as a function of their rate-constant ratio. Curves 1–7 correspond to reaction times between 1 and 4 min (measurements made at 30-sec intervals). $[A]_0 = [B]_0 = 10^{-3}M$; $k_B = 2.0$ l.mole^{-1}.sec^{-1}.

As shown in Fig. 5.5, the relative error also decreases with increasing measurement time for a given rate-constant ratio and $[A]_0 = [B]_0$.

According to Eq. (5.15), the concentration of the faster reactant should not exceed that of the slower by more than a certain factor if some degree of accuracy is to be achieved. Thus, as shown in Fig. 5.6, an error of less than 0.5% for a k_A/k_B ratio of 3.5 will be obtained only if the $[A]_0/[B]_0$ ratio is below 8.

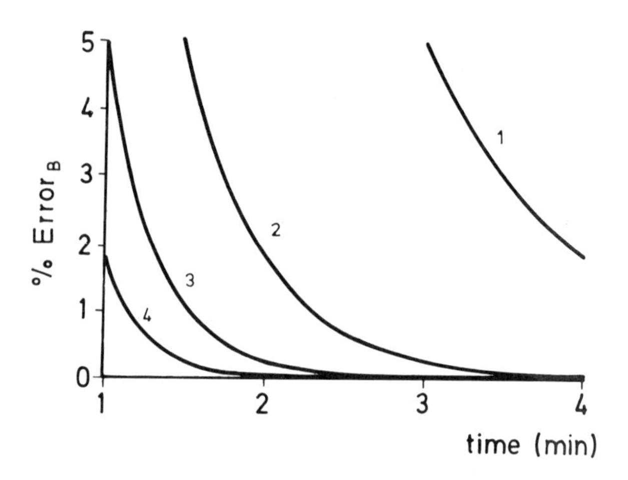

Fig. 5.5 — Variation of the relative error in the determination of B, as a function of the reaction time for different rate-constant ratios. Curves 1–4 correspond to k_A/k_B ratios of 1.5, 2.0, 2.5 and 3.0, respectively. Conditions as in Fig. 5.4.

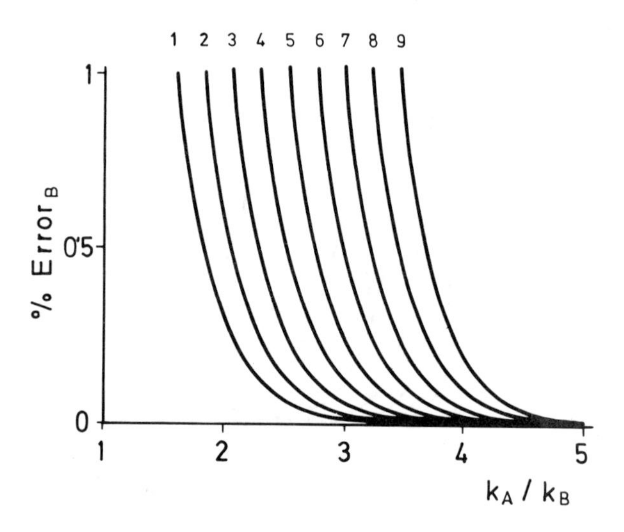

Fig. 5.6 — Influence of the $[A]_0/[B]_0$ ratio on the relative error in the determination of B. Curves 1–9 correspond to ratios of 1/16, 1/8, 1/4, 1/2, 1, 2, 4, 8 and 16, respectively: $k_B = 2.0$ l.mole^{-1}.sec^{-1}.

5.3 APPLICATIONS

As stated at the beginning of this chapter, uncatalysed reactions are rarely used for the determination of organic or inorganic species, because equilibrium methods are faster and just as sensitive. Their use is limited to slow reactions, since there is then no advantage in using an equilibrium method.

5.3.1 Determination of inorganic species

Although there is no restriction on the nature of the determination process, complex-formation and redox reactions have so far been the most frequently used in the determination of both metals and non-metals.

5.3.1.1 Complexation reactions

Complex-formation and ligand-exchange reactions have been applied to the determination of metal ions, particularly Fe(II), which can be determined (by means of a flow-cell) at concentrations of the order of $10^{-4}M$ on the basis of its reaction with ferrozine at pH 5 [6]. The kinetics of this process was studied recently and compared with that of other complexation reactions involving various ligands such as 1,10-phenanthroline and 2,2′-bipyridyl [7]. The Fe(II)/ferrozine and Fe(II)/ferroin systems have also been used for the determination of iron by ligand-exchange reactions with oxalate and citrate, respectively [8]. Iron can also be determined by direct formation of its thiocyanate complex, and a stopped-flow technique [9]. A ligand-exchange reaction is also the basis for the determination of ferrocyanide and ferricyanide: these ions convert 1,10-phenanthroline into ferroin in the presence of ascorbic acid, with the aid of Hg(II) as a catalyst of the exchange reaction [10].

Copper can be determined both by direct formation and ligand-exchange reactions. A representative example of the former is the reaction with 2-methylcyclohexane-1,3-dione bis(4-phenyl-3-thiosemicarbazone) at pH 1–2, which is monitored photometrically at 480 nm and is not very sensitive (linear calibration range 2–10 μg/ml) [11]. A typical ligand-exchange determination for copper involves 2-(2-thiazolylazo)-5-dimethylaminophenol and EDTA, allowing the metal to be determined at the ng/ml level by a stopped-flow technique [12].

Other interesting determinations based on complex-formation reactions are those of aluminium with 8-hydroxyquinoline-5-sulphonic acid [13], rhodium and palladium with sodium azide [14] and thiocyanate with bromine in the presence of pyridine, which yields a yellow complex with the BrCN thus formed [15]. All these reactions are monitored photometrically at the wavelengths of maximum absorption of the corresponding complexes.

Gallium, mercury and nickel can be determined by displacement reactions in the Cu(II)–EDTA [16], Cu(II)–diethyldithiocarbamate [17] and Zn(II)–dimercaptomaleonitrile systems [13]. Nickel can also be determined by displacement from its dimethylglyoxime complex by cyanide [18].

Fluorimetry has been used to determine magnesium at concentrations between 0.4 and 1.2 μg/ml on the basis of initial-rate measurements of the decrease it causes in the green fluorescence of fluoren-2-aldehyde-2-pyridylhydrazone in a micellar medium. Though not very precise, the method is reasonably selective towards other alkaline-earth metals [19]. The decrease in the fluorescence of the morin–Be(II) complex is the basis for the determination of manganese in natural waters in the range 5–50 ng/ml [20].

5.3.1.2 Redox reactions

This type of reaction has been applied to the determination of anionic rather than cationic species. The most representative example of the latter is probably the determination of Fe(II) with chlorate, based on the decolorization of Methyl Orange by the chlorine released [21]. Iron and thallium are determined fluorimetrically at the ng/ml level through their oxidizing action on 1,4-diamino-2,3-dihydroxyanthraquinone [22]. Another determination of interest in this context is that of Fe(III), detectable in amounts between 2 and 10 μmole through the photochemical reduction of its oxalate complex in the presence of 1,10-phenanthroline [23].

The oxidizing power of Ce(IV) has been exploited for the kinetic fluorimetric determination of this species at concentrations between 0.02 and 0.9 μg/ml, based on the oxidation of different organic compounds in acid media [24,25]. In a similar manner V(V) is determined through its oxidation of pyrazoline and various isoxazolines [26], and 1-amino-4-hydroxyanthraquinone [27]. This last reagent is used for the determination of vanadium in crude oil, which entails dry ashing to remove the organic matter and subsequent dissolution of the residue in nitric acid before the initial-rate method is applied.

As stated above, far more applications based on uncatalysed redox reactions are concerned with anions, particularly with halides and halates. Chloride and bromide are determined through their inhibitory effect on the oxidation of 4,4′-dihydroxybiphenyl by Tl(III), which results in the formation of a very stable complex between the halide and the metal ion [28]. Other determinations of bromide involve the BrO_3^-/Br^- system and detection of the liberated bromine with Methyl Orange [29] or Chromotrope 2B (linear range 0.1–20 μg/ml) [30], or its inhibitory effect on the photochemical reaction between Ag(I) and Fe(II), monitored by a fixed-time photometric method [31]. Iodide is determined by the absorptiostat technique on the basis of its reaction with Ce(IV) [32] or its inhibitory effect on a photochemical reaction [33]. Hypochlorite and perchlorate are determined through the inhibitory effect exerted on the reaction between luminol and hydrogen peroxide [34], and by the reduction to chloride by Ti(III) [35] or ruthenium [36], respectively. Lazarou *et al.* [37] have developed a method for determination of 10^{-6}–$10^{-5}M$ perbromate by absorbance measurements of tri-iodide formed by the reaction between this halate and iodide.

Nitrate can be determined in effluents, over the range 10–100 μg, on the basis of absorbance measurements made at 510 nm to follow the photochemical reduction of this anion in the presence of Methyl Orange [38]. Nitrate and nitrite are determined jointly with Fe(II) and sodium iodide, and nitrite selectively with iodide [39], or by reduction to NO with ascorbic acid [40]. Both determinations involve monitoring the luminescence yielded in the reaction between NO and ozone. Nitrite is also determined by the Griess reaction [41,42] or in mixtures with nitrate by prior reduction in a column [42].

Sulphur species such as S^{2-} or SCN^- are also determined by the fixed- and variable-time methods. The former is determined by densitometric measurements made after the sample has been in contact with a photographic plate for a given period of time [43]. The determination of the latter in the range 4×10^{-5}–$5 \times 10^{-4}M$ is based on its inhibitory effect on the reaction between IO_3^- and $H_2PO_2^-$ to yield iodine, which interacts with the thiocyanate ion [44].

The oxidation of luminol by heteropoly acids has been applied to the determination of phosphate and silicate, which are first converted into their corresponding heteropoly acids with ammonium molybdate and vanadate [45].

A recent kinetic study of the reaction between oxalate and permanganate has been the basis for the kinetic determination of the former at concentrations between 0 and 20 μg/ml and of Mn(II) (1×10^{-4}–10×10^{-4} *M*) [46].

Selenite is determined turbidimetrically at 460 nm through its reduction to elemental selenium with ascorbic acid in 2.5*M* hydrochloric acid medium [47].

Finally, hypophosphite can be determined in the range 0.06–4 mg on the basis of its reaction with excess of iodate, which exerts a Landolt effect [48].

A combination of redox and complexation reactions has been used in developing a determination for metal ions such as Fe(II), Co, Ni and Pd, based on the oxidation of benzyl-2-pyridylketone hydrazone by bromate to yield a fluorescent product. The complexation of the metals with this organic compound results in a decrease in the reaction rate that under certain conditions (the same pH for formation of the oxidation product and the complex) can be related to the metal ion concentration [49]. This principle is also the basis for the determination of 20–150 μg/ml Co(II) with the aid of the $KBrO_3$/pyridine-2-aldehyde-2-pyridyl hydrazone system [50].

It is also worth mentioning the determination of cyanide with the KI/MnO_4^- system, in which the iodine released is complexed by cyanide to form ICN, thus retarding the appearance of I_2 (detected by photometric measurements at 575 nm of the blue I_2–starch complex) [51].

5.3.2 Determination of organic compounds

Kinetic methods involving uncatalysed reactions are commonplace in the determination of organic substances. These generally exhibit no catalytic properties, so kinetic methods are a suitable alternative to other conventional equilibrium methods.

The reactions on which these determinations are based are as varied as the nature of the organic compounds themselves, namely: oxidation reactions for compounds bearing reducible groups, bromination reactions for substances possessing aromatic nuclei, condensation or addition reactions for amines and carbonyl compounds, hydrolytic reactions, etc.

5.3.2.1 Oxidation reactions

These have been the basis, among others, for the determination of both aliphatic and aromatic alcohols at the μg/ml level. Thus, ethanol is determined over the range 0.2–2.0 μg/ml through absorbance measurements (at 400 nm) of permanganate, which is reduced to manganate [52]. Up to a fivefold concentration of methanol is tolerated, but other aliphatic alcohols such as 1-propanol, 2-propanol or butanol interfere seriously. Induction period measurements on the oxidation of *tert*-butyl alcohol by xenon tetroxide allow the determination of this alcohol at levels above 22 μg [53]. Methanol, ethanol and 2-propanol are equally determinable on the basis of their effect on this induction period.

Periodate is a common oxidant for organic hydroxyl compounds such as phenols and chlorophenols, which are determined by a fixed-time method [54,55] based on absorbance measurements at 340 nm (the wavelength of maximum absorption of the resulting quinols and quinones). The determination range covered is 50–500 mg/l. Vicinal glycols are also oxidized by periodate (Malaprade reaction). Thus, ethylene, propylene and butylene glycols can be determined over the range 1.4×10^{-3}–$7.0 \times$

$10^{-3}M$ with errors of the order of 0.7% by a variable-time method involving the measurement of the time needed for a preselected amount of periodate to be consumed by the analytes. This is determined automatically with the aid of a perchlorate-selective electrode [56]. A similar procedure is applied to the determination of carbohydrates [57].

Organic compounds of pharmacological interest such as vitamins B_1 and C are determined by an oxidation reaction. Ascorbic acid can be determined in the range $0.1\mu M$–$0.1M$ by a stopped-flow technique through reaction with 2,6-dichlorophenol-indophenol [58,59]. Thiamine is oxidized by Hg(II) to thiochrome, a fluorescent compound on which its determination is based [60]. The procedure is quite sensitive (detection limit $2 \times 10^{-8}M$) and has been applied to the determination of this substance in mineral/multivitamin formulations.

Uric acid can be determined on the basis of its reducing action on the iron(III) complex of 2,4,6-tripyridyl-1,3,5-triazine, which is converted into its iron(II) analogue [61]. The method is suitable for determinations in serum. Catecholamines are oxidized to *o*-benzoquinones with iridium hexachloride or to aminochromes with periodate [62]. In both cases the stopped-flow technique allows the determination of adrenaline and L-dopa in the range 0.2–2.0mM with errors of roughly 2%. The use of periodate as oxidant makes possible the resolution of mixtures.

5.3.2.2 Bromination reactions

Determinations based on bromination reactions usually involve phenols as substrates and the use of bromine generated *in situ* from BrO_3^-/Br^- in an acid medium. At low analyte concentrations, the rate of bromination and bromine generation are practically the same, so Methyl Orange present in the reaction medium will only be decolorized when the bromination process has been completed. The decolorization time will be proportional to the concentration of analyte present. Cresols, xylenols [63], ethyl and phenyl phenols [64,65], paracetamol [66], salicylic and acetylsalicylic acids [67], etc., are among the substances determined in this fashion (variable-time method, detection limit about $10^{-3}M$).

5.3.2.3 Other reactions

The formation of addition compounds of species containing carbonyl groups has been exploited for the determination of organic compounds such as 2-furaldehyde. This reacts with bisulphite in the presence of acetone, methanol or acetic acid [68]. Formaldehyde can be determined in the range 1–6 μg/ml by reaction with cyanide, with the aid of an ion-selective electrode [69]. If the sample also contains hexamine, this can be determined at concentrations between 1 and 5 μg/ml because its hydrolysis yields formaldehyde. Tryptophan reacts with formaldehyde at pH 10.8 to yield a fluorescent product that allows its determination in the range 0–100 nmole/ml [70] by a kinetic method suitable for application to natural and prepared foods.

Amino-compounds can be determined in a variety of ways. Thus, urea is determined at levels between 50 and 1200 mg/l. by reaction with phthalaldehyde and *N*,*N*-dimethyl-*N*′-(1-naphthyl)ethylenediamine, with a coefficient of variation of about 2% [71]. Primary, secondary and tertiary amines can be determined on the basis of the formation of coloured complexes (λ_{max} 450–480 nm) with tetrachloro-*p*-benzoquinone [72]. Sulphonamides are determined in urine in the range $0.5 \times$

10^{-5}–5×10^{-5} by a stopped-flow technique involving the Griess reaction [75].

Steroids such as hydrocortisone [74] and corticosteroids such as betamethasone or valerate [75] are determined by the modified Tetrazolium Blue reaction. The stopped-flow technique applied involves making measurements at 525 nm every 20–30 sec. The application of the method of proportional equations allows the resolution of mixtures of corticosteroids such as the two mentioned above [75].

Nitroglycerine can be determined in tablets by monitoring the absorbance at 336 nm for 20 min, during the course of its hydrolysis in methanol [76]. A picrate-selective electrode makes possible the determination of creatinine in serum by a variable-time method [77]. Finally, some 2-oxohexoses such as fructose, sorbose or psycose can be determined by reaction with cysteine in a sulphuric acid medium, on the basis of the variation of the absorbance as a function of time [78].

REFERENCES

[1] S. Siggia, *Quantitative Organic Analysis Via Functional Groups*, Wiley, New York, 1949, p. 18.
[2] W. J. Blaedel and P. L. Petitjean, *Anal. Chem.*, 1958, **30**, 1958.
[3] A. E. Burgess and J. L. Latham, *Analyst*, 1966, **91**, 343.
[4] J. D. Ingle, Jr. and S. R. Crouch, *Anal. Chem.*, 1971, **43**, 697.
[5] K. J. Laidler, *Chemical Kinetics*, McGraw-Hill, New York, 1965, pp. 19–21.
[6] V. V. S. E. Dutt, A. E. Hanna and H. A. Mottola, *Anal. Chem.*, 1976, **48**, 1027.
[7] J. C. Thomson and H. A. Mottola, *Anal. Chem.*, 1984, **56**, 755.
[8] V. V. S. E. Dutt and H. A. Mottola, *Anal. Chem.*, 1977, **49**, 319.
[9] J. Kurzawa, Z. Kurzawa and Z. Swit, *Chem. Anal. Warsaw*, 1976, **21**, 791.
[10] T. Yogo, G. Nakagawa and M. Kodama, *Bunseki Kagaku*, 1975, **24**, 677.
[11] J. Rodríguez, A. García de Torres and J. M. Cano-Pavón, *Anal. Chim. Acta*, 1984, **156**, 319.
[12] K. Nagakawa, T. Ogata, K. Haraguchi and S. Ito, *Chem. Abstr.*, 1980, 93, 106336z.
[13] R. L. Wilson and J. D. Ingle, Jr., *Anal. Chim. Acta*, 1977, **92**, 417.
[14] N. Gorski and B. Gorski, *Chem. Abstr.*, 1979, 89, 36022r.
[15] J. B. Landis, M. Rebec and H. L. Pardue, *Anal. Chem.*, 1977, **49**, 785.
[16] T. Nozaki and M. Sakamoto, *Bunseki Kagaku*, 1975, **24**, 677.
[17] V. I. Rychkova and I. F. Dolmanova, *J. Anal. Chem. USSR*, 1979, **34**, 1094.
[18] H. Weisz, S. Pantel and R. Giesin, *Anal. Chim. Acta*, 1978, **101**, 187.
[19] J. J. Laserna, A. Navas and F. García-Sánchez, *Microchem. J.*, 1982, **27**, 312.
[20] E. A. Morgan, N. A. Vlasov and L. A. Kozhemyakina, *Zh. Analit. Khim.*, 1972, **27**, 2064.
[21] L. M. Tamarchenko, *Zh. Analit. Khim.*, 1975, **30**, 127.
[22] F. Salinas, C. Genestar and F. Grases, *Anal. Chim. Acta*, 1981, **130**, 337.
[23] R. J. Sukasiewicz and J. M. Fitzgerald, *Anal. Lett.*, 1969, **2**, 159.
[24] F. Salinas, C. Genestar and F. Grases, *Microchem. J.*, 1982, **27**, 32.
[25] A. Navas, F. Sánchez-Rojas and F. García-Sánchez, *Mikrochim. Acta*, 1982 **I**, 175.
[26] F. Grases, C. Genestar and F. Salinas, *Anal. Chim. Acta*, 1983, **148**, 245.
[27] F. Salinas, F. García-Sánchez, F. Grases and C. Genestar, *Anal. Lett.*, 1980, **13**, 473.
[28] E. Mentasti and E. Pelizzetti, *Anal. Chim. Acta*, 1975, **78**, 227.
[29] M. B. Babkin, *Chem. Abstr.*, 1972, 76, 20909a.
[30] M. E. Babkin, *Zavodsk. Lab.*, 1971, **37**, 524.
[31] A. T. Pilipenko, L. V. Markova and T. S. Maksimenko, *Ukr. Khim. Zh.*, 1976, **42**, 1081.
[32] H. Weisz and G. Fritz, *Anal. Chim. Acta*, 1981, **123**, 239.
[33] T. Pérez-Ruiz, C. Martínez-Lozano and J. Ochotorena, *Talanta*, 1982, **29**, 479.
[34] U. Isacsson and G. Wettermark, *Anal. Lett.*, 1978, **11**, 13.
[35] E. Bishop and N. Evans, *Talanta*, 1970, **17**, 1125.
[36] J. F. Endicott and H. Taube, *Inorg. Chem.*, 1965, **4**, 437.
[37] L. A. Lazarou, P. A. Siskos, M. A. Koupparis and T. P. Hadjiioannou, *Anal. Chim. Acta*, 1977, **94**, 475.
[38] E. I. Dodin, L. S. Makarendko, V. F. Tsvetkov, I. P. Kharlamov and A. M. Pavlova, *Zavodsk. Lab.*, 1973, **39**, 1050.
[39] R. D. Cox, *Anal. Chem.*, 1980, **52**, 332.
[40] R. C. Doerr, J. B. Fox, L. Sakritz and W. Fiddler, *Anal. Chem.*, 1981, **53**, 381.

[41] M. A. Koupparis, K. M. Walczak and H. V. Malmstadt, *Analyst*, 1982, **107**, 1309.
[42] M. A. Koupparis, K. M. Walczak and H. V. Malmstadt, *Anal. Chim. Acta*, 1982, **142**, 119.
[43] A. K. Babko, L. V. Markova and M. L. Kaplan, *Zavodsk. Lab.*, 1968, **34**, 1053.
[44] M. I. Karayannis and D. P. Nikolelis, *Microchem. J.*, 1977, **22**, 356.
[45] M. M. Sukovskaya and V. A. Biochenko, *Zavodsk. Lab.*, 1974, **40**, 936.
[46] M. A. Koupparis and M. I. Karayannis, *Anal. Chim. Acta*, 1982, **138**, 303.
[47] K. Nakagawa, T. Ogata, K. Haraguchi and S. Ito, *Chem. Abstr.*, 1980, **92**, 14772q.
[48] D. P. Nikolelis, M. I. Karayannis, E. V. Kordi and T. P. Hadjiioannou, *Anal. Chim. Acta*, 1977, **90**, 209.
[49] F. García-Sánchez, A. Navas and J. J. Laserna, *Anal. Chem.*, 1983, **55**, 253.
[50] F. García-Sánchez, A. Navas, J. J. Laserna and M. R. Martínez de la Barrera, *Z. Anal. Chem.*, 1983, **315**, 491.
[51] S. N. Rai, M. Phull and P. C. Nigam, *Indian J. Chem.*, 1983, **22A**, 482.
[52] S. U. Kreingol'd, L. I. Kefilyan and V. N. Antonov, *Zh. Analit. Khim.*, 1977, **32**, 2424.
[53] R. H. Krenger, S. Vas and B. Jaselskis, *Talanta*, 1971, **18**, 116.
[54] L. R. Sherman, V. L. Trust and H. Hoang, *Talanta*, 1981, **28**, 408.
[55] N. G. Buckman, R. J. Magee and J. O. Hill, *Anal. Chim. Acta*, 1983, **153**, 285.
[56] C. H. Efstathiou and T. P. Hadjiioannou, *Anal. Chem.*, 1975, **47**, 864.
[57] C. H. Efstathiou and T. P. Hadjiioannou, *Anal. Chim. Acta*, 1977, **89**, 55.
[58] M. I. Karayannis, *Anal. Chim. Acta*, 1975, **76**, 121.
[59] K. Hiromi, H. Fujiromi, J. Yamaguchi-Ito, H. Nakatani, M. Ohnishi and B. Tonomura, *Chem. Lett.*, 1977, 1333.
[60] M. A. Ryan and J. D. Ingle, Jr., *Anal. Chem.*, 1980, **52**, 2177.
[61] A. Tabacco, F. Bardelli, F. Meiattini and P. Taril, *Clin. Chim. Acta*, 1980, **104**, 405.
[62] E. Pelizzetti, E. Mentasti, E. Pramauro and G. Giraudi, *Anal. Chim. Acta*, 1976, **85**, 161.
[63] W. Rodziewicz, I. Kwiatkowska and E. Kwiatkowski, *Chem. Anal. Warsaw*, 1968, **13**, 783.
[64] W. Rodziewicz, I. Kwiatkowska and E. Kwiatkowski, *Chem. Anal. Warsaw*, 1968, **13**, 1067.
[65] W. Rodziewicz, I. Kwiatkowska and E. Kwiatkowski, *Chem. Anal. Warsaw*, 1968, **13**, 1305.
[66] M. Elsayed, *Pharmazie*, 1979, **34**, 569.
[67] G. N. Rao, *Z. Anal. Chem.*, 1973, **264**, 414.
[68] V. I. Tikhonova, *Zh. Analit. Khim.*, 1968, **23**, 1720.
[69] M. A. Koupparis, C. E. Efstathiou and T. P. Hadjiioannou, *Anal. Chim. Acta*, 1979, **107**, 91.
[70] H. Steinhart, *Anal. Chem.*, 1979, **51**, 1012.
[71] T. Momose and T. Momose, *Clin. Chim. Acta*, 1981, **114**, 297.
[72] S. U. Kreingol'd, V. N. Antonov and E. M. Yutal, *Zh. Analit. Khim.*, 1977, **32**, 1618.
[73] A. G. Xenakis and M. I. Karayannis, *Anal. Chim. Acta*, 1984, **159**, 343.
[74] R. M. Oteiza, D. L. Drottinger, M. S. McCracken and H. V. Malmstadt, *Anal. Chem.*, 1977, **49**, 1586.
[75] M. Koupparis, K. M. Walczak and H. V. Malmstadt, *J. Pharm. Sci.*, 1979, **68**, 1479.
[76] H. Fung, P. Dalecki, E. Ise and C. I. Rhodes, *J. Pharm. Sci.*, 1973, **62**, 696.
[77] E. P. Diamandis and T. P. Hadjiioannou, *Clin. Chem.*, 1981, **27**, 455.
[78] D. L. Bissett, T. E. Hanson and R. L. Anderson, *Microchem. J.*, 1974, **19**, 71.

6

Differential reaction-rate methods

6.1 INTRODUCTION

Differential reaction-rate methods are an area of recent exploitation in chemical analysis and the answer to a perpetual analytical problem, namely the resolution of mixtures of closely related species.

These methods are based on the different rate at which two or more species interact with a common reagent. Differential reaction-rate methods, though not as sensitive as direct methods for determination of catalysts at the micro or submicro trace level, allow analysis for two or more species without prior separation.

As a rule, these methods are based on uncatalysed reactions, though they are occasionally applied to the determination of mixtures of species acting as catalysts for a given catalytic reaction.

This chapter deals with the simultaneous determination of several species, rather than with analysis for a given species in the presence of others (commented on in the preceding chapter), though due reference will also be made to the use of a separate aliquot of the sample for the determination of each species of interest in a mixture.

Judging by the length at which differential reaction-rate methods are dealt with in the monograph by Mark and Rechnitz [1], the methodology and theoretical background of differential reaction-rate methods are by now completely established, but few analytical applications of these methods have been reported to date. A general classification of these appears in Table 6.1.

6.2 FOUNDATIONS AND METHODOLOGY

6.2.1 Pseudo first-order reactions $[R]_0 \gg [A]_0 + [B]_0$

Let A and B be two substances to be determined in a mixture and R a reagent with which both react to yield two products (P and P′) which, though different, are similar as regards the analytical property to be measured. The chemistry involved can be expressed as

Table 6.1 — Differential reaction-rate methods

Reaction order	Reactant concentrations	Method
Pseudo first-order (in $[A]_0+[B]_0$)	$[R]_0 \gg [A]_0+[B]_0$	Logarithmic extrapolation Linear graphical Worthington and Pardue Connors Single-point (Lee and Kolthoff) Initial-rate Proportional equation
Pseudo first-order (in $[R]_0$)	$[R]_0 \ll [A]_0+[B]_0$	Single-point (Roberts and Regan) Proportional equation
Second-order	$[R]_0 \gg [A]_0+[B]_0$ or $[R]_0 \ll [A]_0+[B]_0$ (50-fold or greater excess of either)	Second-order logarithmic extrapolation Linear extrapolation
	$[R]_0=[A]_0+[B]_0$	Second-order single-point
Pseudo zero-order	$[R]_0 \gg [A]_0+[B]_0$ or $[R]_0=[A]_0+[B]_0$ or $[R]_0 \ll [A]_0+[B]_0$	Single-point Proportional equation

$$A + R \xrightarrow{k_A} P$$

$$B + R \xrightarrow{k_B} P$$

where k_A and k_B are the rate constants for reaction of A and B, respectively. If the two rates are different and correspond to first-order kinetics with respect to each of the components of the mixture (i.e. $[R]_0 \gg [A]_0$ and $[R]_0 \gg [B]_0$), then the sum of the concentrations of A and B at time t is given by

$$C_t = [A]_t + [B]_t = [P]_\infty - [P]_t = [A]_0 \exp(-k_A t) + [B]_0 \exp(-k_B t) \quad (6.1)$$

and the concentration of the sum of the products at time t is given by

$$\begin{aligned} [P]_t &= [A]_0 - [A]_t + [B]_0 - [B]_t \\ &= [A]_0[1 - \exp(-k_A t)] + [B]_0[1 - \exp(-k_B t)] \end{aligned} \quad (6.2)$$

if no product is assumed to be present at the beginning of the reaction. $[P]_\infty$ is the maximum total product concentration.

These expressions are analytically useful provided that k_A and k_B are sufficiently different. $[P]_\infty$ is usually measured after allowing the reaction to proceed for a sufficiently long time, and is in itself a measure of the total initial concentrations of the reactants, $[A]_0 + [B]_0$. [P] can be replaced by a measurable property proportional to the product concentration, such as the absorbance, fluorescence intensity or potential.

If $[R]_0 \approx [A]_0 + [B]_0$ (which is sometimes desirable, as otherwise the reactions involved are too fast to be monitored by conventional methods), the overall kinetics will be second-order (dealt with later on). However, the reaction can also be made pseudo first-order with respect to the reagent, i.e. $[R]_0 \ll [A]_0 + [B]_0$, and the product measured with high sensitivity.

The first-order kinetic equations listed in Table 1.2 and Eqs. (6.1) and (6.2) have been used to develop various graphical and mathematical methods for the differential kinetic resolution of mixtures. The most outstanding are the logarithmic extrapolation, linear graph and proportional equation methods. Obviously, the analytical applicability of these methods is dependent on the relative values of the rate constants, as well as on the ratio between the concentrations of the mixture components.

The simultaneous determination of two or more species in a mixture is a relatively easy task if k_A/k_B is either large or small (e.g. if one of the constants is very small). If this is not the case, the procedure involves selective determination of the more rapidly reacting species either by suitable changes in the reaction conditions (e.g. temperature, solvent, reactant concentrations) or masking the species with a suitable inhibitor. The concentration of the second species can be obtained by a similar approach or by difference from a determination of the sum of the two concentrations.

6.2.1.1 Logarithmic extrapolation method

This is one of the methods most frequently used in differential kinetics [2] and involves taking logarithms in Eq. (6.1) and plotting log C_t or log $([P]_\infty - [P]_t)$ as a function of time (Fig. 6.1). The result is a curve unless $k_A = k_B$, in which case a

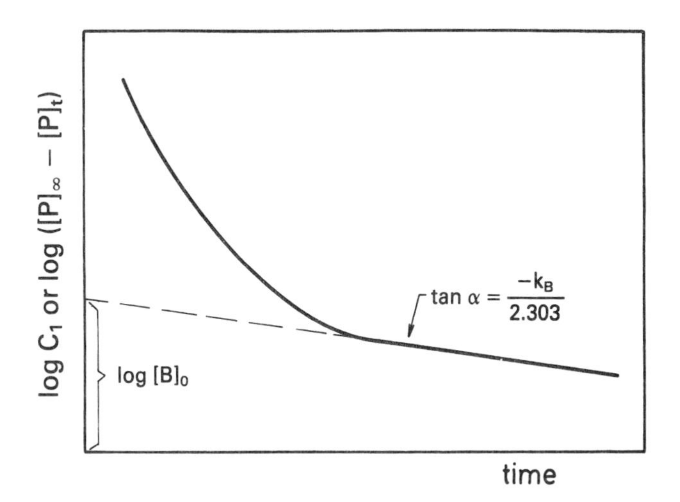

Fig. 6.1 — Logarithmic extrapolation method for first-order reactions.

straight line is obained. If species A disappears at a higher rate than B (i.e. $k_A > k_B$), then $[A]_t \rightarrow 0$ and the curve finally becomes linear (Fig. 6.1), so that

$$\log C_t = \log([P]_\infty - [P]_t) = \log [B]_0 - k_B t/2.303 \qquad (6.3)$$

The intercept on the ordinate allows calculation of the initial concentration of B, $[B]_0$, while that of the other component, $[A]_0$, is obtained by difference after determining $[A]_0 + [B]_0$ from $[P]_\infty$. If a photometric detection technique is applied, then Eq. (6.3) can be written as:

$$\log(D_\infty - D_t) = \log \varepsilon_B [B]_0 - k_B t/2.303 \qquad (6.4)$$

where D_∞ and D_t are the absorbances measured after the reaction has developed to completion and after a time t near the start of the linear region, and ε_B is the molar absorptivity of B (known beforehand).

This method features several advantages, namely: (a) no prior knowledge of the rate constants of A and B is necessary; (b) it is less prone to error as the linear plot is constructed from several points; (c) it can be applied to ternary mixtures, provided that the rate constants of the three components are sufficiently different, and even to reactions subject to synergistic effects, as it is not limited to pseudo first-order reactions. However, at least 99% of the more reactive species should be consumed in the process if valid data are to be obtained, and the total initial concentration of all the species involved should be known, for the method to be applicable. In addition, the method is somewhat laborious as it entails the continuous recording of the measured property and obtaining the required $[P]_\infty - [P]_t$ values from the resulting curve.

6.2.1.2 Linear graph methods

There are by now several linear graph methods available. Thus, Worthington and Pardue [3], assuming $k_A > k_B$ in Eq. (6.2) and a time t at which $\exp(-k_B t) \rightarrow 0$ and $\exp(-k_B t) \neq 1$ (but is not near zero), have derived the expression:

$$[P]_t = [A]_0 + [B]_0[1 - \exp(-k_B t)] \qquad (6.5)$$

In practice, the method involves plotting $[P]_t$ as a function of $\exp(-k_B t)$. The result is a straight line, the intercepts of which with the y-axis at $t=0$ and $t=\infty$ correspond to $[A]_0$ and $[A]_0 + [B]_0$, respectively. $[B]_0$ can obviously be calculated by difference. Alternatively, $[A]_0$ can be determined (less accurately) by plotting $[P]_t$ *vs.* $\exp(-k_A t)$. Figure 6.2 shows examples of these plots for the individual components at the same concentration and for their mixture. The practical work involved entails plotting $[P]_t$ as a function of both $\exp(-k_A t)$ and $\exp(-k_B t)$ and extrapolating to obtain the required data from them.

This method does not call for prior knowledge of the initial total concentration of the mixture, but it does require the rate constants of the different species to be known beforehand. To avoid the laborious point-by-point tracing of the plot, an instrumental technique can be used in which the exponential functions are generated by means of an analogue circuit. The curve shown in Fig. 6.2 was constructed in this manner

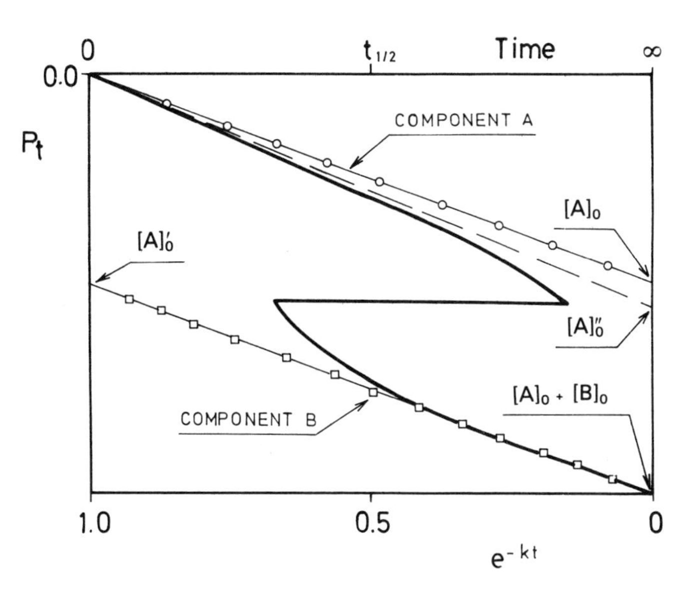

Fig. 6.2 — Linear extrapolation method. $[A]_0$, initial concentration of A in the absence of B; $[A]'_0$, concentration of A in the mixture, obtained by extrapolation of the curve corresponding to the slower reacting component; $[A]''_0$, concentration of A in the mixture, obtained by extrapolation of the curve corresponding to the faster reacting component. (Reproduced with permission, from J. B. Worthington and H. L. Pardue, *Anal. Chem.*, 1972, **44**, 767. Copyright 1977, American Chemical Society.)

and can also be obtained by replacing $[P]_t$ with absorbance values. The system is started at $t=0$ (i.e. as the last reactant is added) and the value of $\exp(-k_B t)$ is recorded on the x-axis as a function of time. When the curve begins to deviate from linearity the circuit is switched to record the variation of $\exp(-k_B t)$ for as long as is required to obtain reliable results from the slower component.

This system has been successfully applied to the determination of lanthanum and neodymium in mixtures [3]. A similar device, designed by Gary and Lagrange, has been used for the analysis of Ca–Mg mixtures [4].

A potential simplification of the method relies on the assumption that $\exp(-k_B t) \approx 1 - k_B t$, which is the case when $k_B t$ is small enough, since $\exp(-k_B t)$ can be expanded as a series,

$$\exp(-k_B t) = 1 - k_B t + (k_B t)^2/2! - (k_B t)^3/3! + \ldots$$

and, under these conditions, only the first two terms in the series are significant. This simplification reduces Eq. (6.5) to:

$$[P]_t = [A]_0 + [B]_0 k_B t \tag{6.6}$$

so the plot of $[P]_t$ as a function of time becomes a straight line with a slope and intercept which allow the calculation of $[B]_0$ (if k_B is known) and $[A]_0$, respectively. This method has been applied to the simultaneous determination of Zn and Cd [5].

The Connors method

The Connors graphical extrapolation method [6] yields satisfactory results when k_A/k_B is rather small or $[A]_0/[B]_0$ is very large. It requires prior knowledge of the rate constants but not of the total initial concentration of the mixture.

With $[P]_\infty = [A]_0 + [B]_0$, and assuming $k_A > k_B$, Eq. (6.1) can be transformed into:

$$([P]_\infty - [P]_t)\exp(k_A t) = [A]_0 + [B]_0 \exp[(k_A - k_B)t] \tag{6.7}$$

In practice, the whole left-hand side of this equation is plotted as a function of $\exp[(k_A - k_B)t]$, as shown in Fig. 6.3a. The slope of the plot yields $[B]_0$. The molar fraction of B in the mixture can be calculated from the slope/intercept ratio since the intercept on the ordinate at $t=0$ (i.e. $\exp[(k_A - k_B)t] = 1$) corresponds to $[P]_\infty = [A]_0 + [B]_0$.

The resolution of a three-component mixture for which $k_A > k_B > k_C$ requires plotting $([P]_\infty - [P]_t)\exp(k_B t)$ as a function of $\exp[(k_B - k_C)t]$. The plot becomes linear once A has reacted completely (Fig. 6.3b), the slope of the straight segment then being $[C]_0$, and its intercept with the ordinate at $\exp[(k_B - k_C)t] = 1$ is equal to $[B]_0 + [C]_0$, so the initial concentrations of the three components can be readily calculated by taking into account that $[P]_\infty = [A]_0 + [B]_0 + [C]_0$.

This method has been applied to the resolution of an ester mixture consisting of *p*-nitrophenyl *p*-chlorobenzoate (A) and *p*-nitrophenyl benzoate (B), which undergo alkaline hydrolysis at different rates ($k_A > k_B$) and cannot be determined satisfactorily by the logarithmic extrapolation method at concentration and rate-constant ratios for A to B of 1.07 and 2.7, respectively. Figure 6.3a is representative of this determination.

Another graphical interpolation method proposed by Connors [7] is of equal use when similar rate constants are involved. It calls for no prior knowledge of the constants, though it does require a knowledge of the overall initial concentration.

On the basis of the $[P]_t$ *vs.* time plots for the unknown mixture and different standard mixtures, the concentrations of the components can be obtained from a reference curve matching that of the sample. In this case $C_t^s = C_t^r$ and hence $[A]_0^r = [A]_0^s$ and $[B]_0^r = [B]_0^s$.

Applying Eq. (6.1) to the sample and reference gives

$$([P]_\infty - [P]_t)^s = [A]_0^s \exp(-k_A t) + [B]_0^s \exp(-k_B t) \tag{6.8}$$

$$([P]_\infty - [P]_t)^r = [A]_0^r \exp(-k_A t) + [B]_0^r \exp(-k_B t) \tag{6.9}$$

Subtraction of Eq. (6.9) from Eq. (6.8) yields:

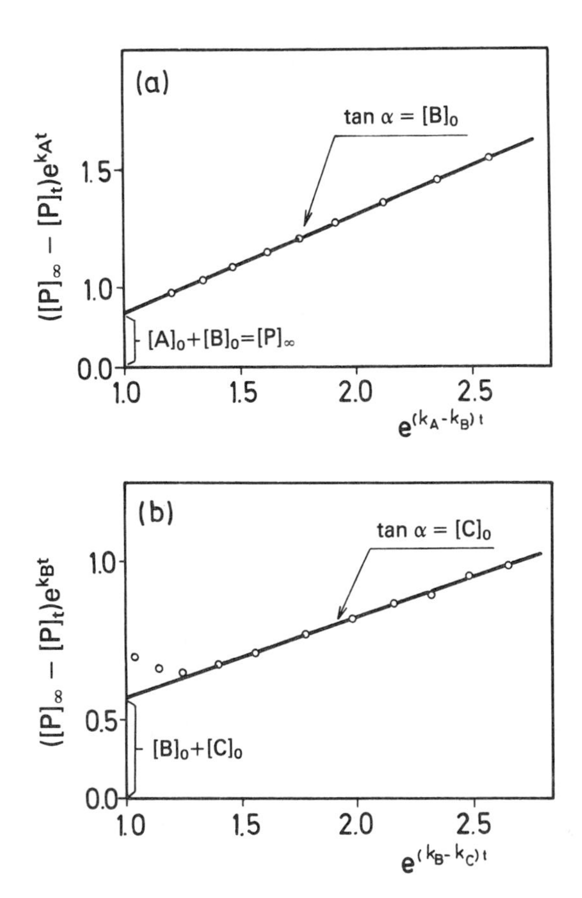

Fig. 6.3 — The Connors method for binary (a) and ternary (b) mixtures. (Reproduced with permission, from K. A. Connors, *Anal. Chem.*, 1976, **48**, 87. Copyright 1977, American Chemical Society.)

$$[P]_t^s - [P]_t^r = ([A]_0^r - [A]_0^s)\exp(-k_A t) + ([B]_0^r - [B]_0^s)\exp(-k_B t) \quad (6.10)$$

Since

$$[A]_0^s + [B]_0^s = [A]_0^r + [B]_0^r \quad (6.11)$$

a combination of Eqs. (6.10) and (6.11) yields:

$$[P]_t^s - [P]_t^r = [B]_0^r[\exp(-k_B t) - \exp(-k_A t)] - [B]_0^s[\exp(-k_B t) - \exp(-k_A t)] \quad (6.12)$$

Plotting $[P]_t^s - [P]_t^r$ *vs*. $[B]_0^s$ for a series of samples with different and accurately known $[B]_0^r$ values gives a straight line with intercept corresponding to $[B]_0^s$ since this is equal to $[B]_0^r$ for $[P]_t^s - [P]_t^r = 0$ (Fig. 6.4). The initial concentration of A, $[A]_0$, is obtained by difference from $[P]_\infty$

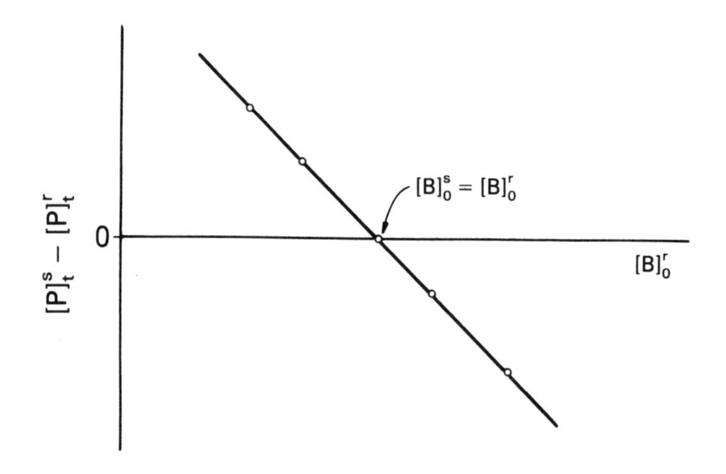

Fig. 6.4 — Modified Connors graphical interpolation method [8].

A modification of this method, requiring prior knowledge of the rate constants, but not of $[P]_\infty$, is based on the use of a further equation for the faster component [8]. This is of special interest when the measurement of $[P]_\infty$ requires an exceedingly long time or when $[P_\infty] \neq [A]_0 + [B]_0$, as is usually the case with the determination of mixtures of catalysts. The equation is derived by assuming that the overall reaction rate of the unknown mixture is equal to the sum of the rates of the individual components, i.e.

$$v_T = v_A + v_B = k_A[A]_0 + k_B[B]_0 \tag{6.13}$$

where k_A and k_B are the pseudo first-order rate constants of A and B, respectively.

Substitution of the $[B]_0$ value obtained by the previous method, into Eq. (6.13), yields the required value of $[A]_0$.

The preparation of the series of standard samples, about which Connors gives no details, is rather troublesome from a practical point of view. The required degree of matching can only be achieved by keeping constant the concentration of one of the components while varying that of the other, in addition to diluting the unknown to an appropriate extent. A detailed description of the procedure involved has been reported [8].

Single-point method

Originally developed by Lee and Kolthoff [9], the single-point method is applicable only to first-order kinetics, which has limited its use to the resolution of binary mixtures. The initial concentrations of the mixture components are calculated from the overall initial concentration and the fraction of it which has reacted after a preselected time, t_{opt}, chosen in terms of the ratio of the two rate constants. Thus, dividing Eq. (6.1) into the total initial concentration, C_{tot}, and expressing $[A]_0$ or $[B]_0$ in terms of $C_{tot} - [B]_0$, results in

$$([P]_\infty - [P]_t)/[P]_\infty = C_t/C_{tot} = ([B]_0/C_{tot})[\exp(-k_B t) - \exp(-k_A t)] + \exp(-k_A t) \quad (6.14)$$

$$([P]_\infty - [P]_t)/[P]_\infty = C_t/C_{tot} = ([A]_0/C_{tot})[\exp(-k_A t) - \exp(-k_B t)] + \exp(-k_B t) \quad (6.15)$$

The calculation of t_{opt} entails taking into account that the steeper the slope of the calibration curve, the greater the accuracy of the analysis. Thus, by differentiation of Eqs. (6.14) and (6.15) with respect to time and equating the derivative to zero, t_{opt} is found to be

$$t_{opt} = \frac{\ln(k_A/k_B)}{k_A - k_B}$$

Plotting C_t/C_{tot} or the $([P]_\infty - [P]_t)/[P]_\infty$ values corresponding to time $t = t_{opt}$ as a function of the molar fraction of A, x_A, gives a straight line with intercepts on the ordinate at $x_A = 0$ and $x_A = 1$, which correspond to $\exp(-k_B t)$ and $\exp(-k_A t)$, respectively, and coincide with those corresponding to $x_B = 1$ and $x_B = 0$ in a plot of C_t/C_{tot} *vs.* x_B (molar fraction of B).

A calibration curve can be constructed by determining the degree to which pure A and B have reacted at t_{opt} and joining these two points, by a straight line, as shown in Fig. 6.5a. These intercepts can also be determined from the rate constants through the expressions $\exp(-k_A t)$ and $\exp(-k_B t)$ (Fig. 6.5a). In practice, the degree of reaction is taken from the calibration curve to determine the molar reactions of A and B, from which their initial concentrations are readily calculated by use of the known value of C_{tot}.

The expressions derived above are not useful in practice, as it is not the concentration, but rather the absorbance, fluorescence intensity, or any other quantity proportional to it that is experimentally measured. Equation (6.1) can be expressed in absorbance terms as:

$$D_\infty - D_t = [A]_0 \varepsilon_A \exp(-k_A t) + [B]_0 \varepsilon_B \exp(-k_B t) \quad (6.16)$$

By assuming the optical-path length to be unity, C_{tot} can be calculated from

$$C_{tot} = D_\infty = [A]_0 \varepsilon_A + [B]_0 \varepsilon_B \quad (6.17)$$

and Eqs. (6.14) and 6.15) can be converted into

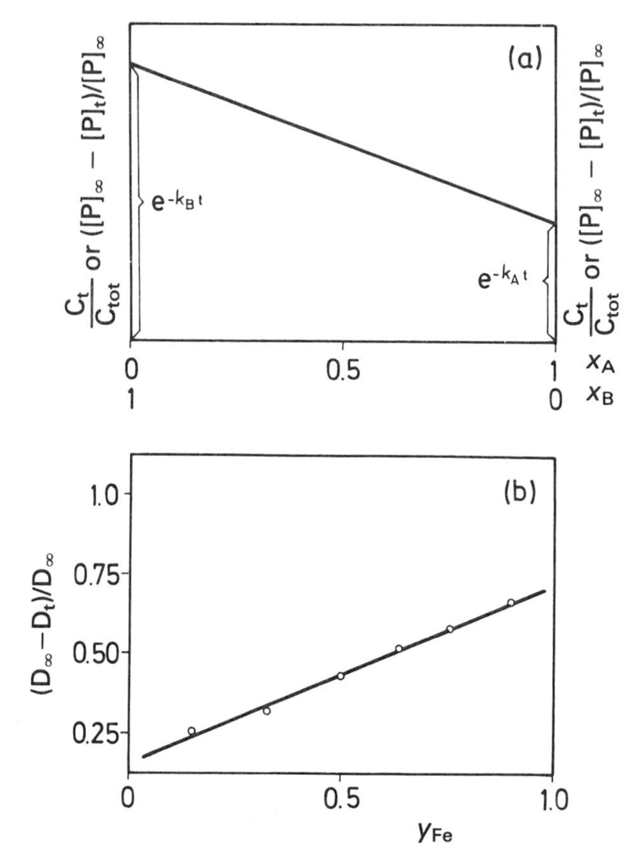

Fig. 6.5 — Single-point method for first-order reactions. (a) Two-component (A, B) mixture; (b) spectrophotometric resolution of Co–Fe mixture.

$$(D_\infty - D_t)/D_\infty = ([B]_0\varepsilon_B/D_\infty)[\exp(-k_B t) - \exp(-k_A t)] + \exp(-k_A t) \tag{6.18}$$

$$(D_\infty - D_t)/D_\infty = ([A]_0\varepsilon_A/D_\infty)[\exp(-k_A t) - \exp(-k_B t)] + \exp(-k_B t) \tag{6.19}$$

Similarly to the molar fraction, the absorbance fraction of each species may be defined as:

$$y_B = [B]_0\varepsilon_B/D_\infty; \quad y_A = [A]_0\varepsilon_A/D_\infty$$

Therefore, by plotting $(D_\infty - D_t)/D_\infty$ as a function of the absorbance fraction of one of the components of the standard sample, a calibration curve is obtained which allows the absorbance fraction of an unknown sample to be readily calculated. From this and D_∞ it is possible to calculate the initial concentrations of both components, by taking into account that $y_A + y_B = 1$ and using the molar absorptivities (known beforehand).

In Fig. 6.5b is shown the calibration curve corresponding to an Fe–Co mixture. Both metals form a complex (though at a different rate) with pyridoxal thiosemicarbazone [10]. The curve shown in the figure corresponds to variation in the absorbance fraction of iron, the slower reactant.

The single-point method requires one of the constants to be at least four times as large as that of the other component. Nevertheless, as the calibration curve is purely empirical, the method is applicable to any type of reaction mechanism.

Reaction-rate method

Also known as the tangent method, this procedure can be applied whenever it is possible to record the variation of $[P]_\infty - [P]_t$ as a function of time.

The method is based on the equation

$$-d([P]_\infty - [P]_t) = (k_A[A]_t + k_B[B]_t)dt \tag{6.20}$$

from which it is evident that $-d([P]_\infty - [P]_t)/dt$ will approach $k_B[B]_t$ as $[A]_t$ tends to zero. Under such conditions, the absorbance measured is exclusively due to B and hence $[B]_t = [P]_\infty - [P]_t$, so Eq. (6.20) can be rewritten as:

$$-d([P]_\infty - [P]_t)/dt = 1/([P]_\infty - [P]_t)k_B$$

from which k_B can be calculated. Thus, at a given point, the product of the slope of the curve and the reciprocal of $[P]_\infty - [P]_t$ reaches a constant value equal to k_B.

Once k_B is known, $[B]_0$ can be calculated from

$$\ln[B]_0 = \ln[B]_t + k_B t = \ln([P]_\infty - [P]_t) + k_B t \tag{6.21}$$

and $[A]_0$ can be obtained from $[P]_\infty$.

This method is similar to the logarithmic extrapolation method, but is subject to more obvious errors in the calculation of the concentrations of the mixture components, as its precision relies on the manner in which the slope of the curve can be determined graphically at a given point.

6.2.1.3 *Proportional-equation method*

This is a mathematical method of wide use in the resolution of mixtures of closely related species [11], usually less time-consuming than those described above (especially if a microcomputer is used), and requiring no prior knowledge of C_{tot}.

A rate-constant ratio of 4 is usually enough for this method, the easy automation of which makes it suitable for fast reactions. However, it cannot be applied to processes subject to synergistic effects, because the measured property must be additive in nature. On the other hand, the rate constants must be carefully calculated; if the reaction involves complex kinetics, the applicability of the method relies on the accurate calculation of the different proportionality constants.

The practical procedure entails measuring C_t at two or more reaction times and formulating two equations similar to Eq. (6.1):

$$[P]_\infty - [P]_{t1} = [A]_0 \exp(-k_A t_1) + [B]_0 \exp(-k_B t_1)$$
$$[P]_\infty - [P]_{t2} = [A]_0 \exp(-k_A t_2) + [B]_0 \exp(-k_B t_2)$$

the resolution of which provides the initial concentration of the different components in terms of their respective rate constants (known beforehand).

Not only the time, but any other experimental variable such as the temperature or one related to the solvent or the physicochemical characteristics of the reactants can be used to formulate the simultaneous equations from which the required concentrations are to be calculated. Thus, the general equations for a binary mixture will be of the form;

$$f_1 = k_{A,1}[A]_0 + K_{B,1}[B]_0$$
$$f_2 = k_{A,2}[A]_0 + K_{B,2}[B]_0$$

where f is any measurable parameter determined under two different sets of conditions, and $K_{A,1}$ and $K_{A,2}$ are empirical or proportionality constants obtained separately for component A under different conditions in the same way as $K_{B,1}$ and $K_{B,2}$ are obtained for component B. Although any experimental variable is usable provided $K_{A,1}K_{B,2} \neq K_{A,2}K_{B,1}$, the optimum results are obtained for $K_{A,1}/K_{B,1} > 1$ for one set of conditions and $K_{A,2}/K_{B,2} < 1$ for the other.

This method has been applied to the simultaneous determination of two catalysts such as ruthenium and osmium by measuring the initial rate of the reaction between Ce(IV) and As(III) (catalysed by both metals) under two different sets of experimental conditions [12].

6.2.2 Pseudo first-order reactions, $[R]_0 \ll [A]_0 + [B]_0$

Methods applied to this type of reaction are less frequent and are normally only applied to fast processes that can be slowed down by decreasing the reactant concentration, though this results in the need for greater accuracy in the determination of the reagent, R.

Roberts and Regan's single-point method and the proportional equation method are by far the commonest alternatives used in the resolution of mixtures involving this type of kinetics.

6.2.2.1 Roberts and Regan's single-point method

This method is based on the general equation:

$$-d[R]/dt = k_A[A]_0[R] + k_B[B]_0[R] = K[R] \quad (6.22)$$

where R is used to monitor the course of the reaction.

The application of this method requires previous knowledge of the respective rate constants k_A and k_B, and also of the initial composition of the mixture, $C_0 = [A]_0 + [B]_0$. A and B are used in large excess relative to R, to make the system pseudo first-order with respect to R.

Integration of Eq. (6.22) between $t_0 = 0$ and $t_1 = t$ and substitution of $k_A[A]_0 + k_B[B]_0$ for K leads to:

$$\ln([R]_t/[R]_0) = -(k_A[A]_0 + k_B[B]_0)t = -Kt \tag{6.23}$$

from which $[A]_0$ and $[B]_0$ can be readily obtained by plotting the left-hand side of the equation as a function of time.

A modified version of this method involves replacing the initial concentration of B by $C_0 - [A]_0$ in Eq. (6.23), which is thus transformed into:

$$\frac{1}{tC_0} = \frac{[A]_0}{C_0}\left(\frac{k_B - k_A}{\ln([R]_t/[R]_0)}\right) - \frac{k_B}{\ln([R]_t/[R]_0)} \tag{6.24}$$

For a given $[R]_t/[R]_0$ ratio, the plot of $1/C_0t$ as a function of the molar fraction of A, $[A]_0/C_0$, is linear and can be used as a calibration curve provided that the time (t_e) at which the error in the determination is minimal is found beforehand. This is derived from the experimental time-dependence of $[R]_t$ and corresponds to the maximum change in R per unit time. Such a time is delimited by the [R] values satisfying an equation derived from Eq. (6.23), namely:

$$[R]_t/[R]_0 = 1/e$$

where e is the base of natural logarithms, or by

$$[R]_2/[R]_1 = 1/e$$

where $[R]_1$ is the concentration of R corresponding to a time t_1 very close to the start of the reaction (Fig. 6.6), when mixing of the reactants takes 2–3 sec.

The derivation of the equality above entails a number of steps which start by formulating Eq. (6.23) in exponential form, i.e.

$$[R]_t = [R]_0 \exp\{-(k_A[A]_0 + k_B[B]_0)t\} \tag{6.25}$$

Differentiation (with respect to time) of the first derivative of this expression with respect to $[R]_t$ at a fixed time, yields

$$\frac{d(d[A]_0/d[R]_t)}{dt} = \frac{1 - (k_A[A]_0 + k_B[B]_0)t}{t^2 \exp[-(k_A[A]_0 + k_B[B]_0)]} \tag{6.26}$$

The minimum error condition for the determination of A is given by t_{opt}, obtained

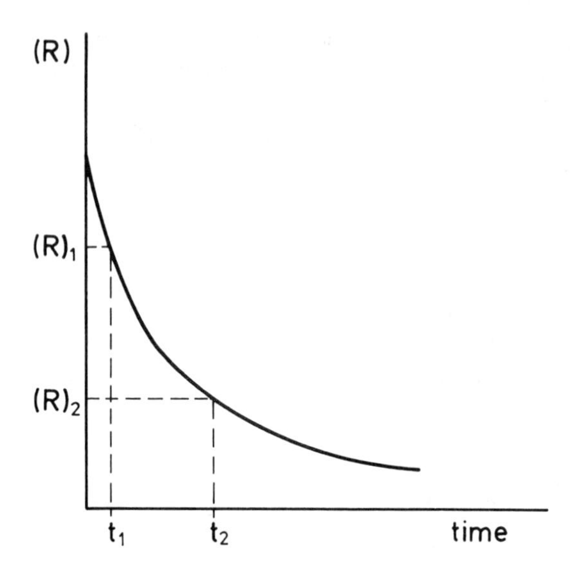

Fig. 6.6 — Determination of the optimum time interval for application of the modified Roberts and Regan method.

when Eq. (6.26) is equated to zero. Under such conditions

$$(k_A[A]_0 + k_B[B]_0)t = 1$$

Substitution of this condition into Eq. (6.25) leads to:

$$[R]_t = [R]_0 \exp(-1) = [R]_0/e$$

and hence

$$[R]_t/[R]_0 = 1/e$$

Another way to calculate $[A]_0$ and $[B]_0$, based on Eq. (6.22), relies on the calculation of K (pseudo rate constant) from the measurement of the variation in [R] with time for different concentrations of R in a mixture of A and B. Rate constants k_A and k_B can be obtained similarly for each of the pure components. Thus, taking into account that $C_0 = [A]_0 + [B]_0$, the concentration of both components can be easily calculated from the expression

$$K = k_A[A]_0 + k_B[B]_0$$

6.2.2.2 Proportional-equation method

This is a modified version of the method described above, proposed by Greinke and Mark [14], which does not require C_0 to be known beforehand and involves the calculation of two values of K (pseudo rate constant) under two given sets of experimental conditions:

$$K_1 = k_{A,1}[A]_0 + k_{B,1}[B]_0$$
$$K_2 = k_{A,2}[A]_0 + k_{B,2}[B]_0$$

The method depends on changing the ratio of the two rate constants (k_A and k_B) by varying the reaction medium or the conditions. This is most easily done by changing the reaction medium. The use of two different solvents yields reliable results [14].

6.2.3 Second-order reactions

In dealing with this type of reaction we shall consider two different situations according to whether the initial concentration of the reagent, $[R]_0$, is equal (linear extrapolation method) or not (logarithmic extrapolation and single-point methods) to the sum of those of both components, $[A]_0$ and $[B]_0$.

6.2.3.1 Logarithmic extrapolation method, $[R]_0 \approx [A]_0 + [B]_0$

The application of the logarithmic extrapolation method to first-order reactions can be extended to second-order processes provided that $[R]_0$ is different from $[A]_0 + [B]_0$ [15].

Consider a second-order irreversible reaction of the type

$$M + R \xrightarrow{k_M} P$$

with rate given by

$$d[P]/dt = k_M([R]_0 - p_M)([M]_0 - p_M) \quad (6.27)$$

where p_M is the concentration of product formed corresponding to the concentration of M or R which has reacted over time t, and k_M is the second-order rate constant.

Integration of Eq. (6.27) between $t_0 = 0$ ($[P]_0 = 0$) and $t_1 = t$ ($[P]_t = p_M$) yields:

$$\ln\frac{[R]_0 - p_M}{[M]_0 - p_M} = k_M([R]_0 - [M]_0)t + K \quad (6.28)$$

If M is assumed to be a mixture of two substances, A and B, which react irreversibly with reagent R at a rate given by constants k_A and k_B, then

$$[M]_0 = [A]_0 + [B]_0 \text{ and } p_M = p_A + p_B$$

If, in addition, A reacts more rapidly with R than does B, by the time that A has reacted completely, $p_A = [A]_0$ and

$$[R]_0 - p_M = [R]_0 - p_A - p_B = [R]_0 - [A]_0 - p_B \tag{6.29}$$

$$[M]_0 - p_M = [A]_0 + [B]_0 - p_A - p_B = [B]_0 - p_B \tag{6.30}$$

Substituting these expressions into Eq. (6.28) gives

$$\ln \frac{[R]_0 - [A]_0 - p_B}{[B]_0 - p_B} = k_B([R]_0 - [A]_0 - [B]_0)t + K$$

i.e. the equation of a straight line representative of the contribution of the slower component. Plotting the left-hand side as a function of time yields the concentration of the faster reacting component, A, which, subtracted from $[M]_0$, gives that of the slower reacting component, B (Fig. 6.7).

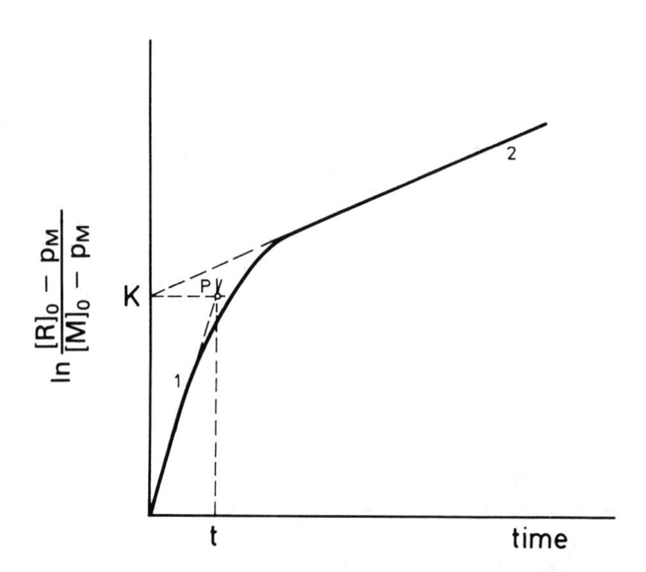

Fig. 6.7 — Logarithmic extrapolation method for second-order reactions. (Adapted by permission, from S. Siggia and J. G. Hanna, *Anal. Chem.*, 1961, **33**, 896. Copyright 1961, American Chemical Society.)

As shown in Fig. 6.7, the initial concentration of A, $[A]_0$, can be calculated in two ways.

(a) By extrapolating line 2 to $t = 0$. In this case,

$$K = \ln \frac{[R]_0 - [A]_0}{[M]_0 - [A]_0}$$

since, at such a time, p_M consists of the contribution of the faster reacting component alone, $[A]_0$.

(b) Drawing a line parallel to the x-axis, from K, and extrapolating line 1 to its intercept with this parallel, gives a point P corresponding to a time t_1 which yields a p value on the kinetic curve $[P] = f(t)$ that is exclusively due to component A, from which $[A]_0$ can be calculated.

A mathematical method based on Eq. (6.28) allows the determination of the faster reacting component, A. It involves obtaining the p_1 and p_2 values corresponding to two times, t_1 and t_2, at which A has reacted virtually completely. These, in turn, allow the formulation of two equations yielding the following solution:

$$[A]_0 = \frac{[R]_0\left\{\left(\dfrac{[M]_0 - p_{M1}}{[R]_0 - p_{M1}}\right)^{t_2}\left(\dfrac{[M]_0 - p_{M2}}{[R]_0 - p_{M2}}\right)t_1\right\}^{1/(t_1 - t_2)} - [M]_0}{\left\{\left(\dfrac{[M]_0 - p_{M1}}{[R]_0 - p_{M1}}\right)^{t_2}\left(\dfrac{[M]_0 - p_{M2}}{[R]_0 - p_{M2}}\right)t_1\right\}^{1/(t_1 - t_2)} - 1}$$

Although this equation contains no rate constants, it is dependent on the temperature, which should be kept constant in order to avoid fluctuations in the experimental values obtained at t_1 and t_2.

Though simpler, (it requires only two measurements) this method is less precise than the graphical method. As a rule, it is applied when the reaction rates of the mixture components are sufficiently different to result in two straight lines (Fig. 6.7).

The method is inapplicable when $[R]_0 = [A]_0 + [B]_0$ since Eq. (6.28) is meaningless in that case.

6.2.3.2 *Linear extrapolation method,* $[R]_0 = [A]_0 + [B]_0$

The starting point of the mathematics of this method is the equation

$$-\mathrm{d}[R]/\mathrm{d}t = k_A[R][A] + k_B[R][B] \tag{6.31}$$

corresponding to an irreversible reaction [16].

Taking into account that $-\mathrm{d}[R]/\mathrm{d}t = \mathrm{d}[P]/\mathrm{d}t$ and $p = p_A + p_B$ (where p_A and p_B are the amounts of product formed from A and B, respectively, or, in other words, the amounts of A and B consumed, and p is therefore the overall amount of product formed after time t), Eq. (6.31) can be rewritten as:

$$\mathrm{d}[P]/\mathrm{d}t = ([A]_0 + [B]_0 - p_A - p_B)\{k_A([A]_0 - p) + k_B([B]_0 - p)\} \tag{6.32}$$

If A reacts more rapidly with R than does B, at the end of the reaction $[A]_0 = p_A$ and Eq. (6.32) becomes equivalent to:

$$\mathrm{d}[P]/\mathrm{d}t = k_B([B]_0 - p_B)^2 \tag{6.33}$$

the integration of which yields:

$$1/([B]_0 - p_B) = k_B t + K \tag{6.34}$$

The integration constant, K, is easily calculated by making $t = 0$ in Eq. (6.34), whence $K = 1/[B]_0$. Substitution of this constant into Eq. (6.34) leads to:

$$\frac{1}{[B]_0 - p_B} - \frac{1}{[B]_0} = k_B t \tag{6.35}$$

and rearrangement results in

$$k_B t = p_B/[B]_0([B]_0 - p_B) \tag{6.36}$$

Since $[R]_0 = [A]_0 + [B]_0$ and $[A]_0 = p_A$, then $[B]_0 = [R]_0 - p_A$ and $p_B = p - [A]_0$, so Eq. (6.36) can be rewritten as

$$p = k_B[B]_0([R]_0 - p)t + [A]_0 \tag{6.37}$$

The plot of p *vs.* $([R]_0 - p)t$ becomes a straight line once A has reacted to completion. The (extrapolated) intercept of such a line yields $[A]_0$, and its slope is equal to the product $k_B[B]_0$ (Fig. 6.8).

The accuracy achieved in the calculation of the initial concentration of A, $[A]_0$, depends on the extent to which the equality $[R]_0 = [A]_0 + [B]_0$ holds, insofar as the concentration of A will diminish proportionally to any excess of reagent present. As can be seen from Eq. (6.37), greater $[R]_0$ values result in straight lines of the same slope but smaller intercepts.

To obtain the concentration of the slower reacting component without prior knowledge of its rate constant, k_B, use can be made of the expression

$$1/([R]_0 - p) = k_B t + (1/[B]_0) \tag{6.38}$$

derived from Eq. (6.35) by making $p - [A]_0 = p_B$ and $[A]_0 + [B]_0 = [R]_0$.

The plot of the left-hand side of Eq. (6.38) *vs.* time is a straight line (for $[A]_t$ values approaching zero), the intercept of which yields $[B]_0$ directly. In addition, its slope gives the rate constant of this component (Fig. 6.9).

A possible simplification of this method lies in obtaining two p values corresponding to two reaction times at which only the slower reacting component contributes significantly to p. Under such conditions, $[B]_0 = [R]_0 - [A]_0$ and Eq. (6.37) becomes:

$$[A]_0 = \frac{p - k_B([R]_0 - p)[R]_0 t}{1 - k_B([R]_0 - p)t}$$

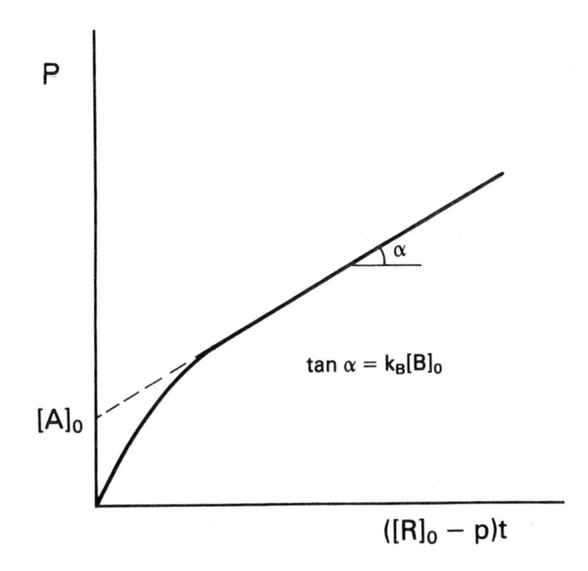

Fig. 6.8 — Linear extrapolation method for second-order reactions. Determination of the concentration of the faster reacting component. (Reproduced with permission, from C. N. Reilley and L. J. Papa, *Anal. Chem.*, 1962, **34**, 801. Copyright 1962, American Chemical Society.)

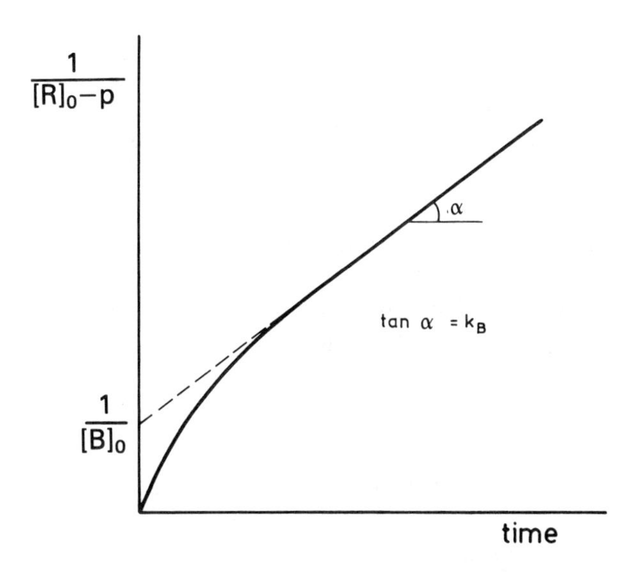

Fig. 6.9 — Linear extrapolation method for second-order reactions. Determination of the slower component. (Reproduced with permission, from C. N. Reilley and L. J. Papa, *Anal. Chem.*, 1962, **34**, 801. Copyright 1962, American Chemical Society.)

which requires only a single reaction time for the calculation of the initial concentration of A. However, it calls for prior knowledge of k_B, which need not be calculated if two reaction times are used, since

$$k_B = \frac{p_1 - [A]_0}{[B]_0([R]_0 - p_1)t_1} = \frac{p_2 - [A]_0}{[B]_0([R]_0 - p_2)t_2} \tag{6.39}$$

which yields the following solution for $[A]_0$:

$$[A]_0 = \frac{p_1 - [([R]_0 - p_1)t_1/([R]_0 - p_2)t_2]p_2}{1 - \{([R]_0 - p_1)t_1/([R]_0 - p_2)t_2\}}$$

Analogous reasoning can be applied to obtain the initial concentration of the slower reacting component from Eq. (6.38) on the basis of one reaction time

$$[B]_0 = \frac{1 - k_B t([R]_0 - p)}{[R]_0 - p}$$

or two reaction times:

$$[B]_0 = \frac{([R]_0 - p_1)([R]_0 - p_2)(t_2 - t_1)}{([R]_0 - p_2)t_2 - ([R]_0 - p_1)t_1}$$

6.2.3.3 Single-point method, $[R]_0 = [A]_0 + [B]_0$

This method is analogous to that described above for pseudo first-order reactions [9] and also involves determining the concentrations of A and B, both at the start of the reaction and at time t_{opt}.

The pertinent calibration curve is run by plotting the fraction of A or B unreacted at t_{opt}, as a function of the molar fraction of the same component in the original mixture. Unlike the pseudo first-order version of the method, the procedure followed to calculate t_{opt} is rather complicated [1] and yields:

$$t_{opt} = (k_B/k_A)^{1/2}/k_B[R]_0$$

from which it is obvious that k_B must be known beforehand.

6.2.4 Pseudo zero-order reactions

If the experimental measurements involved in a differential reaction-rate method are performed at the beginning of the reaction, it may reasonably be assumed that the concentrations of the reactants (A and B) are virtually constant and the kinetics of the process is pseudo zero-order. Such processes have the advantage that the mathematical treatment involved is always the same, irrespective of the reactant concentration in the mixture.

Only those methods involving no mathematical integration, namely the single-point and proportional equation methods, will be dealth with here.

When two substances A and B react with the same reagent, R, the overall initial rate for the mixture is given by:

$$v_{0,\mathrm{A+B}} = k_{\mathrm{A}}[\mathrm{A}]_0 + k_{\mathrm{B}}[\mathrm{B}]_0 \tag{6.40}$$

so the concentration of species A and B can be readily determined from knowledge of the sum of their initial concentrations, $[\mathrm{A}]_0 + [\mathrm{B}]_0$, and of v_0. This is the basis for the development of the corresponding version of the single-point method [17], similar to that described by Roberts and Regan [13] for pseudo first-order reactions. Its chief application is probably to the resolution of sugar mixtures.

When the sum of the initial concentrations of A and B is unknown, it is still possible to resolve the mixture by simply applying Eq. (6.40) under two different sets of conditions, *viz*.

$$v_{0,(\mathrm{A+B})1} = k_{\mathrm{A},1}[\mathrm{A}]_0 + k_{\mathrm{B},1}[\mathrm{B}]_0$$
$$v_{0,(\mathrm{A+B})2} = k_{\mathrm{A},2}[\mathrm{A}]_0 + k_{\mathrm{B},2}[\mathrm{B}]_0$$

Solving this equation system gives $[\mathrm{A}]_0$ and $[\mathrm{B}]_0$.

This method, similar to the proportional equation method for pseudo first-order reactions, has been applied to the resolution of mixtures of alcohols [18].

6.3 SYNERGISM IN KINETIC ANALYSIS

The kinetic methods described so far were developed by assuming the reaction rates of the species concerned to be independent of each other or, in other words, that the presence of one component has no effect on the kinetic behaviour of another. When this is not the case, the mixture is said to be subject to a synergistic effect.

Although synergism is not a frequent phenomenon, it is the chief source of error in differential kinetic analysis, so much so that only the second-order logarithmic extrapolation [15] and Connors graphical extrapolation [8] methods are applicable to the resolution of mixtures affected by synergistic effects; in fact, there are few antecedents dealing with the resolution of these troublesome mixtures, which are only tackled once the synergistic effect has been eliminated.

The presence of synergism is revealed by determining the individual rate constant for one component, first in the absence of the other component and then in the presence of increasing amounts of it. Any increase or decrease in the calculated constant will be indicative of a positive or negative synergistic effect, respectively. Inasmuch as this phenomenon alters the rate of a reaction, it could also be referred to as 'pseudo-activation' or 'pseudo-inhibition'.

6.3.1 Causes of synergism

Synergistic effects may arise from a variety of causes such as (a) interaction between the solvent and the reactants; (b) possible changes in the activity coefficients (effect of the total concentration) and (c) changes in the rate constants resulting from catalytic effects.

The interaction between the solvent and the reactants has been found to result in a synergistic effect on the determination of different types of amine by reaction with methyl and ethyl iodide in the presence of acetone as solvent [19]. The effect is attributed to the formation of Schiff's bases between the amine and the acetone. The extent to which these reactions occur varies according to the mixture composition insofar as the rate of formation of the Schiff's base of one of the components is modified by the presence of the other (competing) component. The addition of 30% water to the reaction medium avoids the interaction.

As stated above, a change in the activity coefficients can be another cause of synergism. Thus, the presence of the faster reacting component has been found to result in a decrease in the reaction rate of the other component. If the concentration of the more slowly reacting species and the temperature are both kept constant, the variation of its reaction rate can only be due to a change in its activity coefficient brought about by the presence of the faster reacting component [20]. This behaviour cannot be generalized, as the rate constants of both components can be affected through their mutual interaction.

The presence of catalytically active impurities in the reaction medium is another possible cause of synergism. Their influence is similar to that of the solvent, their catalytic effect being exerted simultaneously on both rate constants. Thus, the presence of small amounts of water has a catalytic effect on the above-mentioned analysis of the amine mixture [19]. If the catalyst is known, its effect can be avoided by adding it in large excess ('swamping' technique), thereby overcoming its trace effect. This is the procedure to be followed if the Roberts and Regan method [21] is applied, as it requires the rate constants to be known beforehand. However, this is unnecessary if the second-order logarithmic extrapolation method is applied, as this calls for no prior knowledge of the rate constants.

6.3.2 Treatment of synergism

As stated above, mixtures affected by synergistic effects can only be resolved by applying the second order logarithmic extrapolation and Connors interpolation methods. The former is advantageous when the rate constant of the slower reacting component is further decreased by the presence of the other component. As can be seen from Fig. 6.10, the slopes of the plots are more divergent in the presence of synergistic effects, which facilitates obtainment of the intercept (which is unchanged), from which the concentration of the slower reacting component is calculated [20].

The foundation of the Connors graphical interpolation method (fitting of the unknown curve to that of the reference) allows it to be applied to any type of mixture, whatever the effect to which it may be subject.

A novel approach to the simultaneous resolution of mixtures subject to synergistic effects has been reported [22]. It involves assuming the mixture A + B to react with a reagent R according to pseudo first-order kinetics, i.e. $[R]_0 \gg [A]_0 + [B]_0$. The rate of formation of the product will be given by

$$v = d[P]/dt = v_A + v_B = k_A[A]_0 + k_B[B]_0 \quad (6.41)$$

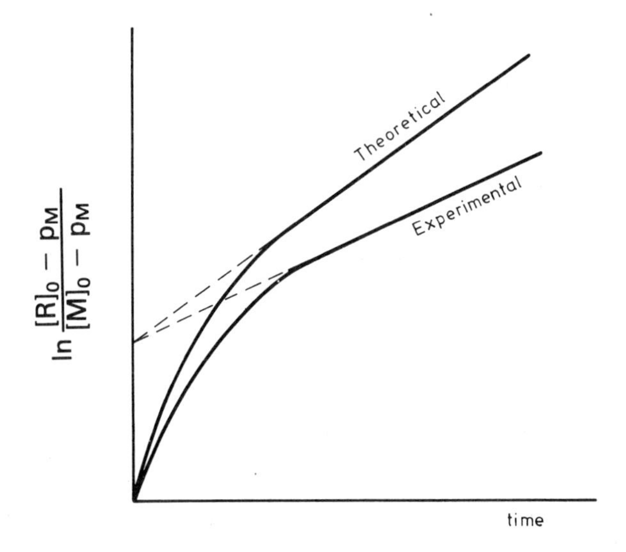

Fig. 6.10 — Influence of synergistic effects on the logarithmic extrapolation method. (Reproduced with permission, from S. Siggia and J. G. Hanna, *Anal. Chem.*, 1964, **36**, 228. Copyright 1977, American Chemical Society.)

If a synergistic effect is present, i.e. if each species influences the kinetic behaviour of the other, then Eq. (6.41) does not hold, as the reaction rates will not be additive as postulated by the expression. However, this equation can be used to take into account the synergistic effect. Thus, the slopes of the calibration curves of the type $v = f([A]_0)$, for component A in the presence of various fixed amounts of B, will differ from those obtained in the absence of B and will be a function of $[B]_0$:

$$\text{slope} = S = C_1[B]_0 + C_2$$

The relationship between the initial concentration and the reaction rate of A will be given by;

$$v_A = S[A]_0 = (C_1[B]_0 + C_2)[A]_0 = C_2[A]_0 + C_1[A]_0[B]_0 \quad (6.42)$$

Similarly, for species B:

$$v_B = S'[B]_0 = (C_3[A]_0 + C_4[B]_0 = C_4[B]_0 + C_3[A]_0[B]_0 \quad (6.43)$$

On comparing Eqs. (6.42) and (6.43) with Eq. (6.41), it is seen that the synergistic effect is represented by the term containing $[A]_0[B]_0$, which makes the reaction rates dependent on the concentration of both components, thus logically taking into account the mutual influence of both species.

The overall rate for the mixture, υ, will therefore be given by:

$$\upsilon = \upsilon_A + \upsilon_B = C_2[A]_0 + C_4[B]_0 + C_5[A]_0[B]_0 \qquad (6.44)$$

where $C_5 = C_1 + C_3$. From Eq. (6.44), $[A]_0$ and $[B]_0$ are easily obtained provided that their sum is known beforehand.

Should the sum of the initial concentrations of both components be unknown, these can still be obtained by formulating Eq. (6.44) under two different sets of experimental conditions in much the same way as with the proportional equation method.

$$f_1 = C_{A,1}[A]_0 + C_{B,1}[B]_0 + C_{AB,1}[A]_0[B]_0$$
$$f_2 = C_{A,2}[A]_0 + C_{B,2}[B]_0 + C_{AB,2}[A]_0[B]_0$$

It is interesting to note that these equations can indeed be formulated on the basis of any parameter proportional to the concentration of the species involved, such as the absorbance, fluorescence intensity, induction period, etc.

This procedure has been satisfactorily applied to the fluorimetric resolution of mixtures of amines on the basis of a C=N− group exchange-reaction between the amines and 2-hydroxybenzaldehyde azine [22].

A recent application of this procedure, though focusing on sequential mixture resolution, is based on the assumption that, over a given concentration range, one of the species can be determined free from interference by the other. Thus, once $[A]_0$ is determined, $[B]_0$ can be obtained from Eq. (6.43). This method, much more restricted in its application, has been used for the resolution of mixtures of two activators such as In(III) and Ga(III), both of which enhance the catalytic effect of copper on a selected reaction [23].

6.4 APPLICATIONS OF DIFFERENTIAL REACTION-RATE ANALYSIS

The graphical or mathematical data-treatment involved in differential reaction-rate methods makes them rather more laborious than other analytical procedures, especially when mixtures of more than two components are dealt with. Hence, computers are an invaluable aid in this area, as will be acknowledged in the chapter devoted to instrumentation.

The relatively small number of applications of these methods reported so far may be a result of the difficulty in finding closely related species (their most promising field of application) reacting at sufficiently different rates with a common reagent. In this regard, metal- and ligand-exchange reactions are the most frequently used with inorganic species, while redox processes are by far the commonest involved in the determination of organic substances.

As a rule, the stopped-flow technique is indispensable in roughly half the applications, and this is probably another reason for the comparative scarcity of applications of differential reaction-rate methods.

The maximum concentration ratios of the two analytes allowed by these methods are usually about 5:1, though some afford ratios of 10:1 or even higher, depending on the rate-constant ratio and the particular method used.

The sensitivity and selectivity of differential reaction-rate methods are similar to those of non-kinetic simultaneous procedures available for the resolution of mixtures of dissolved species. Interferences are numerous, except for catalysed reactions. The most serious limitation of these methods, however, is their poor precision (errors greater than 5% are commonplace).

A better knowledge of the kinetics and mechanism of the reactions involved, the introduction of new systems, the use of computers and the reduction of costs thanks to the incorporation of mixing chambers in conventional detectors (spectrophotometers and spectrofluorimeters) will foreseeably give differential reaction-rate methods a more prominent place in the framework of kinetic analysis.

6.4.1 Resolution of mixtures of inorganic species

Applications to the resolution of mixtures of inorganic species can be classified according to the type of reaction on which the method used is based.

6.4.1.1 Complexation reactions

This group comprises both formation and dissociation reactions, and ligand- and metal-substitution reactions.

The ligands most frequently used are aminopolycarboxylic acids such as EDTA (and especially DCTA), employed either as chelating agents or scavengers. Classical chromogenic reagents such as PAR, PAN or SPADNS are often part of the starting complexes, from which they are displaced by one of the complexones, and are only occasionally themselves used as displacing ligands.

Substitution reactions, by far the commonest in differential kinetic analysis, are dealt with first. Methods based on substitution reactions can in turn be divided according to whether a ligand or a metal is replaced from the starting complex:

$$ML + L' \rightarrow ML' + L \qquad (I)$$

$$ML + M' \rightarrow M'L + M \qquad (II)$$

DCTA is one of the ligands most frequently used on account of the advantages offered by the singular kinetic behaviour of M–DCTA complexes [24]. The cyclohexane ring of DCTA restricts the flexibility of the ligand and this favours retention of metal ions in a cage formed by the nitrogen atoms and carboxylic groups.

The beneficial effects of DCTA can be summed up as follows.

(a) If an excess of metal M′ is present [reaction (II)], each M′–DCTA complex reacts independently of the other complexes or metal ions present in the bulk solution.

(b) Metal-substitution reactions involving DCTA are rather slow compared with those of EDTA, as a result of a more restrictive reaction mechanism.

(c) Suitable pH changes can result in suitable pseudo first-order rate constants and hence in useful reaction half-lives.

Ligand-substitution reactions

These generally involve multidentate ligands such as EGTA and DCTA, which are displaced by chromogenic azo ligands, especially suited to photometric monitoring of the reaction rate. However, the reverse reaction (i.e. the displacement by EDTA and related ligands) has been far more frequently used, probably because EDTA complexes are normally more stable than those formed by chromogenic ligands. The practical procedure commonly used in the reverse reaction involves recording the decrease in absorbance of the solution as a function of time.

The earliest application of this nature was reported by Tanaka *et al.* [25], who used 4-(2-pyridylazo)resorcinol (PAR) to displace EGTA from its Ni and Co complexes and to determine both metals by differential kinetic analysis (EGTA is displaced more rapidly from the cobalt complex than from that of nickel). The substitution reaction is monitored spectrophotometrically at 500 nm (absorption wavelength of the PAR complexes of Ni and Co), and the absorbance readings are treated by the logarithmic extrapolation method. This is one of the few differential kinetic procedures not requiring the application of the stopped-flow technique, and has been applied to the resolution of binary mixtures of Fe, Co, Ni, Cu, Zn, Cd and Pb, and of quaternary mixtures of Co, Ni, Zn and Pb, at the micromolar concentration level.

A comprehensive study of the rate of ligand substitution between the complexes of Co, Ni, Pb and Cu with 2-(4-sulphophenylazo)-1,8-dihydroxy-3,6-naphthalenedisulphonic acid (SPADNS) and EDTA (or DCTA) at pH 8.0–9.3 and 25°C was made by Mentasti [26]. He determined these ions in mixtures by applying the stopped-flow technique (with measurement at 600 nm) to the rate of disappearance of the different complexes according to the mechanistic scheme

$$\text{M–SPADNS} + \text{EDTA} \underset{k_{-1}}{\overset{k_1}{\rightleftharpoons}} \text{SPADNS–M–EDTA}$$

$$\text{SPADNS–M–EDTA} \overset{k_2}{\rightarrow} \text{M–EDTA} + \text{SPADNS}$$

The process involves formation of an intermediate mixed SPADNS–M–EDTA complex, the decomposition of which is the rate-determining step. The application of the logarithmic extrapolation method allows mixtures of Ni and Co in ratios from 1:10 to 10:1 to be resolved with a precision of about 5%. Mixtures of Cu and Co, Ni or Pb; Co and Pb; or of several components such as Zn (Mn, Cd), Co, Pb (Ni) and Cu can also be satisfactorily resolved.

Finally, it is worth mentioning the replacement of an inorganic by an organic ligand, as in the reaction between the ternary complexes formed by molybdophosphoric acid with species such as Ti(IV), V(V), Zr(IV) and Nb(V), and a displacing ligand such as citrate, oxalate or tartrate [27]. The technique uses the logarithmic extrapolation method and requires no stopped-flow instrumentation to resolve binary mixtures of Zr–Nb, V–Zr, Nb–V and Ti–Nb by use of oxalate or tartrate, with acceptable relative errors of ±5%. The highest concentration ratio that can be dealt with is 5:1.

Table 6.2 lists the applications of differential reaction-rate methods involving ligand-exchange reactions.

Metal-exchange reactions

The earliest use of this type of reaction in differential kinetic analysis was due to Pausch and Margerum [24], who proposed the kinetic determination of mixtures of Mg, Ca, Sr and Ba on the basis of their different rates of displacement from DCTA complexes by Pb(II). The displacement reactions are followed by monitoring the ultraviolet absorption of the Pb–DCTA complex at 260 nm. The general mechanism can be formulated as follows:

$$\text{M–DCTA}^{2-} + \text{H}^{+} \underset{k_{\text{M}}^{\text{Y}}}{\overset{k_{\text{H}}^{\text{MY}}}{\rightleftharpoons}} \text{HDCTA}^{3-} + \text{M}^{2+} \tag{III}$$

$$\text{HDCTA}^{3-} + \text{Pb}^{2+} \overset{k_{\text{Pb}}^{\text{Y}}}{\rightarrow} \text{Pb–DCTA}^{2-} + \text{H}^{+} \tag{IV}$$

The step determining the overall rate of reaction of each complex is reaction (III) which is first-order with respect to hydrogen ions. Assuming $k_{\text{Pb}}^{\text{Y}}[\text{Pb}] \gg k_{\text{M}}^{\text{Y}}[\text{M}]$, the exchange rate will be given by $v = k_{\text{H}}^{\text{MY}}[\text{H}^{+}][\text{M–DCTA}^{2-}]$. The rate constants are in the proportion 1:6.5:96:1660 for Mg:Ca:Sr:Ba. The reaction is performed in acetate medium (pH 8) to avoid precipitation of lead without significantly decreasing its rate of reaction with DCTA [reaction (IV)]. In this manner binary and ternary mixtures of the above-mentioned alkaline-earth metals have been resolved in the range 1×10^{-6}–$5 \times 10^{-4}M$, with a precision of between 5 and 10%. The pH chosen, set according to the particular mixture, ranges between 5.5 and 7.6. The concentration ratios suitable for analysis are 0.42:1–1.8:1, 0.4:1–9.3:1 and 0.1:1–1:1 for the Ca/Mg, Sr/Ca and Ba/Sr mixtures. This procedure involves the use of the logarithmic extrapolation method and the stopped-flow technique and also allows the resolution of ternary Mg/Ca/Sr and Ca/Sr/Ba mixtures. Table 6.3 lists the methods published.

Jensen *et al.* [44, 46–48] have used FIA to apply differential kinetic analysis to the resolution of various mixtures of alkaline earths.

The systems proposed by Pausch and Margerum have inspired development of methods for the resolution of binary mixtures of alkaline-earth metals by means of electroanalytical rather than photometric measurements. Thus, Ca–Mg and Ca–Sr mixtures can be resolved with the aid of the M–DCTA/Cd system by recording the changes in the concentration of free, unreacted Cd(II) added as a scavenger, measured by square-wave polarography [45], and arising from the reaction

$$\text{Ca–DCTA} + \text{Mg–DCTA} + 2\text{Cd}^{2+} \rightarrow \text{Ca}^{2+} + \text{Mg}^{2+} + 2\text{Cd–DCTA}$$

The reactivity increases in the order Mg < Ca < Sr, and the rate-constant ratios $k_{\text{Ca}}/k_{\text{Mg}}$ and $k_{\text{Sr}}/k_{\text{Ca}}$ are 6.8 and 12.5, respectively. The procedure uses the linear extrapolation method and therefore requires prior knowledge of the overall initial concentration of both alkaline-earth metals, which is calculated by measuring the square-wave polarography peaks corresponding to Cd(II) released in the reaction

Table 6.2 — Differential reaction-rate determinations based on ligand-substitution reactions

Type of reaction	Chemical system	Mixture: Binary	Mixture: Ternary and multi-component	Technique	Method	Reference
Substitution of organic ligands	M–EGTA+PAR→M–PAR+EGTA	Ni–Co Zn–Pb Fe–Ni Zn–Cd Cu–Pb	Co–Ni–Zn–Pb	$A(\lambda=495\text{ nm})$	a	[25]
	M–DCTA+Arsenazo III→ M–Arsenazo III+DCTA	Th–U Th–Np				[28]
		Th–Pu U–Pu				[29]
	M–TAC+DCTA→M–DCTA+TAC	Zn–Cd		$A(\lambda=600\text{ nm})$ (back-extraction)	a′(2)	[5]
	M–TAC+EDTA→M–EDTA+TAC		Cu–Ni–Co	$A^*(\lambda=610\text{ nm})$ (with persulphate)	a	[30]
	M–TAN+EDTA→M–EDTA+TAN	Zn–Mn		$A^*(\lambda=590\text{ nm})$	a′(2)	[31]
	M–PAN+EDTA→M–EDTA+PAN	Cd–Mn		$A^*(\lambda=554\text{ nm})$	a′(2)	[32]
	M–SPADNS+EDTA (DCTA)→ M–EDTA (M–DCTA)+SPADNS	Co–Ni Cu–Co Cu–Ni Cu–Pb Co–Pb	Zn–Co–Pb–Cu Zn–Co–Ni–Cu Mn–Co–Pb–Cu Cd–Co–Pb–Cu	$A^*(\lambda=600\text{ nm})$	a	[26]
	M–Zincon+DCTA→ M–DCTA+Zincon	Various binary and quaternary Zn–Cd–Hg–Cu mixtures		$A^*(\lambda=620\text{ nm})$	c	[33]
	M–XO+EDTA→M–EDTA+XO	Pr–Yb			a′	[34]
		La–Sm		$A^*(\lambda=570\text{ nm})$	a′	[35]
			Dy–Ho–Yb		b	[36]
Substitution of an inorganic by an organic ligand	M–molybdophosphoric acid+ oxalate/citrate/tartrate→ M–oxalate/citrate/tartrate+ molybdophosphoric acid	Zr–Nb Zr–V Ti–V Ti–Nb		$A(\lambda=400\text{ nm})$	a	[27]

EDTA: ethylenediaminetetra-acetic acid; EGTA: ethyleneglycol bis(2-aminoethyl ether)-*N*,*N*′-tetra-acetic acid; DCTA: 1,2-diaminocyclohexane-*N*,*N*,*N*′,*N*′-tetra-acetic acid; Arsenazo III: sodium 2,7-bis[*o*-arsonophenylazo]-1,8-dihydroxynaphthalene-3,6-disulphonate; PAR: 4-(2-pyridylazo)resorcinol; TAC: 2-(2-thiazolylazo)-4-methylphenol; TAN: 1-(2-thiazolylazo)-2-naphthol; PAN: 1-(2-pyridylazo)-2-naphthol; SPADNS: sodium 2-(*p*-sulphophenylazo)-1,8-dihydroxynaphthalene-3,6-disulphonate; Zincon: 1-(2-hydroxy-5-sulphophenyl)-3-phenyl-5-(2-carboxyphenyl)formazan, sodium salt; XO: Xylenol Orange, *A*: spectrophotometry, *A**: stopped-flow, a: logarithmic extrapolation, a′: graphical extrapolation, a′(2): simplified method, b: proportional equations, c: other methods, including computerized methods.
(Reproduced from [37] with permission of the copyright holders, the Royal Society of Chemistry.)

Table 6.3 — Differential reaction-rate determinations based on metal-substitution reactions

Chemical system	Mixture: Binary	Mixture: Ternary and multi-component	Technique†	Method†	Reference
M–DCTA+Pb→Pb–DCTA+M	Mg–Ca Sr–Ca Sr–Ba	Mg–Ca–Sr	$A^*(\lambda=260$ nm$)$	a	[24]
	Sr–Ca	Sr–Ca–Mg	$A^*(\lambda=260$ nm$)$	c	[38]
	Ca–Mg		B	a, a′(1)	[39]
	Ca–Mg		C	c	[40]
	Ca–Mg		C, D	c	[41]
M–DCTA+Cu→Cu–DCTA+M	Various mixtures of Zn, Mg, Cd and Pb, and of lanthanides		$A^*(\lambda=300$ nm$)$	a′(1)	[42]
	La–Nd		$A(\lambda=300$ nm$)$	a′(1)	[3]
	Mg–Zn		$A(\lambda=310$ nm$)$ (cont. pH change)	c	[43]
	Sr–Mg		$A(\lambda=300$ nm$)$ (FIA)		[44]
M–DCTA+Cd→Cd–DCTA+M	Ca–Mg Ca–Sr		B	a′(2)	[45]
M–cryptand(2.2.1)+Na→Na–cryptand(2.2.1)+M	Ca–Mg		$A(\lambda=586$ nm$)$ (FIA)		[46]
M–cryptand(2.2.2)+K→K–cryptand(2.2.2)+M		Mg–Ca–Sr	$A(\lambda=575$ nm$)$ (FIA)		[47]
	Ca–Mg		$A(\lambda=575$ nm$)$ (SFIA)		[48]

A: spectrometry; *A**: stopped-flow; *B*: polarography; *C*: amperometry; *D*: potentiometry; a′(1) Worthington and Pardue method; FIA: flow-injection analysis; SFIA: stopped-flow injection analysis.
†See footnote to Table 6.2.
(Reproduced from [37] with permission of the copyright holders, the Royal Society of Chemistry.)

$$2\text{Cd–EDTA} + Ca^{2+} + Mg^{2+} \xrightarrow{NH_3\ \text{excess}} 2Cd(NH_3)_4^{2+} + \text{Ca–EDTA} + \text{Mg–EDTA}$$

for the Ca/Mg mixture [45]. The method relies on the assumption that the concentration of cadmium found (which should be equal to the sum of those of the alkaline-earth metals, according to the reaction above) is equal to the amount of DCTA added to form the M–DCTA complexes quantitatively. This is also the amount of Cd to be added to the M–DCTA mixture to start the substitution reaction. The alkaline-earth metals can be resolved at the micromolar level in ratios from 1:0.15 to 1:6.5, but the average error of the method is rather large (±9.5%).

Other electroanalytical determinations of metals in mixtures are also listed in Table 6.3.

Complex-formation and dissociation reactions

These types of reaction are less commonly applied than substitution reactions in differential kinetic analysis. A representative example of use of a dissociation reaction is the decomposition of the NTA complexes of Mo(VI) and W(VI) by attack with hydroxide ions. This reaction takes place at different rates for the two species [49]. The experimental data are generated by applying the stopped-flow technique to absorbance measurements made in the ultraviolet region, where molybdate absorbs strongly, the absorption by tungstate is less intense, and that by NTA takes place at a shorter wavelength. The data are subsequently treated by two mathematical methods applied with the aid of an on-line computer. The method permits the resolution of Mo/W mixtures in ratios from 1:5 to 5:1 in the concentration ranges 8×10^{-4}–$1 \times 10^{-5}M$ and 1×10^{-5}–$2 \times 10^{-6}M$ for Mo and W, respectively, with precisions of 2 and 4%.

Other determinations based on decomposition reactions are summarized in Table 6.4.

There are few methods based on complex-formation reactions. A typical example is the simultaneous analysis for nitric oxide and nitrogen dioxide in air [52] by the formation of the Fe(II)–nitrosyl complexes from NO and NO_2,

$$Fe^{2+} + NO = Fe(NO)^{2+}$$

$$3Fe^{2+} + NO_2 + 2H^+ = Fe(NO)^{2+} + 2Fe^{3+} + H_2O$$

which take place at different rates. At the 0.050*M* Fe(II) level (roughly 50 times that of the analyte) used to ensure pseudo first-order conditions, the rate-constant ratio k_{NO}/k_{NO_2} is 270. The procedure uses the ordinary and stopped-flow spectrophotometric techniques; the latter involves using the logarithmic extrapolation method (the less reactive species, NO_2, is determined graphically first, and the NO concentration is calculated by difference). The method is not very sensitive, the minimum overall concentration that can be determined being $8.6 \times 10^{-4}M$. The samples

Table 6.4 — Differential kinetic determinations based on dissociation and direct complex-formation reactions

Type of reaction	Chemical system	Mixture		Technique†	Method†	Reference
		Binary	Ternary and multi-component			
Complex dissociation	M–NTA + $OH^- \rightarrow$ MO_4^{2-} (MO_3^-) + NTA	Mo–W		$A^*(\lambda = 260$ nm)	c	[49]
		Mo–V	Mo–W–V	$A^*(\lambda = 260$ nm)	c	[50]
	M–EDTA + $OH^- \rightarrow$ MO_4^{2-} (MO_3^-) + EDTA	Mo–V		$A^*(\lambda = 260$ nm)	c	[50]
	M–DTPA + $H^+ \rightarrow$ DTPA + M	Binary and ternary mixtures of La, Ce, Pr, Nd and Sm				[51]
Direct complex formation	X + $Fe^{2+} \rightarrow Fe(X)^{2+}$	NO–NO_2		$A, A^*(\lambda = 450$ nm)	a	[52]
	M + CPCD $\xrightarrow{O_2}$ M–CPCD	Ni–Co		$A^*(\lambda = 324$ nm)	a	[53]
	M + PT $\xrightarrow{Ac^-}$ M–PT	Fe–Co		$A^*(\lambda = 425$ nm)	a,a′(4)	[10]
	M + SQ $\rightarrow$ M–SQ	Al–Ga Mg–Cd		Fluor. decay	a	[54]

NTA: nitrilotriacetic acid; DTPA: diethylenetriamine-*N*,*N*,*N*′,*N*″,*N*″-penta-acetic acid; CPCD: 2-carboxy-1-pyrrolidine carbodithioic acid; PT: pyridoxal thiosemicarbazone; SQ: 5-sulpho-8-quinolinol; a′(4): single-point method. †See footnote to Table 6.2.
(Reproduced from [37] with permission of the copyright holders, the Royal Society of Chemistry.)

assayed usually contain NO_2 at concentrations 2.3–5.5 times that of NO. Obviously a large volume of air has to be sampled.

There are only two published methods for the resolution of mixtures of metals on the basis of different rates of formation of their chelates. One was proposed by Kitagawa and Fugikawa [53] for the determination of Ni and Co in mixtures, with 2-carboxy-1-pyrrolidinecarbodithioic acid (CPCD) as chelating ligand. The formation of the cobalt complex is slower than that of the nickel chelate as Co(II) is simultaneously oxidized to Co(III) by dissolved oxygen. The first-order Ni/Co rate-constant ratio for air-saturated solutions is about 70. The application of the logarithmic extrapolation method allows both metals to be determined at concentrations between 2×10^{-6} and $8 \times 10^{-6}M$.

The other method [10] exploits formation of the pyridoxal thiosemicarbazone (PT) complexes of Fe(III) and Co(II) in an acetate medium. The rate-constant ratio of these complexes, which absorb at 425 nm, is $k_{Co}/k_{Fe} = 10.5$ and no stopped-flow technique needs to be used. The proponents of the method have used and compared the logarithmic extrapolation and single-point methods to resolve Fe/Co mixtures in ratios from 1:4 to 4:1 with a precision of 1%. These and a few other determinations are listed in Table 6.4.

6.4.1.2 Redox reactions

Of the few differential reaction-rate methods relying on redox reactions, most are based on the reduction of the heteropoly acids formed by molybdate with silicate, phosphate and germanate. Thus, Ingle and Crouch [55] have devised a combined redox/complex-formation procedure for the sequential determination of phosphate and silicate, which react with Mo(VI) to yield yellow heteropoly acids that can in turn be reduced to 'molybdenum blue' with ascorbic acid. Under specific conditions, the reduction of phosphomolybdate is much faster than that of silicomolybdate. Thus, the absorbance–time curves recorded at 840 nm (λ_{max} for molybdenum blue) provide initial rate data related to the phosphate concentration and are barely affected by the presence of silicate. Phosphate can be determined in this manner in the range 2.5–10 μg/ml P in the presence of up to 50 μg/ml Si, with a relative error of less than 1%. Both yellow heteropoly acids are formed (at different rates) in the absence of ascorbic acid. Under these conditions, silicate can be determined (2.5–10 μg/ml Si) in the presence of up to 10 μg/ml P, with a precision of 1–3% by means of the corresponding kinetic curve obtained by measurement at 400 nm. The maximum permissible P/Si and Si/P ratios are 4:1 and 20:1, respectively. Higher phosphate concentrations result in negative errors in the determination of silicate.

Mixtures of Ge/Si and Ti/V have also been resolved on the same principle (see Table 6.5, which lists the applications of redox reactions as the basis for differential reaction-rate methods).

6.4.1.3 Catalysed reactions

Inasmuch as most of the catalysed reactions exploited in kinetic analyses are redox in nature, it is not surprising that they have also been the only catalysed reactions used in differential kinetic analyses.

Among these redox reactions, that of Ce(IV) with As(III) occupies a prominent place. This is catalysed by iodide, Os(VIII) and even Ru(VIII). Its first application

Table 6.5 — Differential reaction-rate determinations based on redox reactions

Type of reaction	Chemical system	Mixture: Binary	Mixture: Ternary and multi-component	Technique†	Method†	Reference
Redox	Mo heteropoly acid (yellow) + ascorbic acid $\rightarrow$ Mo heteropoly acid (blue)	Si–P		$A(\lambda = 650$ nm$)$	a′(1)	[55]
	Mo heteropoly acid (yellow) + ascorbic acid $\rightarrow$ Mo heteropoly acid (blue)	Ge–Si		$A(\lambda = 700$ nm$)$	b	[56]
	Mo heteropoly acid (yellow) + $SnCl_2 \rightarrow$ Mo heteropoly acid (blue)	Ti–V		$A(\lambda = 700$ nm$)$		[57]
	$XO_3^- \ (XO_4^-) + I^- \rightarrow I_2$	IO_3^-–BrO_3^- IO_4^-–BrO_3^-		Continuous flow	a	[58]
Redox + complex-formation	XO_4^{n-} + Mo(V)/Mo(VI) $\rightarrow$ Mo heteropoly acid (blue)	Si–P		$A(\lambda = 840$ nm$)$	a′(1) and tangent method	[59]
Redox + complex-dissociation	M(IV)–DCTA + $H^+ \rightarrow$ M(IV) + DCTA M(IV) + ascorbic acid + Arsenazo III $\rightarrow$ M(III)–Arsenazo III	Pu–Np	Pu–Np–U	A	a′	[60]

†See footnote to Table 6.2.
(Reproduced from [37] with permission of the copyright holders, the Royal Society of Chemistry.)

was reported by Rodríguez and Pardue [61], who used it to resolve a mixture of I^- and Os(VIII) by masking the iodide with Hg(II) or Ag(I). The reaction is monitored spectrophotometrically through the decrease in the absorbance of Ce(IV) at 407 nm. After determination of the initial rate for the mixture, the inhibitor is added to the spectrophotometer cell, followed by the catalyst mixture. The iodide concentration is calculated from the corresponding rate law, and that of osmium is found from the initial rate found in the presence of Hg(II) and the pertinent rate law in which the rate constant (as well as that of iodide) is calculated beforehand. This method has been applied to the determination of the two species in mixtures in I/Os ratios from 100:1 to 10:1, at concentrations as low as $10^{-10}M$ for osmium. The system has also been applied to the resolution of iodide/iodate mixtures in ratios from 1:10 to 10:1, based on the fact that iodate is not a catalyst for the reaction, but is slowly reduced by As(III) to an oxidation state with a catalytic activity equivalent to that of iodide.

A third application of the Ce(IV)/As(III) system is the resolution of Os–Ru mixtures [62]. The method is based on the different kinetic dependence of the reaction rate on the concentration of Ce(IV) and As(III) in each of the catalysed reactions. The procedure uses the proportional equation method and the establishment of two rate equations, one for the conditions under which the reaction catalysed by ruthenium takes place preferentially and the other corresponding to the conditions favouring the catalytic effect of osmium. This method has been applied to the determination of both metal ions at roughly equimolar concentrations (at the $10^{-8}M$ level) and containing more Ru than Os, with errors ranging between 0.5 and 4%.

Another redox system of frequent use in this field is that of iodide and hydrogen peroxide, which allows the resolution of mixtures such as Zr/Hf, Mo/W and Nb/Ta, as shown in Table 6.6, which lists the salient applications of catalysed reactions in differential kinetic analysis.

The use of indicator systems consisting of organic reagents is a novel contribution to this type of raction. Thus, mixtures of Fe(III) and Mn(II) can be resolved on the basis of the different kinetic dependence on these ions in the oxidation of salicylaldehyde thiosemicarbazone by H_2O_2 [68]. The reaction is monitored spectrofluorimetrically at $\lambda_{em} = 440$ nm, $\lambda_{ex} = 365$ nm, where the oxidation product of the thiosemicarbazone shows maximum fluorescence. This procedure, which uses the proportional equation method to process the data obtained, permits the resolution of mixtures with Fe/Mn ratios from 8:1 to 1:2, respectively, with error and precision better than 3 and 1%. The method has been applied to various types of sample with excellent results.

6.4.2 Resolution of mixtures of organic compounds

The earliest applications of differential reaction-rate methods involved organic compounds. Since most of the organic reactions exploited by these methods were of second-order, special differential methods were developed for use with them, simultaneously with those for first- and pseudo first-order processes.

The original applications involving simple organic compounds such as acids, alcohols, amines and ketones have been extended to more complex substances in the domain of pharmacological and clinical chemistry.

As with inorganic species, the applications of differential reaction-rate methods are discussed here according to the type of reaction involved, and only those

Table 6.6 — Differential reaction-rate determinations based on catalysed reactions

Chemical system	Binary mixtures	Technique†	Method†	Reference
$Ce(IV) + As(III) \xrightarrow{X \text{ or } M} Ce(III) + As(V)$	I^-–Os	$A(\lambda = 407$ nm)	initial-rate	[61]
	IO_3^-–I^-			
	Ru–Os	$A(\lambda = 407$ nm)	b	[62]
$I^- + H_2O_2 \xrightarrow{M} I_2$	Zr–Hf	A	b	[63]
	Mo–W	A (with citric acid)	tangent method	[64]
$I^- + BrO_3^- \xrightarrow{M} I_2$	Nb–Ta			[65]
	Mo–W	C	b	[66]
	Mo–Cr			
I^- + isopolymolybdic acid $\xrightarrow{M} I_2$	Ge–P	A		[67]
$SAT + H_2O_2 \xrightarrow{M} SAT_{ox}$	Fe–Mn	spectrofluorimetry	b	[68]

SAT: salicylaldehyde thiosemicarbazone. †See footnotes to Table 6.2.
(Reproduced from [37] with permission of the copyright holders, the Royal Society of Chemistry.)

applications of special relevance reported in the 1970s and 80s are dealt with (earlier applications are described in detail in Chapter 6 of the monograph by Mark and Rechnitz [1]). The applications are based on redox reactions, followed in importance by hydrolysis and condensation reactions.

6.4.2.1 Redox reactions

These are based on the oxidation of organic compounds with strong oxidants such as permanganate, periodate, etc. Thus, tartaric and formic acid react at a different rate with $KMnO_4$ to yield carbon dioxide and water. The reaction is performed at pH 4 and room temperature, at which formic acid reacts more rapidly than tartaric acid. The process is stopped at various times by adding Mn(II); this reacts with unreacted permanganate to yield MnO_2, which is dissolved and converted into the Mn(III) pyrophosphate complex which is then titrated potentiometrically with hydroquinone. The overall process is second-order and calls for the application of the logarithmic extrapolation method. The errors encountered increase with increasing concentration of the faster reacting component (formic acid), and the method affords the resolution of mixtures with formic/tartaric ratios from 1:10 to 4.5:1 [69].

Mixtures of carbohydrates such as glucose and fructose can be readily resolved in a few minutes with a precision of 1% on the basis of the different rate of oxidation of the sugar by periodate. The reaction is followed enthalpimetrically with the aid of a reaction cell fitted with a thermistor. The highest pseudo first-order rate-constant ratio (3) is obtained at pH 5.2, and the determination involves the application of the proportional equation method to equimolar concentrations of both sugars [70].

Mentasti *et al.* [71] have proposed the determination of binary mixtures of carboxylic acids (formic, acetic, propionic, butyric, isobutyric and pivalic) based on their oxidation by the Ag(II) obtained by anodic oxidation of the Ag(I) at a Pt electrode:

$$2\mathrm{Ag(II)} + \mathrm{RCOOH} \rightarrow 2\mathrm{Ag(I)} + 2\mathrm{H}^+ + \mathrm{CO}_2 + \text{products}$$

The release of the first electron gives rise to a radical ·R, which is then oxidized by another Ag(II) ion to yield alkenes, ketones or esters (with the obvious exception of formic acid, which yields only CO_2).

The rate law of the process is

$$-\mathrm{d[Ag(II)]}/\mathrm{d}t = k\mathrm{[Ag(II)][RCOOH]}$$

In the presence of excess of Ag(II), the reaction is pseudo first-order provided that the remaining conditions are kept constant.

The process is monitored photometrically at $\lambda = 470$ nm ($\varepsilon = 138 \pm 2$ $\mathrm{l.mole^{-1}.cm^{-1}}$) by following the disappearance of Ag(II), and applying the stopped-flow technique and the Roberts and Regan method.

The principle has also been applied to the resolution of binary mixtures of aliphatic and aromatic alcohols by the single-point and integration methods, with detection limits $\geq 5 \times 10^{-4} M$ [72].

The coupling of a redox reaction with one of a different nature has allowed the resolution of mixtures of aromatic phenols and amines. It is widely known that the oxidation of *p*-diamines by ferricyanide or a similar oxidant yields unstable coloured semiquinone radicals (in two steps), which can be used as indicators of the degree and rate of reaction:

$$RR^1N{-}C_6H_4{-}NH_2 \xrightarrow{-e^-} R^+R^1N{-}C_6H_4{-}NH_2 \xrightarrow{-e^-} R^+R^1N{=}C_6H_4{=}NH + H^+$$

p-diamine SQ QDI

where QDI and SQ denote a quinonedi-imine and a semiquinone, respectively. If a QDI such as that from *N*,*N*′-dimethyl-*p*-phenylenediamine is reacted with an aromatic amine (e.g. aniline), the result is an indamine,

$$(H_3C^+)(H_3C)N{=}C_6H_4{=}NH + C_6H_5NH_2 \xrightarrow{\text{slow}} (CH_3)_2N{-}C_6H_4{-}NH{-}C_6H_4{-}NH_2$$

QDI leucoindamine

$$\xrightarrow{\text{fast}} (CH_3)_2N{-}C_6H_4{-}N{=}C_6H_4{=}NH$$

indamine

a process which can be used to resolve a large variety of mixtures of aromatic amines [73]. These determinations are seriously interfered with by phenols, which can also be determined in mixtures [74] by use of *N*,*N*-diethyl-*p*-phenylenediamine in the presence of ferricyanide. The QDI formed reacts with the phenol to form a dye (indaniline) instead of the indamine, in a reaction that is monitored photometrically by the stopped-flow technique.

Substances of biological interest such as betamethasone or its valeric ester (at position 17) have been determined through oxidation by Cu(II) of the hydroxyl group at position 21 of the steroid to give an aldehyde which is then condensed with 3-methylbenzothiazole-2-one hydrazone to yield an azine that absorbs at 394 nm and is used to monitor the development of the reaction. The proportional equation method is used [75].

Mixtures of organic peroxides can also be resolved by reduction with various sulphides, and application of the proportional equation method [76].

Uric and ascorbic acids have been determined in mixtures at an overall concentration of $10^{-5}M$ by using the 2,2′-bipyridyl/Fe(II) complex instead of an oxidant. The stopped-flow technique applied to monitor the reduction of the complex results in large errors ($\sim 10\%$) [77] and does not allow the single-point method to be used; instead, $(A_\infty - A_t)/A_\infty$, where A_∞ and A_t denote the absorbances at equilibrium and at a reaction time of 0.5 sec, respectively, is plotted as a function of the molar fraction of uric acid. Though non-linear, the resulting curve is useful for estimating the mixture composition.

Few catalysed redox reactions have been used in the resolution of mixtures of organic compounds. Among them is the analysis for sulphophthaleins in binary and ternary mixtures, proposed by Mottola and Ellis [78], based on selective oxidation by periodate, catalysed by Mn(II). The reactions involved are pseudo first-order and the data are treated by the proportional equation method.

Organophosphorus pesticides have also been determined in mixtures through their accelerating effect on the rate of oxidation of *o*-dianisidine by H_2O_2, although the method is mainly used for individual determinations and is subject to large errors (~ 19%) [79].

6.4.2.2 Hydrolysis reactions

These have been used in the determination of penicillins and their derivatives in mixtures by degradation.

Thus, the logarithmic extrapolation method allows the resolution of binary mixtures of penicillins and 6-aminopenicillinic acid (6-APA) on the basis of their different rates of degradation, which are highly influenced by pH. By selection of suitable pH values, the difference in the rates of the hydrolysis reaction catalysed by acids can be maximized. In this manner mixtures of cloxacillin/6-APA, ampicillin-G/6-APA, cloxacillin/ampicillin and penicillin-G/cloxacillin have been resolved, the last named yielding the most accurate results [80].

The simultaneous determination of carbenecillin and the drugs carindacellin and carfecillin from which it is formed by hydrolysis, has been performed by the single-point (Lee and Kolthoff) and logarithmic extrapolation methods on the basis of the different rates of degradation of carbenecillin and its carboxy esters in acid solution [81]. The reaction is monitored spectrophotometrically by the imidazole test. Carindacellin and carfecillin are hydrolysed quantitatively to carbenecillin in 0.05*M* phosphate buffer of pH 7.4 at 37°C, with half-lives of 5.5 and 2.5 hr, respectively.

On the basis of this principle binary mixtures of ampicillin and its parent drugs pivampicillin and bacampicillin (as hydrochlorides) have also been resolved by the proportional equation method, with acceptable precision (≈5%) over the range 10–200 μg/ml [82].

The decomposition of the π-complexes of tetracyanoethylene with tertiary amines has been employed for the determination of mixtures of the latter by tristimulus colorimetry [83].

6.4.2.3 Other reactions

Primary amines can be determined in mixtures in the presence of secondary and tertiary amines, by means of their condensation with salicylaldehyde in chloroform medium. The differences in the rate of reaction of secondary amines with methyl acrylate in methanol have been used for the resolution of binary and ternary mixtures of these amines in the presence of tertiary (but not primary) amines. The reactions involved are second-order in both cases and are followed by titration in non-aqueous medium. The results are processed by the second-order version of the logarithmic extrapolation method [84].

The different rates of reaction of 6(R)- and 6(S)-mecillinam with glycine to yield a 4-aminoethyleneimidazol-5(4H)-one derivative that is monitored spectrophotometrically at 330 nm is the basis for the determination of the two epimers of this derivative of penicillinic acid by application of the proportional equation method at two different times. The 6(S)-epimer reacts more slowly, and the kinetic information obtained points to the occurrence of an intermediate compound in the formation of the imidazole [85].

Also based on the formation of a coloured compound is a recently reported simultaneous determination of amino-acids in binary or ternary mixtures through reaction with ninhydrin. Two multi-point curve-fitting methods (linear and non-linear regression) were applied and gave similar results in the resolution of histidine/lysine/isoleucine, leucine/isoleucine and glycine/isoleucine mixtures at concentrations between 5 and $50 \mu M$ [86].

A bromination reaction is the basis for the differential determination of salicylic acid and paracetamol in mixtures with caffeine by use of the bromate/bromide system in acid medium [87]. The procedure entails measuring the time required for the complete bromination, which is marked by the presence of free bromine, which decolorizes Methyl Orange. This time is directly proportional to the concentration of the brominated species. The method requires both components of the mixture to be determined independently.

Ligand-exchange reactions have also been applied to the resolution of mixtures of aminopolycarboxylic acids on the basis of the reaction of their nickel complexes with cyanide. The stopped-flow technique and a linear regression method are used in the procedure [88].

The novel reaction of C=N– group exchange is the basis for resolving mixtures of hydrazine and hydroxylamine, and of ammonia with either of these two compounds. Both compounds can exchange C=N– groups with 2-hydroxybenzaldehyde azine, giving rise to the corresponding hydrazone, a fluorescent product by means of which the development of the reaction can be monitored. These mixtures are subject to synergistic effects that can be overcome by the method described in Section 6.3 [22].

REFERENCES

[1] H. B. Mark, Jr. and G. A. Rechnitz, *Kinetics in Analytical Chemistry*, Wiley, New York, 1968.
[2] S. Siggia, J. G. Hanna and N. M. Serencha, *Anal. Chem.*, 1963, **35**, 362.
[3] J. B. Worthington and H. L. Pardue, *Anal. Chem.*, 1972, **44**, 767.
[4] A. M. Gary and P. Lagrange, *Bull. Soc. Chim. France*, 1974, 1219.
[5] K. Haraguchi and S. Ito, *Bunseki Kagaku*, 1975, **24**, 405.
[6] K. A. Connors, *Anal. Chem.*, 1976, **48**, 87.
[7] K. A. Connors, *Anal. Chem.*, 1977, **49**, 1650.
[8] A. Garrido, *Minor Thesis*, University of Córdoba, 1984.
[9] T. S. Lee and I. M. Kolthoff, *Ann. N. Y. Acad. Sci.*, 1951, **53**, 1903.
[10] L. Ballesteros and D. Pérez-Bendito, *Analyst*, 1983, **108**, 443.
[11] R. G. Garmon and C. N. Reilly, *Anal. Chem.*, 1962, **34**, 600.
[12] J. B. Worthington and H. L. Pardue, *Anal. Chem.*, 1970, **42**, 1157.
[13] J. D. Roberts and C. McG. Regan, *Anal. Chem.*, 1952, **24**, 360.
[14] R. A. Greinke and H. B. Mark, Jr., *Anal. Chem.*, 1966, **38**, 340.
[15] S. Siggia and J. G. Hanna, *Anal. Chem.*, 1961, **33**, 896.
[16] C. N. Reilley and J. J. Papa, *Anal. Chem.*, 1962, **34**, 801.
[17] H. B. Mark, Jr., L. M. Backes and D. Pinkel, *Talanta*, 1965, **12**, 27.
[18] H. B. Mark, Jr., *Anal. Chem.*, 1964, **36**, 1668.
[19] R. A. Greinke and H. B. Mark, Jr., *Anal. Chem.*, 1966, **38**, 1001.
[20] S. Siggia and J. G. Hanna, *Anal. Chem.*, 1964, **36**, 228.

[21] L. J. Papa, J. H. Patterson, H. B. Mark, Jr. and C. N. Reilley, *Anal. Chem.*, 1963, **35**, 1889.
[22] A. Ríos, M. Silva and M. Valcárcel, *Z. Anal. Chem.*, 1985, **320**, 762.
[23] A. Marín, M. Silva and D. Pérez-Bendito, *Anal. Chim. Acta*, 1987, **197**, 77.
[24] J. B. Pausch and D. W. Margerum, *Anal. Chem.*, 1969, **41**, 226.
[25] M. Tanaka, S. Funashi and D. Shirai, *Anal. Chim. Acta*, 1967, **39**, 427.
[26] E. Mentasti, *Anal. Chim. Acta*, 1979, **111**, 177.
[27] Y. Nagaosa and T. Yonekubo, *Bull. Chem. Soc. Japan*, 1973, **46**, 1667.
[28] A. V. Stepanov, M. A. Nemtsova, S. A. Nikitina and T. A. Dem'yanova, *Radiokhimiya*, 1978, **20**, 906.
[29] A. V. Stepanov, S. A. Nikitina and T. A. Dem'yanova, *Radiokhimiya*, 1979, **21**, 34.
[30] S. Ito, K. Haraguchi, K. Nagagawa and K. Yamada, *Bunseki Kagaku*, 1977, **26**, 554.
[31] K. Haraguchi, K. Nagagawa, T. Ogata and S. Ito, *Bunseki Kagaku*, 1980, **29**, 809.
[32] K. Haraguchi, K. Nagagawa, T. Ogata and S. Ito, *Bunseki Kagaku*, 1981, **30**, 149.
[33] G. R. Riddler and D. W. Margerum, *Anal. Chem.*, 1977, **49**, 2090.
[34] K. B. Yatsimirskii, L. I. Budarin and A. G. Khachatryan, *Dokl. Akad. Nauk. SSSR*, 1970, **195**, 898.
[35] K. B. Yatsimirskii, L. I. Budarin and A. G. Khachatryan, *Zh. Analit. Khim.*, 1971, **26**, 1499.
[36] K. B. Yatsimirskii, L. I. Budarin and A. G. Khachatryan, *Dokl. Akad. Nauk. SSSR*, 1973, **211**, 1139.
[37] D. Pérez-Bendito, *Analyst*, 1984, **109**, 891.
[38] B. G. Willis, W. H. Woodruff, Jr., Frysinger, D. W. Margerum and H. L. Pardue, *Anal. Chem.*, 1970, **42**, 1350.
[39] A. M. Gary and J. P. Schwing, *Bull. Soc. Chim. France*, 1975, 441.
[40] A. M. Gary and J. P. Schwing, *Bull. Soc. Chim. France*, 1976, 1609.
[41] A. M. Albrecht-Gary, J. P. Collin, P. Jost, P. Lagrange and J. P. Schwing, *Analyst*, 1978, **103**, 227.
[42] D. W. Margerum, J. P. Pausch, G. A. Nyssen and G. F. Smith, *Anal. Chem.*, 1969, **41**, 223.
[43] J. B. Kloosterboes, *Anal. Chem.*, 1974, **46**, 1143.
[44] J. H. Dahl, D. Espersen and A. Jensen, *Anal. Chim. Acta*, 1979, **105**, 327.
[45] M. Kopanica and V. Stará, *Collection Czech. Chem. Commun.*, 1976, **41**, 3275.
[46] D. Espersen and A. Jensen, *Anal. Chim. Acta*, 1979, **108**, 241.
[47] H. Kagenow and A. Jensen, *Anal. Chim. Acta*, 1980, **114**, 227.
[48] H. Kagenow and A. Jensen, *Anal. Chim. Acta*, 1983, **145**, 125.
[49] J. P. Collin and P. Lagrange, *Bull. Soc. Chim. France*, 1976, 1309.
[50] J. Lagrange, G. Lagrange and Z. Zare, *Bull. Soc. Chim. France*, 1978, **I**, 7.
[51] A. V. Stepanov, T. P. Makarova and A. M. Fridkin, *Zh. Analit. Khim.*, 1979, **34**, 2337.
[52] J. F. Coetzee, D. R. Bayla and P. K. Chattopadhyay, *Anal. Chem.*, 1973, **45**, 2266.
[53] T. Kitagawa and K. Fugikawa, *Nippon Kagaku Kaishi*, 1977, **7**, 998.
[54] K. Hiraki, K. Morishige and Y. Nishikawa, *Anal. Chim. Acta*, 1978, **97**, 121.
[55] J. D. Ingle, Jr. and S. R. Crouch, *Anal. Chem.*, 1971, **43**, 7.
[56] T. Yonekubo, Y. Nagaosa and Y. Nakahigashi, *Nippon Kagaku Kaishi*, 1974, **2**, 269.
[57] Y. Nagaosa, T. Yonekubo, M. Satake and R. Seto, *Bunseki Kagaku*, 1972, **21**, 215.
[58] J. B. Hanna and S. Siggia, *Anal. Chem.*, 1964, **36**, 2022.
[59] K. Ohashi, H. Kawaguchi and K. Yamamoto, *Anal. Chim. Acta*, 1979, **111**, 301.
[60] A. V. Stepanov, T. P. Makarova and A. M. Fridkin, *J. Radioanal. Chem.*, 1979, **51**, 385.
[61] P. A. Rodríguez and H. L. Pardue, *Anal. Chem.*, 1969, **41**, 1376.
[62] J. B. Worthington, P. A. Rodríguez and H. L. Pardue, *Anal. Chem.*, 1970, **42**, 1157.
[63] K. B. Yatsimirskii and L. P. Raizman, *Zh. Analit. Khim.*, 1963, **18**, 829.
[64] I. I. Alekseeva, L. P. Ruzinov, E. G. Khachaturyan and L. M. Chemysova, *Zh. Analit. Khim.*, 1980, **35**, 60.
[65] I. I. Alekseeva, L. P. Ruzinov, E. G. Khachaturyan and L. M. Chemysova, *Izv. Vyssh. Uchebn. Zaved. Khim. Khim. Tekhnol.*, 1973, **16**, 1145.
[66] C. M. Wolff and J. P. Schwing, *Bull. Soc. Chim. France* 1976, 679.
[67] I. I. Alekseeva and I. I. Nemzer, *Zh. Analit. Khim.*, 1970, **25**, 1118.
[68] A. Moreno, M. Silva and D. Pérez-Bendito, *Anal. Chim. Acta*, 1984, **159**, 319.
[69] A. Berka and J. Korečková, *Anal. Lett.*, 1973, **6**, 1113.
[70] W. A. de Oliveira and A. A. Rodella, *Talanta*, 1979, **26**, 965.
[71] E. Mentasti, E. Pelizzetti and G. Saini, *Anal. Chim. Acta*, 1976, **86**, 303.
[72] E. Mentasti and C. Baiocchi, *Anal. Chim. Acta*, 1980, **119**, 91.
[73] R. Rawa and S. Hirose, *Chem. Pharm. Bull.*, 1980, **28**, 2136.
[74] E. Pelizzetti, G. Girandi and E. Mentasti, *Anal. Chim. Acta*, 1977, **94**, 479.
[75] J. Hansen and H. Bundgaard, *Int. J. Pharm.*, 1981, **8**, 121.
[76] J. P. Hawk, E. L. McDaniel, T. D. Parish and K. E. Simmons, *Anal. Chem.*, 1972, **44**, 1315.
[77] E. Pelizzetti and E. Mentasti, *Anal. Chim. Acta*, 1979, **108**, 441.
[78] G. L. Ellis and H. A. Mottola, *Anal. Chem.*, 1972, **44**, 2037.

[79] *G. Kh. Shapenova, Sh. T. Talipov and I. A. Orlik, Uzb. Khim. Zh.*, 1978, **4**, 8; *Chem. Abstr.*, 1978, **89**, 141700c.
[80] L. Koprivc, E. Polla and J. Hranilovic, *Acta Pharm. Suec.*, 1976, **13**, 421.
[81] H. Bundgaard, *Arch. Pharm. Chemi Sci. Ed.*, 1979, **7**, 95.
[82] H. Bundgaard, *Arch. Pharm. Chemi*, 1979, **86**, 607; *Arch. Pharm. Chemi Sci. Ed.*, 1979, **7**, 81.
[83] R. Tawa, S. Hiroshi and K. Adachi, *Chem. Pharm. Bull.*, 1982, **30**, 1872.
[84] I. L. Shresta and M. N. Das, *Anal. Chim. Acta*, 1970, **50**, 135.
[85] H. Bundgaard, *Int. J. Pharm.*, 1980, **5**, 257.
[86] Y. R. Tahboub and H. L. Pardue, *Anal. Chim. Acta*, 1985, **43**, 173.
[87] M. A. Elsayed and O. S. Ogbonnia, *Pharmazie*, 1980, **35**, 474.
[88] L. C. Coombs, J. Vasiliades and D. W. Margerum, *Anal. Chem.*, 1972, **44**, 2325.

7

Instrumentation

7.1 INTRODUCTION

The degree of development attained by kinetic methods compared with static (equilibrium) methods owes much to major advances in instrumentation over the last decade. On the one hand, the incorporation of microelectronics and microcomputer science into analytical instrumentation has provided a means of treating data in a completely or partly automatic manner. On the other hand, there is the degree of perfection reached in the design of detectors, patent in the stabilization of the light-source and photodetectors in photometric techniques, the advent of novel techniques using image detectors, the design of the high-speed recorders for use with optical techniques, the development of differential pulse polarography, and the popularization of selective electrodes in electroanalytical techniques.

All this has led to the accuracy and precision of measurement with which kinetic methods are endowed today and accounts for their present popularity. More and more kinetic methods are being incorporated into teaching manuals and used in routine analyses (e.g. clinical analyses) and an increasing number of commercially available instruments are marketed, with special devices for kinetic measurements.

With the exception of a relatively recent review by Malmstadt [1], most of the literature on analytical kinetic instrumentation was published in the early 1970s in reviews such as those by Malmstadt [2,3], Reich [4], Crouch [5] and Mark *et al.* [6]. Neither of the two monographs on kinetic methods published so far [7,8] deals with instrumentation at length. In addition, the chapter devoted to instrumentation in the work by Kopanica and Stará [9] is quite brief. Finally, it is worth mentioning the attention paid to this aspect in some overviews of kinetic methods [10,11] and the biennial reviews published in *Analytical Chemistry* in the last few years.

7.2 GENERAL CONSIDERATIONS

Kinetic methods do not require measurement of the absolute value of the parameter chosen to monitor the course of the reaction (absorbance, fluorescence, potential etc.), but rather its variation as a function of time, so that measurements are free from the influence of factors contributing to error in absolute values (turbidity, liquid-junction potential, presence of other absorbing or fluorescent substances, provided that these do not take part in the reaction of interest or modify the response of the parameter). This is a clear advantage over static methods, but the inherent nature of dynamic systems calls for high precision and sensitivity in the measurement of the instrumental parameter, as well as strict control of experimental conditions and the accurate timing of measurements, requisites that were not satisfactorily met until modern instrumentation became available. Thus, the strict control of time and temperature, the relative nature of the signal measurements made and the use of systems for data collection and treatment in some cases (fast kinetics) are essential features of kinetic methods, in contrast to equilibrium methods.

Time is one of the variables to be strictly controlled and measured in every operation involved in a kinetic method; not only measurements and recording periods, but also operations such as sample treatment should be accurately timed. In addition, kinetic methods require knowledge of the influence of this variable on the performance of the data-processing system. The transduced signal to be processed should be collected accurately with regard to time.

The temperature is a critical variable in kinetic methods insofar as it affects the kinetics much more strongly than the equilibrium of a reaction (provided the equilibrium constant is large enough). While not so relevant to static methods, this variable has to be strictly controlled ($\pm$0.01–0.1°C) in the application of kinetic methods if reproducible results are to be obtained (a temperature change of 1°C may result in a concomitant change of up to 10% in the reaction rate at normal laboratory temperatures).

The method used to measure the rate of reaction is dictated by the half-life. The instrumentation used to monitor slow reactions (half-lives longer than 10 sec), is simpler than that employed with fast reactions (half-lives shorter than 10 sec) both as regards mixing of the reactants and the measurement and collection of the analytical signals, so much so that the particular system involved dictates the instrumentation to be used according to whether the external experimental conditions remain constant (closed systems) or not (open systems). Closed systems are not suitable for fast reactions; conversely, open systems are especially suited to reactions with fast kinetics and have contributed significantly to their use for analytical purposes.

7.3 BASIC COMPONENTS OF KINETIC INSTRUMENTATION

Kinetic methods make use of instruments of different complexity, from conventional elementary systems (e.g. a titrator) to sophisticated set-ups (e.g. completely automatic instruments capable of collecting the sample, transporting it to the detector and processing the results obtained). The analytical procedure involved in the measurement of the reaction rate comprises the following stages (Fig. 7.1).

(1) Preparation, measurement, transport and mixing of the reactants (i.e. mixing of sample and reagents). This first stage, common to other analytical methods, is a key step in kinetic methods, as regards measurement of the parameter by which the process is monitored.

(2) Signal monitoring and transduction at constant temperature. This is per-

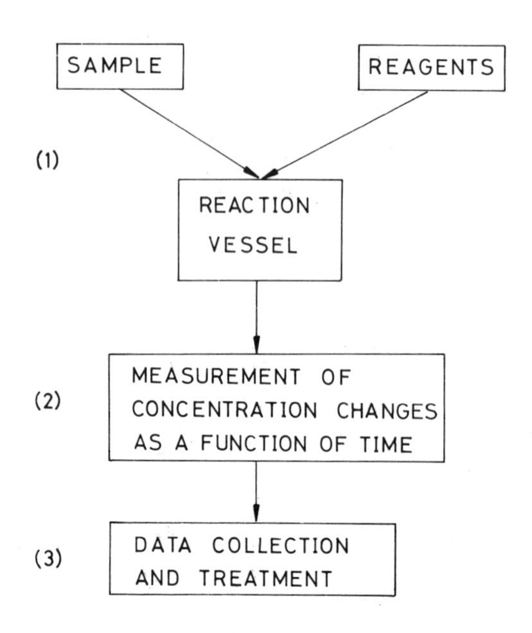

Fig. 7.1 — Basic stages of reaction-rate determinations.

formed by the measuring system (photometer, potentiometer) which follows the course of the reaction by determining the changes in a property of one of the reactants or products.

(3) Timed collection of the data, which are treated simultaneously or sequentially either by manual computation or with the aid of a computer.

Completely automated systems execute all three stages, while partly automated instruments usually perform stages (2) and (3) without simultaneously treating the data generated.

The evolution of automation in the context of kinetic methods is illustrated in Fig. 7.2. At the lowest level of automation, the operator performs most of the steps involved (Fig. 7.2a). In more elaborate systems, the instrument performs time measurements (Fig. 7.2b). Finally, completely automatic systems (Fig. 7.2c) only require the operator to act on a feed-back computer controlling and regulating all the stages involved in the analytical procedure.

Each of the three stages described above is related to one of the basic components of instrumentation and will be described in greater detail later on. The first stage, mixing of sample and reagents, is done differently according to whether an open or closed system is used. In dealing with it, we shall refer only to closed systems. The control of the temperature, compulsory in stages 1 and 2 irrespective of the type of system used, will also be dealt with at length.

7.3.1 Control of temperature

As stated above, the temperature must be strictly controlled in kinetic methods as its fluctuations affect the accuracy and precision of the analytical results. Not only must

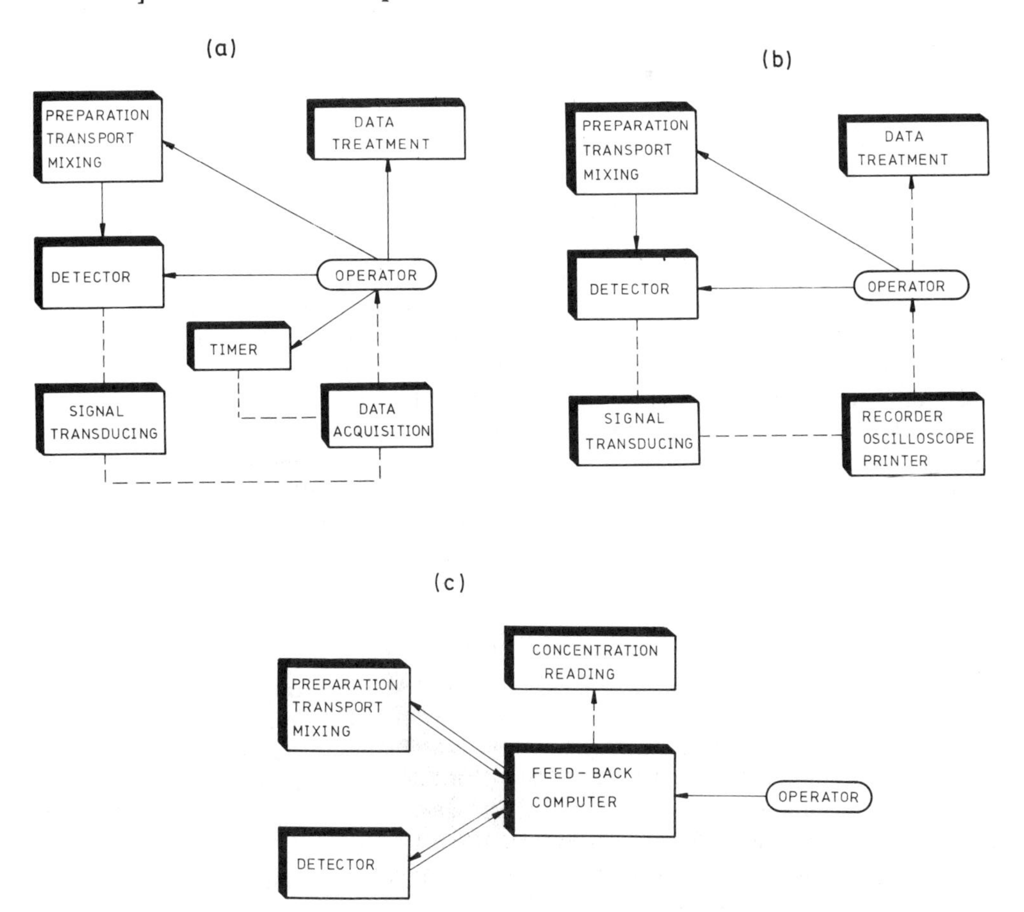

Fig. 7.2 — Automation potential of kinetic instrumentation: (a) manual procedure; (b) automation of time measurements; (c) complete automation by means of a computer with feed-back.

the cell compartment and the mixing chamber (if it is independent of the former) be strictly thermostatically controlled, but the absorption or release of heat by the reaction of interest and the consequent temperature change in the solution until thermal equilibrium with the thermostat is reached, should also be allowed for.

There are several ways to control the temperature in manual closed systems, namely [12]: (a) by placing all the sample and reagent solutions separately in the thermostat prior to their mixing; (b) by use of a thermostatic measuring cell containing all the components of the reaction except the one initiating it, which should be added in a small volume (10–50 μl) once the system has reached the desired temperature; (c) by transferring the reaction mixture to the measuring compartment and waiting until it reaches the desired temperature, which is only feasible when very slow reactions are dealt with.

There are two general mechanisms for controlling the temperature: by recirculation of a liquid kept at constant temperature, and by electrical heating. The first method is the commoner of the two and relies on the circulation of a stream of water or another solvent from a thermostatic bath through a jacket surrounding the connecting tubing and/or reaction cell. The material of which the jacket is made is of great importance, as it should ensure fast exchange of heat between the two liquids it separates.

An alternative is heating or cooling by the Peltier effect; the temperature is then often shown as a digital display. Many commercial spectrophotometers are supplied with fixed-temperature heating systems, commonly operating at temperatures of enzymatic significance (25, 33 and 37°C). The different commercial thermoelectric systems developed recently for this purpose afford fast temperature changes and are capable of establishing positive and negative temperature gradients in the course of a reaction. Their one disadvantage is probably their high cost.

Working temperatures below room temperature result in a further difficulty with optical methods, namely, clouding of the cuvettes through condensation of atmospheric moisture. This may lead to anomalous results and should be avoided at all costs by passing a stream of a dry inert gas (e.g. nitrogen) through the cuvette compartment.

7.4 CLOSED SYSTEMS

Whenever a kinetic method is applied, the sample must be thoroughly mixed with the reagent(s) at constant temperature before its transfer to the detection system. This mixing operation can be done manually (the usual situation with closed systems which are applied to slow reactions) or automatically. Automation of the mixing of sample and reagents calls for special mechanical devices when fast reactions, which require the use of open systems, are involved.

For manual mixing it is usually sufficient to mix the reactants in the reaction vessel, which may be the spectrophotometer cuvette or the electrode cell. The reagents are usually injected from a syringe into the reaction vessel, the contents of which are under continuous agitation. The instant at which the last reactant is added is taken as the time for the start of the reaction ($t = 0$) and must therefore be known precisely. If the reaction takes place in the measuring cell itself, the solution must be mixed manually or magnetically, and some spectrophotometers accommodate magnetic stirrers in their sample compartments.

Clinical kinetic analysis often involves the use of straightforward photometers accommodating a flow-cell wherein sample and reagents are thoroughly mixed after transfer by means of a peristaltic pump. This reduces sample consumption in routine kinetic enzymatic determinations. The flow-cell is thermostatically controlled, so the volume of reaction mixture with which it is filled (between 0.05 and 0.1 ml) reaches the desired temperature in a very short time.

The reactants can be incorporated into the measuring system automatically in order to improve the reproducibility and throughput of closed or batch systems. One such system is the computer-controlled Technicon RA 1000. This consists of three discs: the central disc or reaction tray can hold 100 disposable transparent plastic (light-path 7 mm) cuvettes; the sampling tray has thirty 0.5-ml cups intended to hold the samples and the reagent tray possesses forty 25-ml cavities containing the different reagents. The system includes high-precision Hamilton pipettes for transfer of the sample (3–20 μl) and reagents (300 μl) to the central tray. Carry-over is avoided by use of the novel 'discretional access inert fluid', TRAF. The detection system is a single colorimeter which allows the instrument to make a preliminary measurement to check the cuvette cleanliness and reagent purity. The spinning of the trays makes any agitation redundant. The computer programs and controls every operation and its results: number of samples analysed, parameters measured per sample, type of analysis (end-point, zero-order or first-order kinetics, etc.), transfer to and operation of the detector, signal collection, and data-processing and presentation.

Centrifugal analysers are one type of automatic instrument widely used in clinical laboratories. The best known is probably the Centrifichem, marketed by Union Carbide Ltd. It consists of two separate instrumental modules, the doser and the analyser. The doser consists of two concentric rings, the outer of which holds the sample microvials. The inner ring or transfer disc is the key part in the system and consists of thirty units arranged radially from the centre in three rows of sample, reagent and transfer cavities. The dosing module is turned 12° at a time to place the different units sequentially under the pipetting zone. The sample- or standard-pipette transfers a given volume from a microvial on the outer ring to the sample cavity, with an intermediate wash. Other pipettes transfer the reagents from their cuvettes to their corresponding cavities. Once the transfer operation has been concluded, the transfer disc is placed manually in the rotor of the analyser module, and covered with a plastic lid. A hydropneumatic system evacuates the chamber both for the mixing of the ingredients of the reaction and for the subsequent washing and drying. The disc is spun at 960 rpm and the centrifugal force ensures the mixing and sweeping of the liquids from their radial unit into a measuring cuvette placed on a disc spinning jointly with the transfer disc and having a transfer port which must be aligned with the cuvette port. The sample is held in the cuvette (the top and bottom of which are made of transparent fused silica) by the centrifugal force. The single optical system used, located in a fixed position, operates continuously and measures the absorbance of each sample traversing its light path. It can make measurements at a preset time or signal level, or be programmed for kinetic measurements (absorbance increment over a given Δt). A printer provides the final result of each parameter measured, and a bar graph showing the absorbance changes in each of the thirty cells is continuously displayed on a screen.

7.5 OPEN SYSTEMS

As stated above, the kinetic methodology used to monitor the reaction of interest involves two basic types of systems, closed and open, the first of which was dealt with in the previous section.

As a rule, the open systems more commonly used in the study of fast reactions can be divided into continuous or flow systems, and batch systems (Table 7.1).

Other ways of mixing the sample and reagents include automatic continuous procedures such as those implemented by air-segmented flow analysis (SFA) with the aid of commercial Technicon AutoAnalyzers or by flow-injection analysis (FIA) [13,14], a description of which will not be attempted as it is beyond the scope of this chapter.

Table 7.1 — Open systems

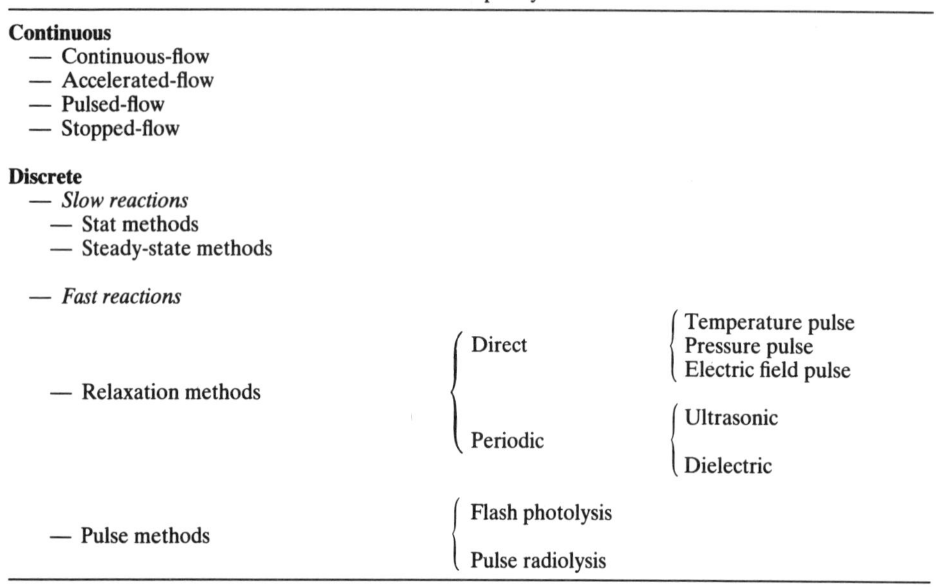

Continuous		
— Continuous-flow		
— Accelerated-flow		
— Pulsed-flow		
— Stopped-flow		
Discrete		
— *Slow reactions*		
— Stat methods		
— Steady-state methods		
— *Fast reactions*		
— Relaxation methods	Direct	Temperature pulse Pressure pulse Electric field pulse
	Periodic	Ultrasonic Dielectric
— Pulse methods	Flash photolysis Pulse radiolysis	

7.5.1 Continuous-flow technique

This was the first flow technique to be developed. It was conceived by Hartridge and Roughton [15], who applied it in 1923 to the study of biochemical processes. Paneth used it in 1929 to establish the existence and reactivity of gas-phase free radicals.

This technique affords the study of reactions with half-lives of a few msec without the need for a fast detection system.

The principle of the continuous flow technique is illustrated in Fig. 7.3, which also shows the variation of the measurement time as a function of the flow-rate. By gas or hydrostatic pressure, or with the aid of a peristaltic pump, the solutions of the reactants are conveyed to a mixing point and thence at a linear flow-rate v_0 through the observation tube, where the signal is measured at a distance d along the tube (observation point) that can be varied to provide for several observations at different reaction times.

The time $t_m = d/v_0$ is that taken for the reacting mixture to travel from the mixing point to the observation point and is therefore a measure of the dead time of the system. Naturally, this time should be much shorter than the reaction half-life, so $t_{1/2} > d/v_0$. Thus, for a flow-rate of 10 msec and a half-life of 1 msec, the separation between the different observation points, d, should be less than 10 mm. The different readings made (kinetic curve) are the basis for the application of the various kinetic determinative methods known.

The considerations above rely on the assumptions that the mixing is virtually instantaneously complete and that the flow-rate profile in a cross-section of the observation tube is flat. The mixing efficiency depends on factors such as the geometry of the mixing chamber, the flow-rate and the viscosity of the solution, and the last two of these determine the flow-rate profile. Both assumptions are effectively valid under turbulent flow conditions. When the Reynolds number Re ($= v_0 l/\nu$, where l is the internal diameter of the observation tube and ν is the kinematic

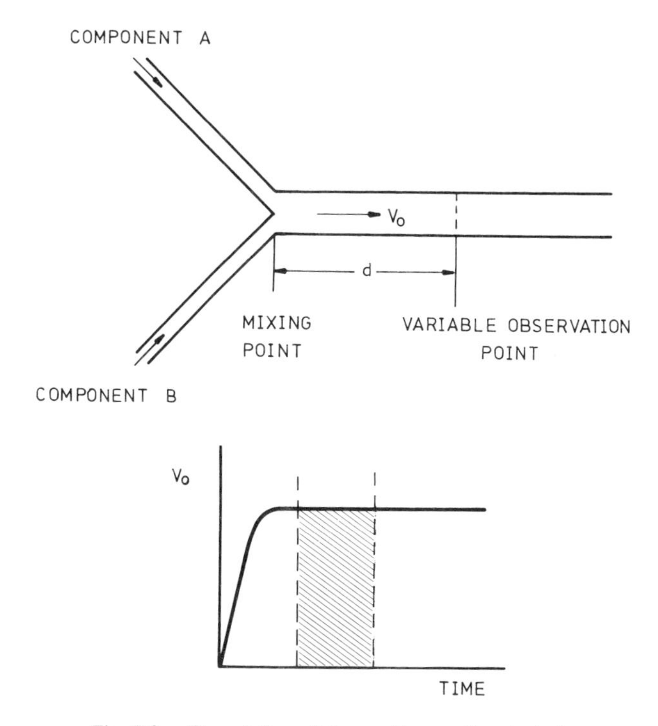

Fig. 7.3 — Foundation of the continuous-flow technique.

viscosity) is > ~2500, the flow becomes turbulent, mixing is rapidly complete, and the flow-profile is flat.

The most serious disadvantage of this technique is its high sample and reagent consumption. This is not so important in work with gas systems, but poses problems with biochemical reactions, as the amount of sample available is usually rather small. This shortcoming has fostered research aimed at reducing the sample volume required, which in turn has resulted in the development of other flow techniques, commented on below.

The detection systems most commonly employed in this technique are spectrophotometric, followed by conductimetric and thermometric. The last-named allows the detection of temperature changes of about 0.03°C by means of a thermocouple or thermistor placed at the centre of the tube through which the flow circulates, the whole system being in a constant-temperature enclosure.

A suitable alternative to this continuous-flow technique is the so-called stirred-flow technique, in which the reactants are led to a mixing chamber where they are rapidly stirred, and a stationary state is reached when the incoming flow of reactants equals the outgoing flow of products. In addition, if the outgoing flow is sensed at a point close enough to the mixing chamber, the concentrations determined will reproduce faithfully those inside the chamber itself. The applicability of such a simple technique is limited by the mixing efficiency, as even stirring at speeds as high as 15000 rpm does not ensure a residence (dead) time shorter than 1 sec.

7.5.2 Accelerated-flow technique

This technique was developed by Chance [16] and is similar to the continuous flow technique described above, with the exception that the flow velocity of the reaction mixture, v_0, varies during the process, and only one observation is made, at a fixed distance, d, from the mixing point (Fig. 7.4). The experimental set-up therefore results in lower reactant consumption.

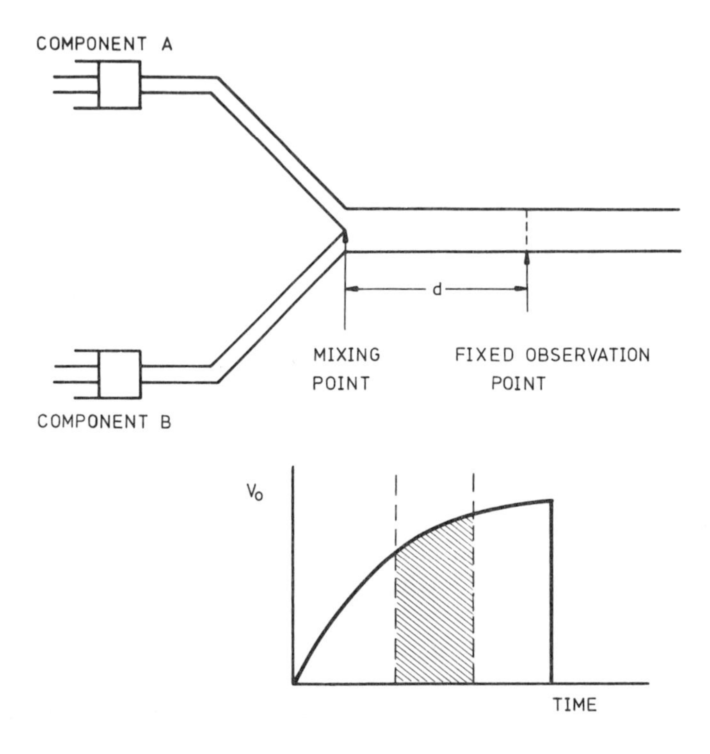

Fig. 7.4 — Foundation of the accelerated-flow technique.

The sample and reagent solutions, held in two hypodermic syringes, are discharged simultaneously and are accelerated from rest to a maximum velocity. As the position of the detector is fixed with respect to the observation point, the reaction time, t, corresponding to the flow reading observation point is inversely proportional to the flow velocity at that point. For a distance d of 7 mm and a maximum velocity of 10 m/sec, an observation tube 1 mm in bore will allow measurement at reaction times as short as 1 msec.

The measurements (usually absorbances) are recorded by a computer or an oscilloscope as a function of the flow velocity. By use of a fast detection system, reagent consumption can be lowered considerably (volumes as low as 100 μl are not unheard of). It is interesting to note that, as with the continuous-flow technique, the scope of application of this technique is limited to reactions involving species with high molar absorptivities, since the detection system is placed perpendicular to the observation tube (optical path length 1–3 mm, depending on the tube diameter). According to Chance [16], this shortcoming can be readily circumvented by monitoring the development of the reaction in the direction of the flow. In this regard, it is worth mentioning the system reported by Stehl *et al.* [17], a mixing chamber coupled to a reaction cell. As can be seen in Fig. 7.5, the flow is generated by two mechanically-actuated 5-ml hypodermic syringes. Once mixing in the chamber is complete, the flow (containing virtually no dead volumes) is passed on to a reaction cell (4 mm diameter, 10 cm length). Absorbance measurements are made as the light beam from a fast-recording spectrophotometric detector crosses the cell lengthwise.

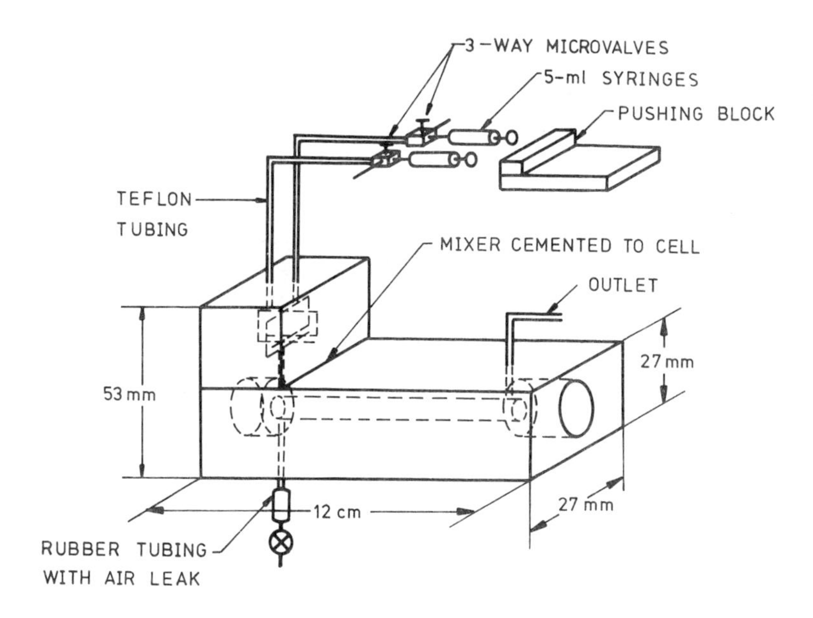

Fig. 7.5 — Mixing system and cell assembly. (Reprinted with permission, from R. H. Stehl, D. W. Margerum and J. J. Latterell, *Anal. Chem.*, 1967, **39**, 1346. Copyright 1967, American Chemical Society).

7.5.3 Pulsed-flow technique

According to Chance, this technique can be considered as a special accelerated-flow technique involving an extremely short circulation time (flow pulse). The principle of this technique is illustrated in Fig. 7.6. Two essential differences from the parent accelerated-flow technique are immediately apparent, namely that measurements are made at a constant flow velocity and throughout the observation tube, so the progress of the reaction can be monitored from the very beginning.

The pulsed-flow technique, described by Gerischer and Heim [17a] and called CFMIO (*c*ontinuous *f*low *m*ethods with *i*ntegrating *o*bservation) is characterized (Fig. 7.6) by the use of three syringes, two for sample and reagent dispensing and the third for halting the flow. This assembly allows the production of very short pulses and hence reduces reactant consumption dramatically ($\approx$ 4 ml). The maximum flow velocity afforded ranges between 2 and 9 m/sec. Its chief innovation is the so-called

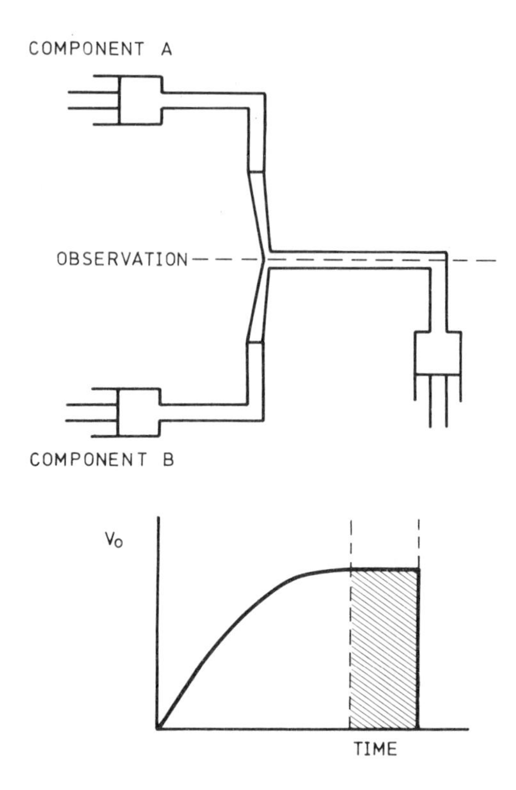

Fig. 7.6 — Foundation of the pulsed-flow technique.

'integrating observation', which involves placing the whole length of the observation tube (including the mixing zone) in the optical path of the sensing system. This results in considerably decreased dead time and allows the use of reactants with low molar absorptivities, thanks to the long optical path.

The most serious limitation of this technique is no doubt the efficiency of the mixing process, which must be completed in a very short length of the observation tube, and this is not feasible merely by use of two conduits leading the reaction components to the mouth of the observation tube, where they merge. Margerum and co-workers [18,19] have tackled this problem by developing a new pulsed-flow spectrophotometer especially suited to fast reactions (Fig. 7.7), incorporating a radially arranged mixing and observation cell (Fig. 7.8). Seven channels are available for each of the two reactants, which are discharged alternately into the mouth of the observation tube. The reactant solutions flow through two circular conduits connected to the seven radial channels. This results in turbulent, highly efficient mixing of the reactants. As the two syringes are actuated, the reactants flow through the channels until they converge in the observation tube. This assembly affords pulses less than 1 sec in length for flow velocities greater than 9 m/sec, and low sample and reagent consumption (3–4 ml).

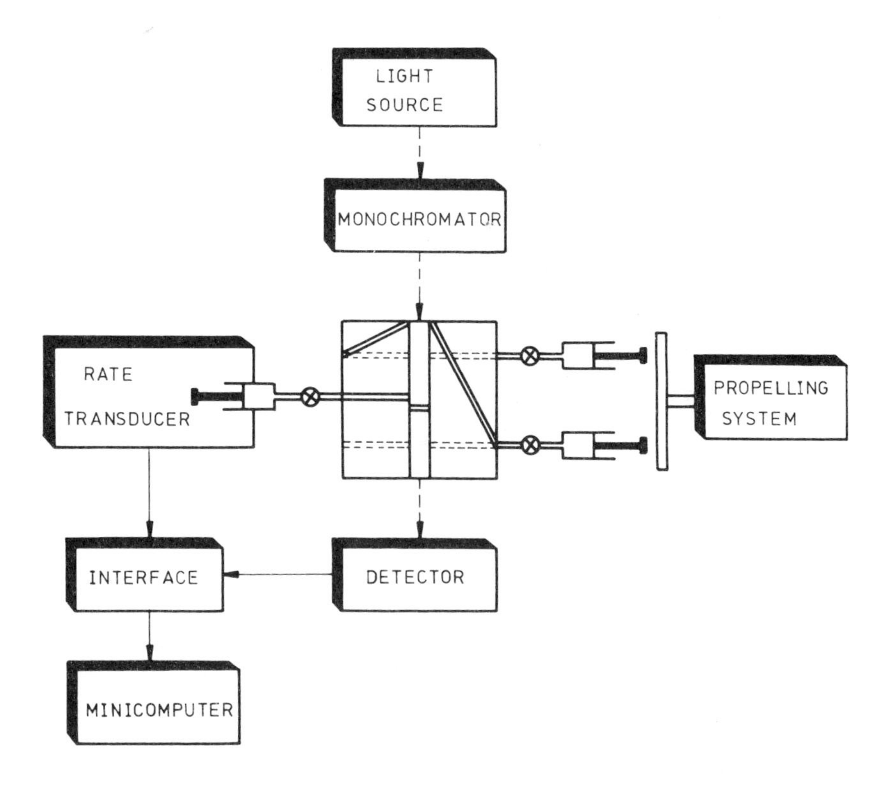

Fig. 7.7 — Block diagram of pulsed-flow spectrometer.

The determination of the rate constant of an irreversible first-order reaction with this type of instrumentation is based on the expressions:

$$(A - A_\infty)/(A_0 - A_\infty) = (1 - e^{-x})/x$$

$$k = xv_0/l$$

where A_∞ is the equilibrium absorbance of the mixture, A_0 the initial absorbance of the reactants, A the measured absorbance (a function of the flow velocity, v_0), x the fraction of monitored species that has reacted, l the optical path length (length of the observation tube) and k the rate constant.

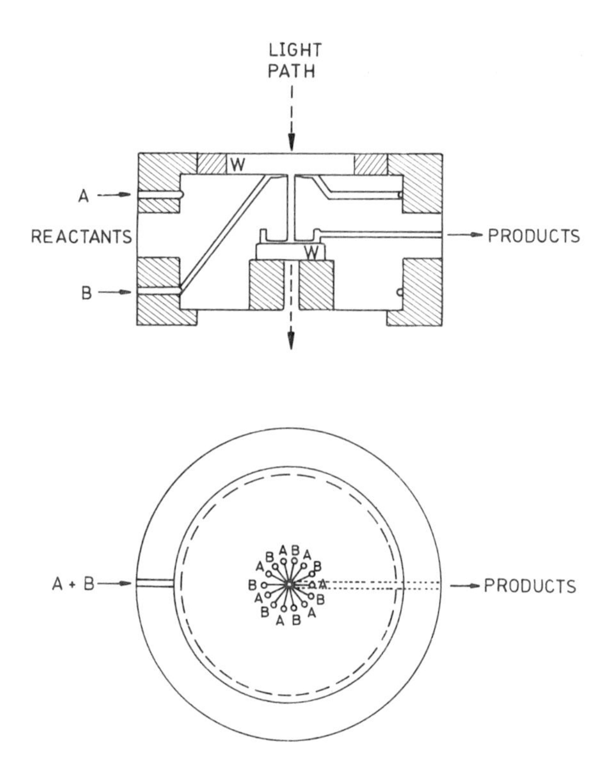

Fig. 7.8 — Top and end-on view of mixing chamber, cell A (14 jet, radial mixer); W, windows; A and B, reactants. (Reproduced with permission from G. D. Owens, R. W. Taylor, T. Y. Ridley and D. W. Margerum, *Anal. Chem.*, 1980, **52**, 130. Copyright 1980, American Chemical Society).

These equations were derived by assuming the mixing process to be faster than the chemical reaction of interest. The mathematical treatment involved is rather complex, particularly for second-order kinetics, and usually requires the aid of a computer. In addition, A_0 and A_∞ must be known beforehand for each concentration and set of experimental conditions under which the experiments are conducted.

One of the requisites of this pulsed-flow instrumentation is a constant velocity at each pulse, as the kinetic determination calls for a large number of data and hence for the application of numerous pulses. This, in turn, is time-consuming and results in increased reactant consumption. More recently, Jacobs *et al.* [20] have implemented the accelerated-flow technique on a pulsed-flow instrument, thereby avoiding the problems described above and allowing the monitoring of very fast reactions in solution. The design of the mixing chamber determines the feasibility of the determination — the model proposed by these authors is similar to that described for the pulsed-flow technique. This instrumentation allows the determination of pseudo first-order rate constants of the order of $1.2 \times 10^4 \text{ sec}^{-1}$ (half-life 60 μsec) by use of relatively low volumes (6 ml) of reactants, for electron-transfer systems such as $Ru(bipyridyl)_3^{3+}/Fe(II)$, $Ce(IV)/Fe(CN)_6^{4-}$, $IrCl_6^{2-}/Fe(II)$ or $IrCl_6^{2-}/Fe(CN)_6^{4-}$.

7.5.4 Stopped-flow technique

The stopped-flow technique [21,22] is by far one of the most popular and frequently used of flow techniques in the study and application of fast reactions with half-lives between some msec and a few seconds. Several European and American firms offer specific instruments, or accessories to be fitted to commercially available spectrophotometers and spectrofluorimeters.

As shown in Fig. 7.9, this technique uses the same components as the pulsed-flow technique, although the time during which the fluid is circulating is somewhat longer. It is thus based on acceleration of the flow by means of two syringes actuated manually or automatically (by a pneumatic device) and its subsequent stoppage. Unlike the pulsed-flow technique, measurements are made when the fluid is stationary ($v_0 = 0$) as can be seen from the plot of the flow velocity as a function of time, shown in Fig. 7.9. The reaction time at the observation point will be given by the ratio d/v_{max}, where d is the distance between the mixing and observation points and v_{max} is the maximum velocity reached prior to stoppage of the flow. In contrast to the continuous-flow technique, a single experiment is enough for the whole kinetic curve to be obtained.

The light path can be perpendicular to the observation tube as with the continuous or accelerated flow techniques, or along its axis as in the pulsed-flow technique.

An important factor to be considered in this technique is the dead time, the length of which is determined by the mixing efficiency (mixing time), the transport time and the stop time. As with other techniques, the efficiency of the mixing process depends on the configuration of the mixing chamber and on the flow velocity. The transport time is that taken for the reaction mixture to travel from the mixing point to the observation area and logically depends on the flow velocity and the distance between the two locations. It can be considerably shortened by building the mixing chamber and the observation cell in a single block. The stop time, or delay between the instant that the syringe plunger meets its retainer and the effective halting of the flow (formation of shock waves) must be much shorter than the mixing and transport times if too long an overall dead time is to be avoided. During the dead time, the flow velocity increases from zero at the moment of impulsion to a maximum value immediately prior to the stopping of the flow, after which the analytical signal is recorded as a function of time.

The dead time can be determined experimentally by a well-known procedure based on the equation [23]:

$$t_m = V/f$$

where V is the volume of fluid held between the mixing point and the centre of the observation cell and f is the mixture flow-rate in ml/sec. When the mixing chamber

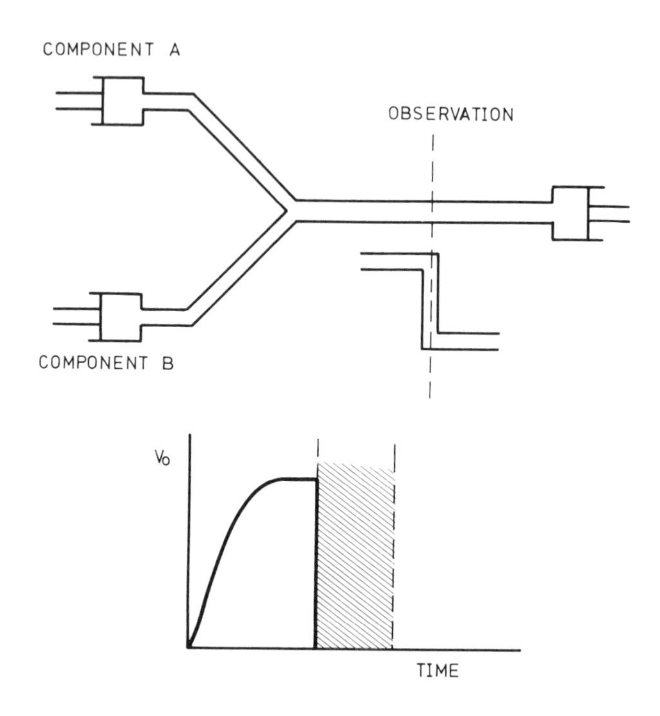

Fig. 7.9 — Foundation of the stopped-flow technique.

and the observation cell form one unit, V is taken as half the volume of the shared chamber and hence the determination of the dead time requires only the measurement of the flow-rate of the mixture from the two propelling syringes. Holler *et al.* [24] have devised an optical electronic system to be fitted to the propelling block to control its output and hence the flow-rate of the reactants dispensed.

Figure 7.10 is a schematic diagram of a stopped-flow spectrophotometer (or spectrofluorimeter) coupled to an automatic analyser controlled by a computer which is also in charge of the data collection and processing operations.

The mechanical system aspirates (loads) and discharges (unloads) the two liquids into the mixing chamber with the aid of the two three-way valves used, which are actuated by the computer. An automatic sampler places a fresh sample in position after each injection. The flow-cell lies in the light-path of the sensing system (spectrophotometer or spectrofluorimeter). The stop syringe receives the mixed solution and a retainer regulated by a micrometer stops the plunger run. As soon as the plunger meets the retainer, the electrical signal generated triggers data collection by the computer. A drain valve sets the instrument ready for a new injection. The retainer should always be fitted with an electrical switch to trigger the signal recording, which cannot be done by ordinary recorders on account of their sluggish response.

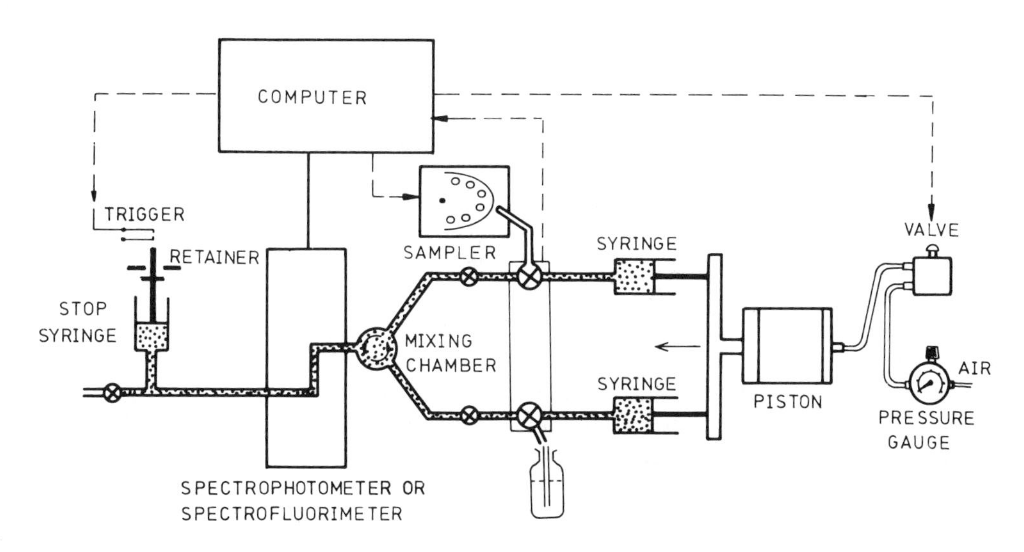

Fig. 7.10 — Scheme of a stopped-flow spectrophotometer or spectrofluorimeter.

Various authors have devised their own stopped-flow spectrophotometers. As a rule, such instruments possess three essential features: (a) they incorporate new developments in the flow system; (b) they use their own optical sensing systems and (c) they are normally automated by use of a computer. Thus, Beckwith and Crouch [25] have developed an automated stopped-flow spectrophotometer featuring solenoid-actuated pneumatic valves for directing the liquid flow and removing waste solutions, and a pneumatic syringe drive system for rapid mixing. The flow system used is vertical to minimize interferences due to bubble formation and can be readily interfaced with a light-source, monochromator and photomultiplier tube for photometric observation. It allows automatic measurement of the initial rate by means of a fixed-time digital read-out system [26]. Recently, Malmstadt and co-workers [27,28] developed a microcomputer-controlled stopped-flow analyser (SFA) sufficiently versatile to be used with a variety of analytical procedures. The most outstanding features of this instrument are the absence of a stop syringe (replaced by a check valve) and the incorporation of a storage cell between the mixer and the observation coil, to make the instrument applicable for slow reactions. The assembly is especially suited to the automatic determination of albumin in serum [29].

As stated above, the stopped-flow technique is one of the more frequently used in the study and application of fast reactions in analytical chemistry. Hence the continual appearance in analytical chemical publications of research papers exploring its vast potential or introducing further improvements in instrumental design. This technique has experienced remarkable developments since the first high-performance stopped-flow apparatus was introduced by Gibson and Milnes [30] in 1964, namely: (a) automation of propulsion and manipulation of sample and reagents, (b) introduction of more efficient mixing chambers and temperature-control units; (c) incorporation of systems for data collection and treatment; (d) design of mixing systems for use with different conventional detectors.

The stopped-flow technique makes use of a variety of means to operate the syringes, which can be driven manually or with the aid of pneumatic (the commonest), hydraulic or electromagnetic devices [31–33], occasionally under the control of a computer. Among the different pneumatic systems used it is worth mentioning the one developed by Morelli [34], which has popularized the use of this technique for teaching purposes. The system is based on the expansion of a previously compressed spring. On release of its lock, the spring impinges on a piston which in turn acts on the syringe plunger. The procedure, though still manual to some extent (squeezing of the spring), provides excellent results.

The automation of the manipulation of sample and reagents prior to their mixing is of great relevance to the stopped-flow technique. In an interesting paper, Sanderson *et al.* [35] describe a computer-controlled system requiring a minimum of human intervention in the preparation of sample and reagent solutions. Figure 7.11 depicts the scheme of the sampling system. It consists of two lines connected to two syringes (1 and 2). The line to the left carries a standard solution of the reagent, which is introduced into it by syringe 2 through valve 4, while the line to the right carries the sample, which is introduced into the coil through valve 3 as a result of evacuation of valve 2. Valve 1, also in the line to the right, introduces another reagent into the system. A 90°-turn of the valves propels the contents of both syringes as far as the point of merging in the mixing chamber. Possible cross-contamination between successively injected samples is avoided by filling the loop twice before monitoring the fresh sample — this operation and those mentioned above are controlled by the computer. This highly automated system can be further improved by automating the sample and/or reagent preparation with the aid of a peristaltic pump and a sampler [36]. A straightforward alternative to this reagent preparation system, developed by Stieg and Nieman [37], involves diluting the reagent completely automatically with the aid of a three-way valve and a small stirred mixing chamber.

The design of mixing chambers has also received much attention from researchers on the stopped-flow technique. The most recent trends are towards the design of units containing both the mixing chamber and the observation area, in order to reduce dead times to a minimum [38]. Such is the case with multi-mixing systems, made up of mixing chambers into which eight channels or more may converge. Figure 7.12 shows the scheme of a three-line multi-mixing system consisting of two mixing chambers. Reactants 1 and 2 are mixed rapidly in the first chamber and the resultant solution is subsequently merged with reactant 3 in a second chamber prior to entering the observation zone.

A detailed description of all the mixing chambers reported is beyond the scope of this monograph. The chamber depicted in Fig. 7.13, which provides an optimum turbulent flow regime, is a representative commercial example. Finally, it is interesting to mention a novel mixing/observation chamber devised by Thompson and Crouch [39] to monitor enzymatic reactions. The chief innovation of the system is the immobilization of the enzyme within the chamber (Fig. 7.14), so that the flow system requires a single line.

Temperature control is one of the major sources of error in stopped-flow systems. The problem, naturally, lies in the occurrence of a temperature gradient between the different components of the flow system. This shortcoming has been circumvented

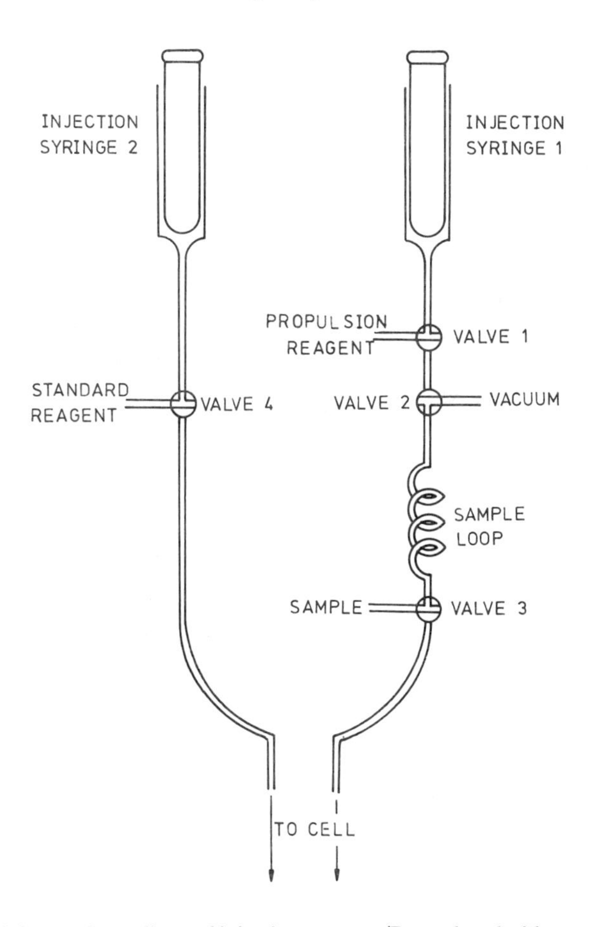

Fig. 7.11 — Scheme of sampling and injection systems. (Reproduced with permission, from D. Sanderson, J. A. Bittikofer and H. L. Pardue, *Anal. Chem.*, 1972, **44**, 1934. Copyright 1972, the American Chemical Society).

by some firms by using manifolds made of stainless steel to give faster heat transfer, but this limits the range of chemicals that can be used. Hence, research in this field has been aimed at developing useful thermostatic systems which generally involve immersing the flow system and the mixing chamber in a constant-temperature bath [40]. The liquid used to fill the bath will vary according to the working temperature range. Thus, Hanahan and Auld [41] have developed a stopped-flow spectrophotometer operating in the range from − 55 to 55°C with a precision of 0.1°C. As can be seen from Fig. 7.15, the temperature is regulated by use of two nitrogen streams, one cooled by liquid nitrogen at − 170°C and the other heated by an oil-bath at 100°C. The optical fibre system fitted to the observation chamber simplifies its design considerably and enables the user to make both photometric and fluorimetric measurements. This instrument is of especial use for ultrafast enzymatic reactions, which are slowed down (and hence suited to the characteristics of the stopped-flow technique) by working at temperatures below 0°C.

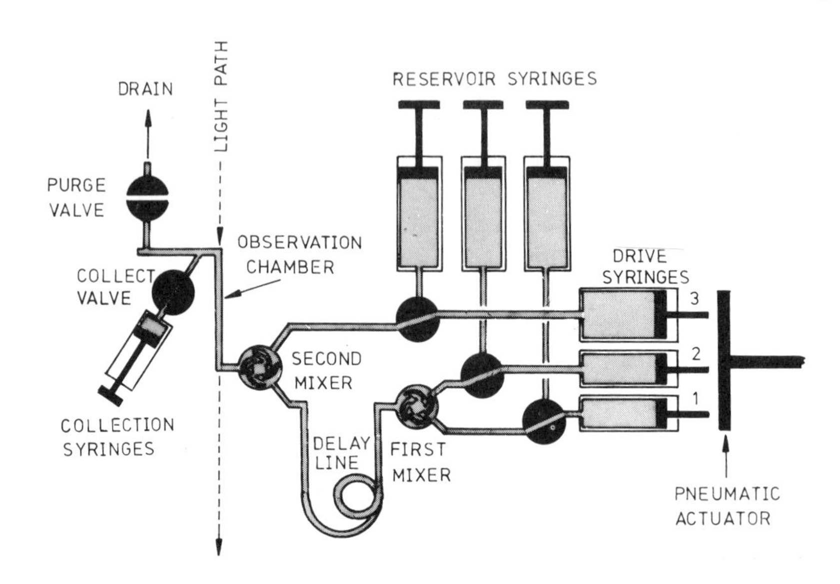

Fig. 7.12 — Multi-mixing system. (By courtesy of Dionex).

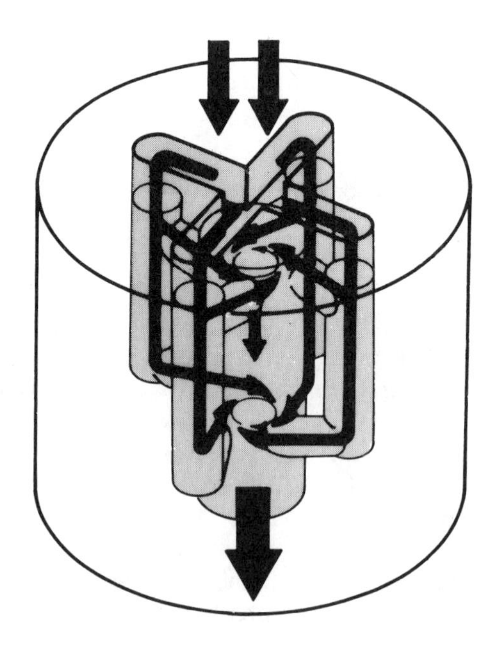

Fig. 7.13 — Flow diagram of mixing jet. (By courtesy of Dionex).

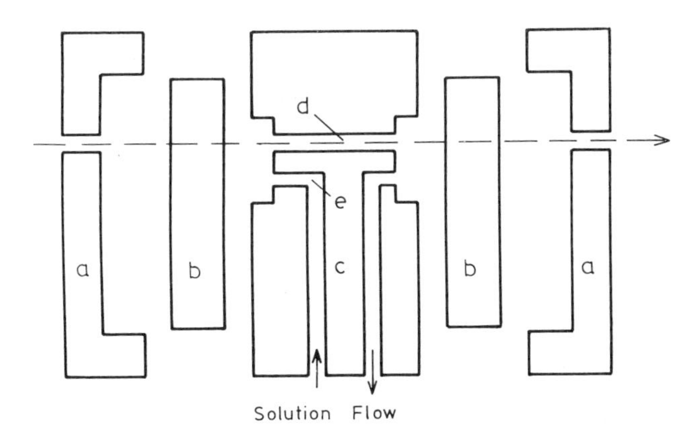

Fig. 7.14 — Cross-sectional view of the GCA McPherson stopped-flow mixer/observation cell unit. The outer aluminium housing (a), quartz windows (b) and the main Kel-F body (c) are press-fitted with three bolts. Mixing occurs at (e), where the two reagent streams meet at 90° to each other; one stream is in the plane of the figure and the other perpendicular to it. The immobilized enzyme reactor is placed inside the observation cell (d). With the reactor in place, the observation cell is 1.75 cm in length and 0.1 cm in diameter. The dashed arrow represents the light-path inside the cell. (Reproduced by permission, from R. Q. Thompson and S. R. Crouch, *Anal. Chim. Acta*, 1982, **144**, 155. Copyright 1982, Elsevier Science Publishers).

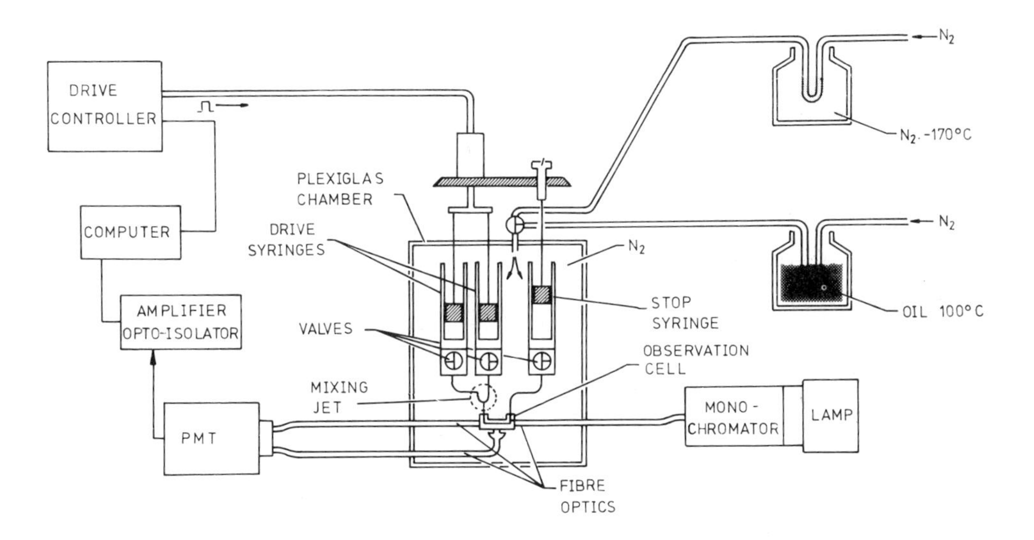

Fig. 7.15 — Schematic diagram of low-temperature stopped-flow apparatus. (Reproduced from D. Hanahan and D. S. Auld, *Anal. Biochem.*, 1980, **108**, 86, by permission. Copyright 1980, Academic Press Inc.).

Because of its nature, the stopped-flow technique has benefited from the growing use of computers for data collection and processing. Computers have superseded the oscilloscopes originally used for recording the kinetic curve. In addition, they are responsible for the whole process of data treatment [42–46] and have facilitated the complete automation of the technique [35,36,47]. A block diagram illustrating the essential parts of a completely automatic stopped-flow assembly is displayed in Fig. 7.16. As can be seen, a microcomputer controls both the reactant preparation unit and the stopped-flow system itself. The signal generated in the system is transmitted through an interface to a minicomputer which treats and displays it as required. In addition, the minicomputer controls the functioning of a microcomputer.

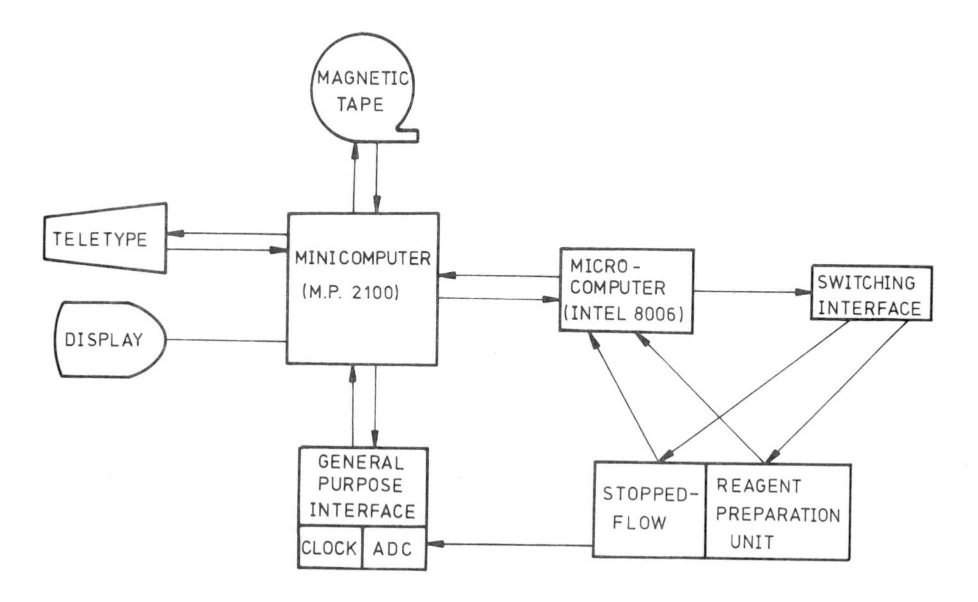

Fig. 7.16 — Block diagram of computer-controlled stopped-flow system. (Reprinted with permission, from G. E. Mieling, R. W. Taylor, L. G. Hargis, J. English and H. L. Pardue, *Anal. Chem.*, 1976, **48**, 1686. Copyright 1976, American Chemical Society).

In connection with the design of mixing modules to be fitted to commercially available spectrophotometers or spectrofluorimeters it is worth mentioning the modular stopped-flow unit for use with commercial spectrofluorimeters proposed by Bartels *et al.* [48]. The system has a dead time of about 10 msec and can be readily adapted to the fluorimeter, and is thus made much more flexible. Other salient features of this set-up are the presence of a temperature-control unit and the absence of the conventional stop-syringe. Hi-Tech Scientific Ltd. [49] have developed a horizontal mixing chamber for use with spectrophotometers and spectrofluorimeters. A novel design featuring improved characteristics was reported recently [50]. It is coupled vertically to the photometric or fluorimetric detector, thereby avoiding bubble formation in the flow system. It used a water-recirculation thermostat system and has a dead time shorter than 15 msec.

Finally, it is also worth commenting on the association of the stopped-flow technique with direct relaxation methods, especially temperature-pulse methods [51] (see Section 7.5.7), in fast ligand-exchange reactions. Their applications were described at length in the preceding chapter. Although the use of this combination is limited to the study of reversible reactions because of the nature of the temperature-pulse methods, it is of great interest inasmuch as it allows the reduction of dead times to a few μsec. In practice, the chemical system is allowed to progress until the flow is halted (pseudo steady-state), whereupon the subsequent relaxation process is monitored.

7.5.5 Stat methods

As stated in Section 7.2 (Table 7.1), slow reactions can be studied by two types of batch method, namely stat methods and steady-state methods.

Stat methods [52] involve the addition of a reagent to the reaction vessel at a given speed in such a way that a characteristic signal of the reaction to be monitored is kept constant. The essential difference from other methods lies in the fact that the kinetic data are not obtained by measuring the variation over time of a property reflecting the concentration of one of the components of the reaction, but rather by controlling the speed of addition of a reagent so as to maintain the system in a given fixed state. This method has been chiefly applied to the study of catalysed reactions for quantitative purposes.

In practice, the method involves addition of a small amount of one of the reaction components until a given value of the monitoring parameter (pH, absorbance, luminescence, etc.) is reached. Any deviation from this state as a result of development of the reaction is immediately compensated for by the automatic addition of this component.

According to the reaction

$$A + B \xrightarrow{C} P$$

if $[B]_0 \gg [A]_0$, then

$$-d[A]/dt = k_1[A][C]_0 + k'$$

If A is the component added to keep the signal constant and it is added at the same rate as it is consumed, then the reaction is pseudo zero-order with respect to A and hence the equation above can be written as:

$$-d[A]/dt = k_1'[C]_0 + k'$$

from which it follows that the speed of signal restoration, $d[A]/dt$, which should be the inverse of the reaction rate, is proportional to the catalyst concentration. Therefore, according to this criterion, these could be categorized as differential (initial-rate) methods.

As stated above, any measurable property proportional to the concentration of one of the reaction components can be used to regulate the speed of reagent addition. Typically, the protons consumed or released in a protolytic reaction are continuously replaced or neutralized by the automatic addition of a standardized acid or base solution, as the case may be. The property by which the reaction is monitored, pH, is continuously measured with the aid of a combined glass–calomel electrode. A block diagram of the instrumental set-up used to monitor the reaction is depicted in Fig. 7.17. A central unit automatically measures the variation of pH as a function of time and traces the kinetic curve as a plot of the speed of titrant addition against time. The unit also actuates the microburette, which is commanded to dispense the appropriate titrant volume. This rather inexpensive system, which allows the pH to be measured with a precision of ± 0.002 [53], has provided outstanding results in enzymatic analysis. The use of a coloured acid–base indicator and the photometric monitoring of the absorbance is a good (though less often used) alternative to the measurement of the pH with the combined glass–calomel electrode [54].

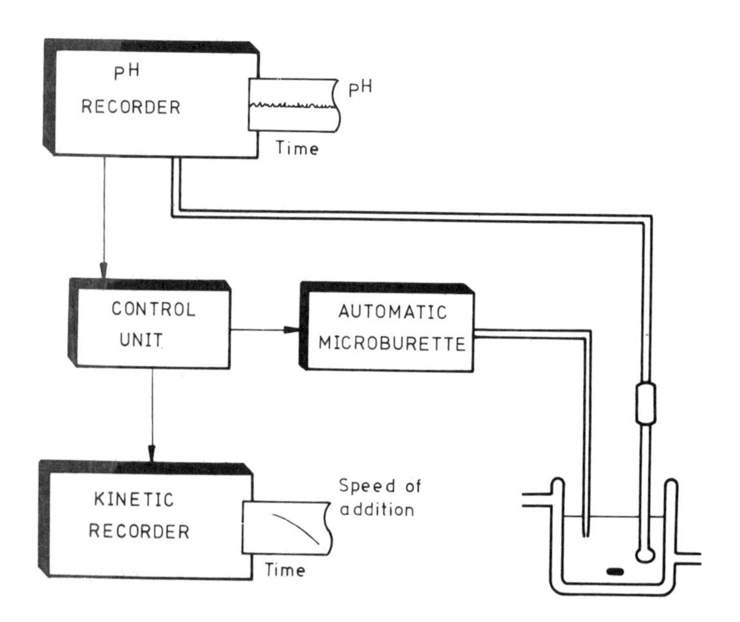

Fig. 7.17 — Block diagram of the typical instrumentation used in pH-stat methods.

This system requires the user to introduce corrections in the slope of the kinetic curve to compensate for the dilution resulting from the increase in solution volume as the stat-titrant is added. This shortcoming can be circumvented by using more concentrated acid or base solutions, though this may result in an uneven speed of titrant addition or in local concentration gradients, which are particularly undesirable in enzymatic reactions. On the other hand, very slow reactions call for use of very dilute solutions, the handling of which also poses serious problems. All these drawbacks can be overcome, at least partly, by using a coulometric titrant (acid or base) generator [55] consisting of a grounded Pt electrode, a Pt generating electrode and an auxiliary Ag/AgCl electrode, plus the usual glass–calomel electrode.

The so-called absorptiostat methods, used by Weisz at the University of Freiburg [56], are based on the monitoring of the absorbance of a coloured reagent. The instrumentation used is schematically depicted in Fig. 7.18. The signal from the light-source (L) traverses the sample and falls on a phototube (P) which amplifies it to a suitable degree. This transduced signal (mV) is measured and compared by the analogue controller with a preselected potential (corresponding to a given absorbance), which determines the reactant concentration to be kept constant. The difference between the preselected potential and each potential measured in the course of the reaction gives rise to a signal controlling the delivery of the coloured reagent from the burette.

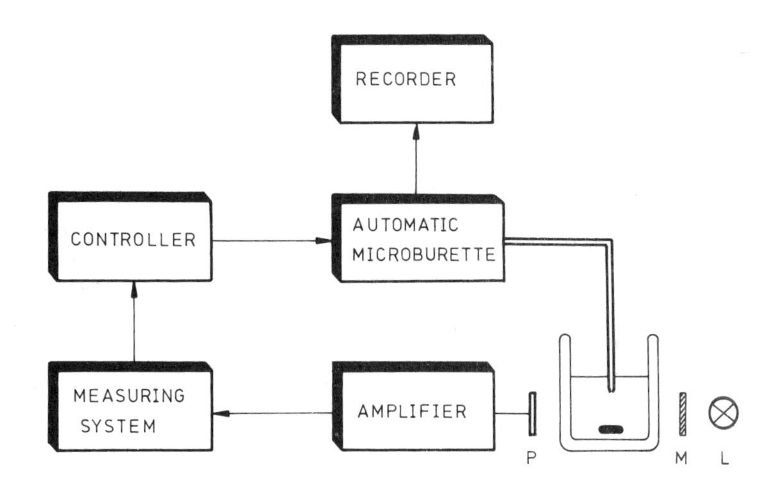

Fig. 7.18 — Block diagram of the instrumentation used in absorptiostat methods.

This system has been successfully applied by Weisz and Rothmaier [56] to the kinetic catalytic determination of iodide, manganese and molybdenum. Thus, iodide is determined through its catalytic effect on the oxidation of As(III) by Ce(IV) (Sandell–Kolthoff reaction). The reaction vessel contains the catalyst and excess of As(III), which is titrated with the Ce(IV) (the indicator), added until a given absorbance is reached, after which the change in signal resulting from the development of the catalysed reaction is compensated for by addition of an appropriate volume of Ce(IV).

Weisz and Pantel have developed a novel system for the biamperometric monitoring of reactions [57,58]. The instrumentation used is similar to that employed in the constant-pH methods, except that the signal (mV) originated in the potentiostat is multiplied by an operational amplifier. This technique has provided excellent results in the determination of copper on the basis of its catalytic effect on the decomposition of hydrogen peroxide, as well as in the study of various enzymatic reactions.

Finally, it is worth noting that Weisz and Pantel have also developed constant-luminescence methods [59] with the aid of an instrumental assembly similar to that used in constant-absorbance methods.

7.5.6 Steady-state methods

Unlike the stat methods, these involve keeping the speed of addition of the species A constant throughout the process, thereby giving rise to a steady state in which the amount of A added is equal to that consumed in the reaction over a given interval of time. Hence the concentration of A in the system remains constant throughout.

Since $d[A]/dt$ is a constant in the kinetic equation corresponding to the reaction,

$$-d[A]/dt = k_1[A][C]_0 + k'$$

then

$$1/[A] = k_1'[C]_0 + k''$$

where $k_1' = k_1/(-d[A]/dt)$ and $k'' = k'/(-d[A]/dt)$. The concentration of A in the equation above can be substituted by $[A]_0 - [P]$ if it is the reaction product rather than a reactant which is monitored. As [P] cannot be neglected with respect to $[A]_0$, steady-state methods can be classed as integral kinetic methods. The plot of the steady-state concentration of A as a function of the catalyst concentration constitutes the calibration curve from which the catalyst is determined.

Because of the small initial concentration of A, the reaction rate is usually very small at the beginning and increases as A is added, as does the consumption of A in the catalysed reaction, until a steady state is attained, after which its concentration in the bulk solution remains constant. Weisz and Ludwig [60] have applied this method to the determination of iodide, based on its catalytic effect on the As(III)/Ce(IV) system. The reaction vessel contains the iodide and As(III) in 0.5*M* sulphuric acid medium, which is titrated by addition of Ce(IV) at a constant speed, the redox reaction being monitored photometrically at 420 nm. Osmium(VIII), a catalyst, and mercury and silver, inhibitors for this reaction, can also be determined by this method.

7.5.7 Direct relaxation methods

Relaxation, or rather, chemical relaxation, methods are an effective tool in the study of fast reactions. They are based on the generic principle that a rapid action on a system in equilibrium alters this significantly. The process of restoring the disturbed equilibrium or establishing a new one (relaxation) is monitored by high-speed recording techniques [61–63].

These methods can be classified as direct or periodic according to the type of perturbation applied. As the former are better suited to the study of reactions in solution, they are of greater relevance to this area of analytical chemistry.

In direct relaxation methods, the system undergoes only a slight shift in its equilibrium and the speed at which the equilibrium is restored or a new one is established is directly proportional to the magnitude of the shift. This concept can be illustrated mathematically by the expression:

$$-\,\mathrm{d}\Delta x/\mathrm{d}t = C\Delta x$$

where Δx represents the equilibrium shift and C is a proportionality constant. Integration of this equation yields:

$$\Delta x = \Delta x_0 \exp(-Ct) = \Delta x_0 \exp(-t/\tau)$$

where Δx_0 is the Δx value corresponding to $t = 0$ and $\tau = 1/C$. This last parameter is known as the 'relaxation time' and is directly proportional to the reciprocal of the overall first-order rate constants of the process. Thus, the relaxation time for a simple reaction such as

$$\mathrm{A} \underset{k_{-1}}{\overset{k_1}{\rightleftharpoons}} \mathrm{B}$$

will be given by $\tau = 1/(k_1 + k_{-1})$.

Multi-stage reactions have as many relaxation times as stages, all of which make up the so-called 'relaxation spectrum', which is resolved by relaxation spectroscopy.

A relaxation process can be induced by altering one of three different properties, namely temperature, pressure and electric field intensity. Each is the origin of a different analytical technique.

7.5.7.1 Temperature-jump method

This is probably the technique most frequently used in the context of relaxation methods. It involves altering the equilibrium of a given chemical system by applying a rapid temperature pulse, the effect on the system being monitored through another physical property of the system. The success of this method lies in the fact that most chemical reactions are either endothermic or exothermic, i.e. they involve changes in enthalpy, so their equilibria can be altered by heating, according to the van't Hoff isobar

$$\left(\frac{\partial \ln K}{\partial T}\right)_{\mathrm{P}} = \Delta H/RT^2$$

which shows the temperature-dependence of the equilibrium constant.

An abrupt increase in the solution temperature can be readily produced by passing an electric current through it for a short interval (Joule heating) or by flash-irradiation with a powerful light-source. The former method is somewhat commoner than the latter and is implemented by discharging the electric energy accumulated in a capacitor through the solution, which should contain an inert electrolyte ($\mu =$

0.1*M*). The electrolyte ions act merely as current carriers and are not involved in the basic chemical reaction. The higher the electrolyte concentration, the lower the cell resistance and the faster the energy transfer. This method allows the heating of a solution from 2 to 10°C in 1–20 μsec. The heating time can be considerably reduced (to 50 μsec) by replacing the capacitor by a coaxial line [64].

Heating by flash irradiation requires the efficient absorption of light by the solvent molecules or those of another substance, known as a 'photochemically inert absorber'. The use of flash-lamps allows the application of light pulses as short as a few μsec [65] or even shorter if microwave or laser sources are used [66].

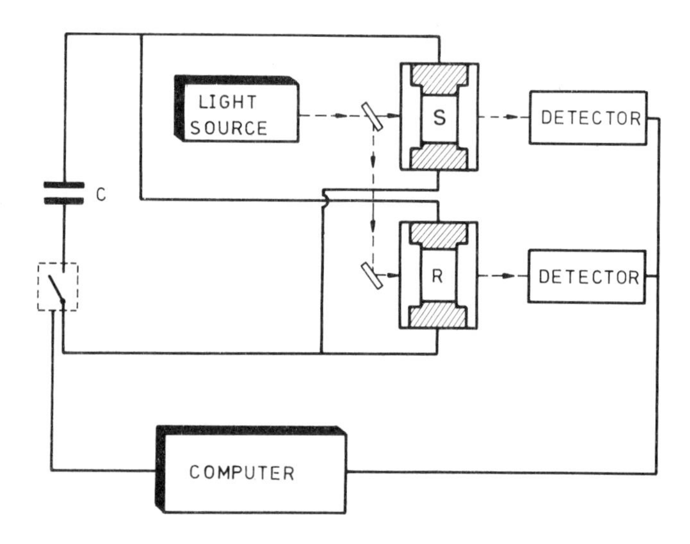

Fig. 7.19 — Block diagram of temperature-jump instrument.

The relaxation process is usually monitored by conductimetry (most of the reactions dealt with are ionic) or spectrophotometry. Figure 7.19 shows the block diagram of a typical temperature-jump instrument with photometric detection. It uses a capacitor, C, as heating source, both for the sample cell (S) and the reference cell (R). A monochromatic light-beam traverses both cells, the readings being collected by two detectors and the differential signal being handled by a computer. It is vital that the heating be uniform, as temperature gradients may give rise to distortions in the optical path as a result of changes in the refractive index.

These temperature-jump methods have been widely used in the study of one of the essential chemical reactions, namely the neutralization of protons by hydroxide ions, and also of complex-formation and enzymatic reactions.

7.5.7.2 Pressure-jump method

This method is based on the equilibrium shift induced in a reversible reaction by a pressure-dependent change in volume, ΔV. At a constant temperature, the relationship between the equilibrium constant and the pressure can be expressed by means of:

$$\left(\frac{\partial \ln K}{\partial P}\right)_T = -\Delta V/RT$$

Marked pressure changes can be induced as follows. The reaction cell is placed in an autoclave at a pressure of 50 atm (Fig. 7.20). Breaking the thin metal diaphragm results either in an expansion or in the formation of a shock wave [67,68]. This system allows the introduction of pressure changes of the order of 1 atm in about 50–100 μsec.

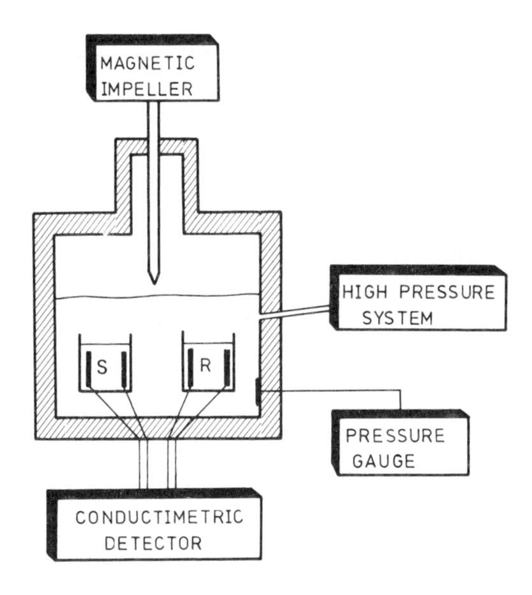

Fig. 7.20 — Block diagram of pressure-jump instrument.

The relaxation process is generally monitored through the changes in the sample conductivity, but other analytical detection techniques are equally suitable. This pressure-jump method has a more limited use than its temperature counterpart, on account of the smaller effect of pressure on chemical equilibrium (e.g. a change of 10°C in the temperature is equivalent to a pressure change of over 50 atm).

7.5.7.3 Electric field-jump method

This method is based on displacement of the equilibrium of a chemical reaction involving ions, dipoles or polarizable species, by application of an electric field, which can have three basic effects: it can (a) increase the ionic conductivity (first Wien effect), (b) distort the ionic atmosphere (second Wien effect) and (c) increase the degree of dissociation. If the first two effects are minimized, the change in the degree of dissociation of an electrolyte [69] on application of an electric field will be given by:

$$(\lambda_x - \lambda_0)/\lambda_0 = (1-\alpha)b\Delta E/(2-\alpha)$$

where λ_x is the electrical conductivity after application of the electric field, ΔE, λ_0 is that measured in the absence of the field, α denotes the degree of dissociation and b is a proportionality coefficient. Thus, the application of an electric field of 200 kV/cm increases the degree of dissociation of acetic acid by 12% [70].

The experimental set-up is similar to that employed for the temperature-jump method. Thus, it is again possible to use the discharge of a capacitor, though the cell resistance should be rather high in order to minimize the circulation of current (negligible in the case of Joule heating) and to ensure the uniform application of the electric field.

This method is generally suitable for the study of weak electrolytes and affords relaxation times of a few μsec, although considerable technical complications make times shorter than 0.1 msec inadvisable. This relaxation method has been primarily used to study proton-transfer processes and the hydrolysis of tervalent metal ions.

7.5.7.4 pH-jump method

This is a recently introduced method [71–73] based on the instantaneous pH change brought about by laser excitation of aromatic hydroxylated compounds such as 8-hydroxypyrene-1,3,6-trisulphonate and 2-naphthol-3,6-disulphonate, which have pK values that are considerably different in the ground and excited states.

7.5.8 Periodic relaxation methods

These are based on the equilibrium shift caused by an oscillating (periodic) perturbation [74]. Their operational principle is related to the relaxation time. Thus, if the perturbation is applied sufficiently slowly, the chemical equilibrium will 'follow' it throughout the cycle. On the other hand, if the perturbation oscillates very rapidly (i.e. if its period is shorter than the relaxation time), the chemical reaction will fail to accommodate to it before recovering its equilibrium state. The net effect is then that the equilibrium position is 'frozen'. Thus, the relative value of the relaxation time and the perturbation period is of utmost importance in periodic relaxation methods.

There are two basic types of periodic relaxation method: (a) ultrasonic relaxation and (b) dielectric relaxation methods.

Ultrasonic relaxation methods, the commoner, are based on the generation of ultrasonic waves by means of piezoelectric quartz crystals and the use of the pulse technique (radiofrequency oscillator) to study the propagation of the waves through the liquid. This technique affords very short relaxation times (of the order of 0.1 μsec) and has been preferentially applied to the study of proton-transfer, association and hydrogen-bond formation reactions.

Dielectric relaxation methods are based on the polarization of a dielectric (capacitor) by application of an alternating electric field. The change in the permittivity of the capacitor allows measurement of relaxation times as short as 0.01 μsec. One of the chief uses of this method is the study of the different rotation rates of the repeat units in a polymer [74].

7.5.9 Pulse methods

Pulse methods involve exposing the system to a short light pulse (flash photolysis) or electron pulse (pulse radiolysis), which starts the reaction of interest. Unlike relaxation methods, they are suited both to reversible and irreversible reactions.

7.5.9.1 Flash photolysis

This methodology relies on the photoexcitation resulting from a discharge through an inert gas (xenon, argon) [75]. The electric circuit used to generate the flash (Fig. 7.21) consists of a thyratron voltage generator and a high-voltage capacitor which provides an amount of energy given by $E = CV^2/2$, where C is its capacity (farads) and V its voltage. Thus, a capacitor of 5–10 μF charged at 10 kV will typically provide a discharge of 1–3 kJ in a time of approximately 5–10 μsec.

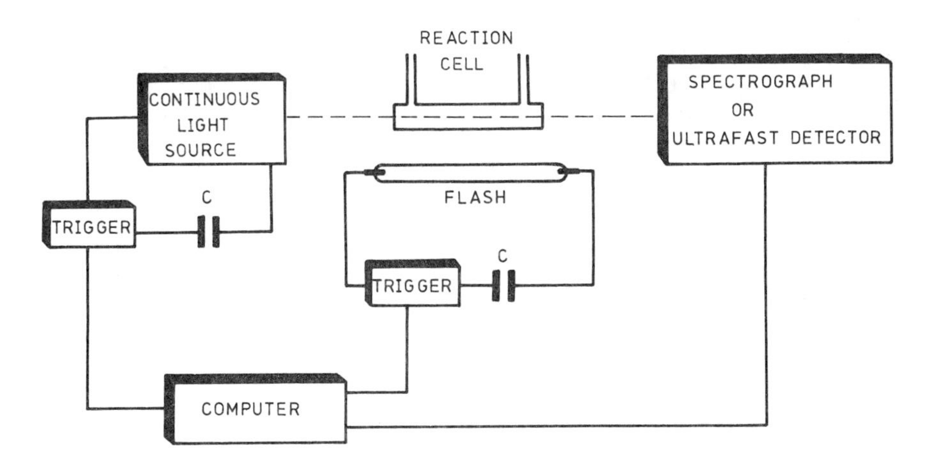

Fig. 7.21 — Block diagram of flash photolysis instrument.

Flash photolysis methods usually make use of absorption spectroscopy or mass spectrometry. Recording of the absorption spectrum with a spectrograph and/or ultrafast detector after each photoexcitation poses problems related to the signal-to-noise ratio if no baseline or continuous spectrum is available. The continuous spectrum should be recorded over a preset time interval once the photoexcitation has been applied. This entails using a second, continuous-spectrum lamp between the reaction cell and the sensing system (Fig. 7.21). The radiation from this lamp yields a reference signal (baseline) at the detector, which does not interfere with the photochemical signal.

This type of instrumentation allows the study of first-order and second-order reactions with rate constants up to 10^6 sec^{-1} and 10^{11} l.mole^{-1}.sec^{-1}, respectively.

One of the most recent advances in the development of the flash photolysis technique is the use of lasers as excitation sources [61]; these allow the generation of extremely intense light pulses (up to 10^{13} W/cm^2) of duration between 5 and 30 nsec in the ultraviolet and infrared regions of the spectrum.

7.5.9.2 Pulse radiolysis

As stated above, this technique involves irradiating the sample with short pulses of high-energy electrons. The experimental foundation of this method is essentially identical with that of flash photolysis, but there are some differences worth commenting on. Thus, the photons used by the flash photolysis technique are absorbed selectively by molecules with an appropriate spectral response and the conversion of a molecule into an excited state is associated with the absorption of a photon from the incident light-beam. The energy of an electron flux, which is much higher than that of a light-beam, is not absorbed selectively, so the net result is the formation of a variety of ions and excited species. These considerations are of importance in the development of instrumentation in this area.

The commonest source of the electron flux used is the linear accelerator, which yields electron pulses of 1–5 MeV energy and 0.5–5 μsec duration [76]. Other sources used for this purpose are the Van der Graaff [77] and the Febetrons accelerators [61]. The high cost of these sources is a major limitation to their application.

Although this methodology permits the detection of transient species by electron spin resonance or electric conductivity measurements, optical sensing techniques are much more commonly used.

Among the salient applications is the study of both inorganic and organic free radicals, molecular ions, excited aromatic molecules (in a triplet state), and, especially, the solvated electron. The passage of ionizing radiation through water leaves positive H_2O^+ ions and secondary electrons. These species react with water molecules to yield a variety of ions and neutral species:

$$e_{aq}^- + e_{aq}^- + 2H_2O \xrightarrow{k_1} H_2 + 2OH^-$$

$$e_{aq}^- + H_3O^+ \xrightarrow{k_2} H + H_2O$$

$$e_{aq}^- + H_2O \xrightarrow{k_3} H + OH^-$$

where $k_1 = 10^{10}$ l.mole^{-1}.sec^{-1}, $k_2 = 2.3 \times 10^{10}$ l.mole^{-1}.sec^{-1} and $k_3 = 16$ l.mole^{-1}.sec^{-1}. The solvated electron shows a well-defined absorption spectrum ($\varepsilon_{720} = 1.5 \times 10^4$ l.mole^{-1}.cm^{-1}), which allows its reactions to be monitored by fast photometric techniques. The hydrated electron is a strong reductant, with rate constants as great as 10^{10} l.mole^{-1}.sec^{-1} for its reactions with organic substances.

7.6 DETECTION SYSTEMS

Although the course of a reaction can be monitored by chemical (titration) or even visual techniques, kinetic methods rely much more frequently on instrumental techniques, the only possibility available when fast reactions are dealt with. In this sense, any instrument capable of transducing the varying chemical information into electrical signals (ultimately converted into analytical data) will be adequate for the monitoring of the kinetics of a reaction. Instrumental techniques permit the continuous measurement of a physicochemical parameter proportional to the concentration of one of the reaction components related to the analyte.

The instrumentation employed in implementing kinetic methods should be designed bearing in mind the problems associated with measurements in a dynamic system, i.e. the measurements should be made relative to one made at a selected reference time. This calls for data processors different from those typically used for equilibrium measurements.

Although the high technological degree of current instrumentation makes it quite suitable for kinetic measurements, the incorporation of microcomputers has further advanced the development of kinetic methods. However, there is very little a computer can do to improve poor input, and the reliability of the data obtained depends much more on the signal sensing and transducing systems.

Another important factor in the development of kinetic methods is knowledge of the mechanism, as this determines the most suitable experimental working conditions. High-speed spectral recorders such as vidicon and diode array detectors have assisted the development of a number of automated kinetic methods.

In selecting a particular instrument for a given purpose, the user should take into account the nature of the analyte, the most suitable species for measuring its concentration, the type and half-life of the reaction involved and the length of the measurement interval. Other aspects such as the cost or availability of the reagents should also be paid due attention.

Spectroscopic (photometric) and electroanalytical detection techniques (particularly those based on the use of ion-selective electrodes) are by far the commonest in this area.

7.6.1 Characteristics of detection systems

Accuracy and precision are essential for both equilibrium and kinetic measurements. However, kinetic methods involve a series of instrumental and experimental factors that can seriously affect the measured parameters.

Most transducers yield non-linear responses as a result of non-proportionality between the property measured (e.g. electrode potential, optical transmission) and concentration. As a rule, this lack of linearity is corrected by instruments dedicated to equilibrium measurements. However, the typical kinetic plot of response *vs.* concentration is more difficult to correct and may result in significant errors in the analytical results. Thus, some devices used to convert transmittance into absorbance are not accurate enough for certain analytical purposes.

The variable-time kinetic method is of advantage whenever a non-linear response is obtained, as the reaction rate measured in the process is directly proportional to the analyte concentration whether or not the effective transducer response is proportional to the concentration. This is a result of measuring the time required for a given concentration to be reached, thereby avoiding the need for a linear relationship between response and concentration. Therefore, it is the instrument itself rather than the chemical reaction which dictates the method to be applied.

Any sensing system employed in kinetic methodology should have the attributes commented on below.

7.6.1.1 Sensitivity and stability

Many kinetic methods are based on measurement of the analytical signal at the beginning of the reaction. This involves instrumental complications arising from the need to measure small changes with high precision, and to ensure that successive measurements are all reproducible and reliable. Fortunately, these requirements are met by the highly sensitive and stable instruments commercially available today.

7.6.1.2 Background noise and drift

Obviously, the sensitivity and stability of an instrument depend to a great extent on its noise level. Broadly speaking, the analytical precision depends on the intensity and nature of the noise associated with the signal, the intensity of the signal itself, and the evaluation method. Kinetic methods are especially sensitive to background noise (positive and negative signal oscillations) when measurements are made at very short time intervals.

Background noise can be said to be of high or low frequency according to whether the signal due to the reaction rate can be distinguished from the perturbation or not. High-frequency noise can be readily eliminated or minimized by using a simple electronic filter which retains the noise, the frequency of which is much higher than that of the rate of change of the signal. Conversely, low-frequency noise, also known as drift (signal trend), which varies at a rate similar to that of the monitored reaction, is a serious perturbation as it cannot be distinguished from the analytical signal. This drawback can only be eliminated by avoiding the perturbation altogether. The problem does not seem to affect fast kinetics to the same degree.

7.6.1.3 Response time

The instrument response time should be short enough for the signal to be as close as possible, in time, to the event giving rise to it, especially when fast reactions or initial-rate methods are used. This requires the so-called *rise time* to be reproducible over all the operative ranges defined by the initial concentration determining the reaction rate. The relevance of this aspect lies in its relationship to electronic filtering; thus no fixed time-constant can be applied throughout a wide concentration range, since the reaction rate corresponding to a given time is a function of the initial concentration. Therefore, the start time for the instrument must be adjusted for a given background noise (damping), and the results compared with the maximum rate constants measured for each series of filter time-constants.

7.6.2 Optical detectors

Most kinetic methods of analysis rely on absorptiometric measurements made in the ultraviolet–visible region. Hence spectrophotometric techniques are discussed in greater detail here than the less frequent fluorimetric or nephelometric techniques. Atomic spectrometry, rarely used in kinetic methods, will not be dealt with in this section.

7.6.2.1 Ultraviolet–visible molecular absorption spectrometry

The popularity of this technique is due to the availability of reactions involving substances absorbing in the near ultraviolet or visible region. Spectrophotometric sensing, unlike fluorimetric detection, allows the simultaneous recording of absorbance readings from the unknown and a blank, by use of a dual-beam spectrophotometer. This is not suitable for fast reactions, as it may provide delayed responses, detracting from the precision of the measurements.

The photometric technique is useful not only for reaction monitoring, but also for simplification purposes. Thus, methods based on the Landolt effect can be made faster and automated to a certain extent by using pairs of solutions, containing different catalyst concentrations, in the two cuvettes of a dual-beam spectrophotometer. The instrument will detect no difference in absorbance at the beginning of the reaction; as soon as the reaction has started, however, the absorbance in the cuvette containing the greater catalyst concentration changes rapidly and then keeps constant until the reaction has developed to completion in the other cuvette, and then the absorbance difference decreases rapidly to zero. A suitable calibration graph can be prepared by plotting the absorbances obtained by keeping the catalyst concentration constant in one of the cuvettes and varying that in the other, during the period from the increase to the decrease of absorbance. This procedure is most readily implemented with the aid of a dual-outlet pipette [78], which allows the simultaneous injection of different concentrations of the catalyst species, [e.g. Mo(VI), a catalyst for the reaction between iodide and hydrogen peroxide] into both cuvettes; molybdenum can thus be determined from the calibration curve plotted from the difference in the absorbance readings taken. These 'differential' methods also make occasional use of other detection techniques such as conductimetry or thermometry. Their precision relies on the incorporation of the second reactant (generally the oxidant), which should be added simultaneously to both reaction cells from the dual-outlet pipette.

As stated above, kinetic methods are more demanding than their equilibrium counterparts as regards some instrumental components, namely the light-source, which must be as stable as possible, and the sensing system, the background noise of which must be minimal.

The use of a dual-beam system is in itself a way of improving the stability of the light-source. However, the rather high cost of dual-beam spectrophotometers and the reasons given above compel the user to employ single-beam instruments and to resort to electronic stabilization of the light-source.

There are two main procedures to improve the stability of the light-source: an indirect one based on stabilization of the voltage or intensity of the source or elimination of the variation arising from the heat flux around the lamp [6], and a direct or automatic one. The latter, much more effective than the former, involves the use of a feed-back system that keeps the lamp intensity constant, and is itself controlled by a second photodetector giving a signal which is fed back to the light-source [79]. This system has been coupled successfully to a Bausch & Lomb Spectronic 20 instrument [80]. The stability of the light-source is always of great importance, as the variation in it can be comparable to the changes in the analytical signal yielded by the reaction monitored.

The particular sensing system used determines the signal-to-noise ratio and the drift. There is no general agreement as to whether it is vacuum phototubes (photodiodes) or photomultiplier tubes which provide the higher signal-to-noise ratio. For a given level of light intensity, the photomultiplier tube yields a high signal-to-noise ratio resulting from its high internal amplification factor (about 10^6), which minimizes the Johnson noise from the load resistor. Thermal noise, which inhibits external amplification, is the one limiting factor of phototubes working at low light levels. For a given anode current, the phototube has a higher signal-to-noise ratio, resulting from the need for a much higher light-intensity to yield an equivalent output. The phototube is therefore recommended when the operator can control the light level and use a powerful light-source; otherwise, the photomultiplier tube must be used.

Ingle and Crouch [81,82] claim that photomultipliers have higher signal-to-noise (S/N) ratios than photodiodes, except at high light intensities; other authors [83] however, state that photodiodes provide acceptable ratios at low light intensities. According to Malmstadt [1], photomultiplier tubes should be used to measure narrow bands, while photodiodes should be reserved for wide bands (colorimetry).

Various electronic differentiation devices coupled to conventional spectrophotometers [6] allow the direct measurement of rates in initial-rate methods. These, however, have currently been superseded by microcomputers coupled on-line to the spectrophotometer.

7.6.2.2 Molecular fluorescence spectrometry

Ingle and Ryan [84] have studied in depth spectrofluorimetric detection in kinetic methods and have come to the conclusion that it represents "a synergistic combination of the features of kinetic methods and fluorescence measurements." These fluorimetric kinetic methods have been reviewed [85].

The fluorimetric technique has a number of advantages over its photometric counterpart. (a) It is more sensitive: detection limits are lower by a factor of 100–1000. This results in greater precision in the reaction rate measurements and avoids the need for preconcentration steps. (b) It is more selective on account of the small number of fluorescent substances available. In addition, its selectivity can be further improved by appropriate selection of the excitation and emission wavelengths. (c) Fluorescence–concentration plots have much wider linear ranges so dilution to bring the sample concentration within the linear range is unnecessary. (d) Temperature can have a self-compensating effect, as it increases the reaction rate and decreases the fluorescence quantum yield.

Notwithstanding its advantages, the fluorimetric technique has also some disadvantages: (a) the instrumentation required is expensive and complex; (b) it is subject to parasitic radiation (both from the excitation radiation and the self-absorption of emitted radiation) and photolytic phenomena; (c) absolute fluorescence intensity values are difficult to obtain (a hindrance in routine enzymatic analysis); (d) background changes from sample to sample, which in principle have no effect on absorbance or fluorescence measurements, are a source of imprecision and result in altered detection limits — their effect on the spectrophotometric technique is much less marked.

The stability of the light-sources and sensing systems used affects molecular fluorescence spectroscopy in much the same way as it does absorption spectrometry. Photomultiplier tubes are used preferentially, as they are better at transducing low light levels. The incorporation of lasers as light-sources has resulted in remarkable improvements, since the fluorescence signal is a function of the incident light intensity. In 1977, Wilson and Ingle [86] reported a laser instrument designed for the fluorimetric monitoring of reaction rates. The use of rapid scanning devices is dealt with below.

7.6.2.3 Rapid scanning spectrometry

This is a promising area in the context of kinetic methods of analysis [83], and is a major breakthrough in dealing with the time-dependent spectra of intermediates or products, recorded with the aid of conventional monochromators. It should be distinguished from so-called kinetic spectroscopy [87], a more general term, which comprises all the instrumentation used to monitor fast reactions and is not necessarily associated with multidetection.

Rapid scanning spectrophotometers are based on various operational principles [88]; thus, image detectors such as the well-known vidicon [89] bear some relationship to television, whereas photodiode–array detectors consist of a large number (from a few hundred to several thousand) of individual detectors, each of which deals with a very narrow band of the transmitted spectrum. Some commercial firms (Pye-Unicam, Hewlett Packard, Perkin-Elmer) already market instruments equipped with diode-array detectors. Figure 7.22 compares schematically a conventional spectrophotometer and a diode-array spectrophotometer.

Unlike conventional spectrophotometers, the instruments incorporating diode-array detectors irradiate samples with a scattered polychromatic light beam. The light-scattering unit is fixed and the light is transduced into an electric signal by every photodiode, each of which works at a given wavelength. The most significant innovation of diode-array spectrophotometers is that the sample is irradiated with light of all wavelengths — hence its classification as an inverse optical technique. The spectral resolution afforded by the diode coverage is determined by the number of elements used. Thus, an array of 250 diodes, covering a wavelength range of 200 nm, will provide a resolution of slightly better than 1 nm. The photometric resolution of the instrument is determined by the number of bits of information provided per diode, so a 12-bit analogue-to-digital converter can distinguish between 4096 (2^{12}) intensity levels.

Diode-array configurations allow the simultaneous kinetic determination of two or more dissolved species, giving rise to two or more different reactions with the same or different reagents, provided that the species monitored absorb at different wavelengths. The oscilloscope or computer display can present spectra repetitively at intervals of 10–100 msec. The computer, provided with suitable software, can

a)

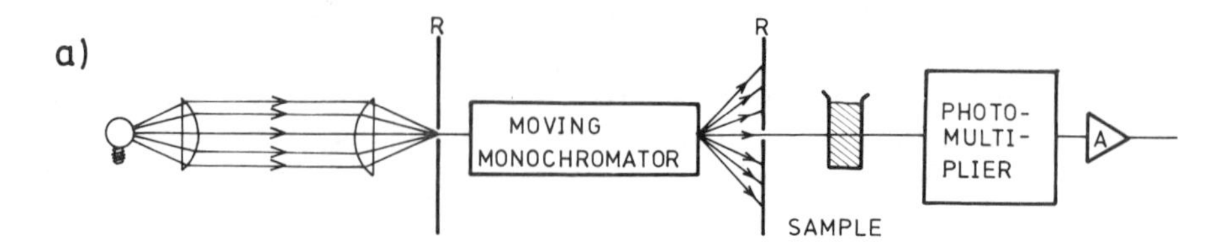

b)

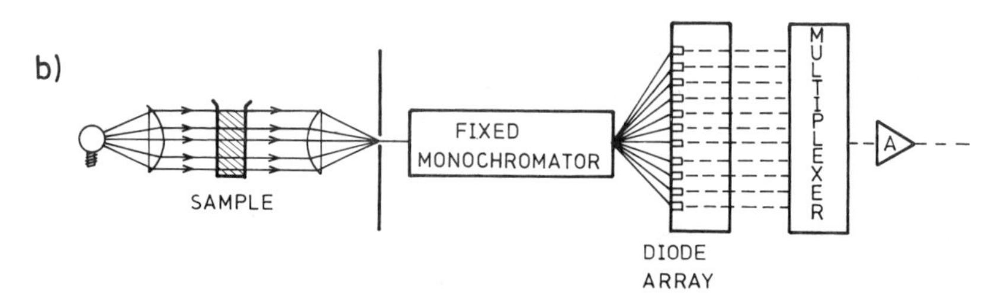

Fig. 7.22 — Schematic comparison of the different components of a conventional spectrophotometer (a) and a rapid-scanning spectrophotometer with diode-array detector (b).

record simultaneously kinetic absorbance–time curves at two different wavelengths. This is the procedure followed in the simultaneous determination (in serum) of the enzymes lactate dehydrogenase (LDH) and alkaline phosphatase [90], which catalyse the reaction between lactate and NAD, and the hydrolysis of phenolphthalein monophosphate, respectively, in the same solution. In practice, the spectra of the reaction products, NADH and phenolphthalein, are recorded as a function of time. Figure 7.23a shows such changes for five different intervals, and Fig. 7.23b presents the variation of the absorbance of each component (LDH at 350 nm and alkaline phosphatase at 550 nm) along the whole cycle. The slopes of the plots obtained vary linearly with enzyme activity.

This kind of spectrometry is also useful for the study of intermediates formed in fast reactions, for mechanistic elucidation purposes. Such is the case with the monitoring of fast reactions of pharmacokinetic interest such as the acid-catalysed hydrolysis of penicillin [91]. Penicillin derivatives such as potassium benzylpenicillin (BP), an oral antibiotic, are subject to acid hydrolysis by gastric juices which convert them into inactive products such as penicillenic acid (**I**) (λ_{max} = 322 nm), which is responsible for allergenic reactions to penicillin, penicillinic acid (**II**) (λ_{max} = 240 nm), penamaldic acid (**III**) (λ_{max} = 240 and 290 nm) and other rearrangement products (**IV**) (λ_{max} = 240 nm). All these species can be identified, and their fraction in the mixture (usually 33% **II**, 5.8% **III** and 61% **IV**, the result of the rapid degradation of BP to **I** and the conversion of the latter into **II**) determined by monitoring the degree of hydrolysis simultaneously at various wavelengths.

Rapid-scanning spectrometry has been used in conjunction with the stopped-flow mixing system designed for fast reactions. Ridder and Margerum [92] have studied

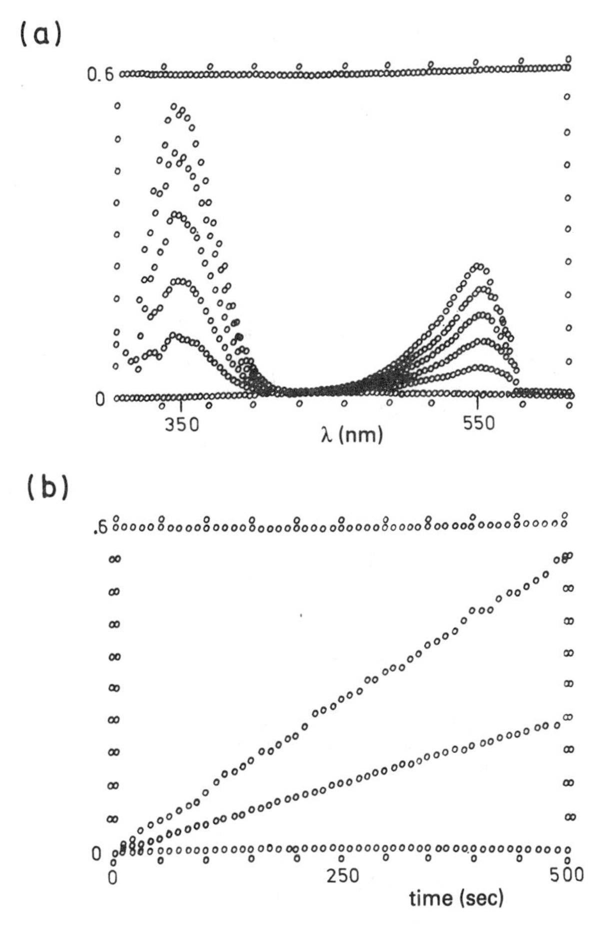

Fig. 7.23 — (a) Time-dependent spectra for LDH and alkaline phosphatase reactions proceeding simultaneously. (b) Absorbance *vs*. time plot for each of the individual reactions, recorded at 350 nm (LDH) and 550 nm (alkaline phosphatase) [90a].

performance of this combination by applying multiwavelength regression analysis to first-order reactions, taking repetitive spectra of reactants or intermediates and applying regression analysis to two parallel first-order reactions by means of the response surface, i.e. the absorbance measured simultaneously as a function of time and wavelength. In their study, these authors used the dissociation of the complexes of Hg(II) and Zn(II) with Zincon and worked in such a manner that the rate constants and spectral changes were of roughly the same magnitude, so as to obtain a three-dimensional response and thus improve the accuracy and precision of the determination of both metal ions. Figure 7.24 shows the schematic diagram of a rapid scanning spectrophotometer coupled to a microcomputer-assisted stopped-flow system.

This novel principle has also been applied to molecular fluorescence spectroscopy and kinetic fluorimetric determinations [89,93].

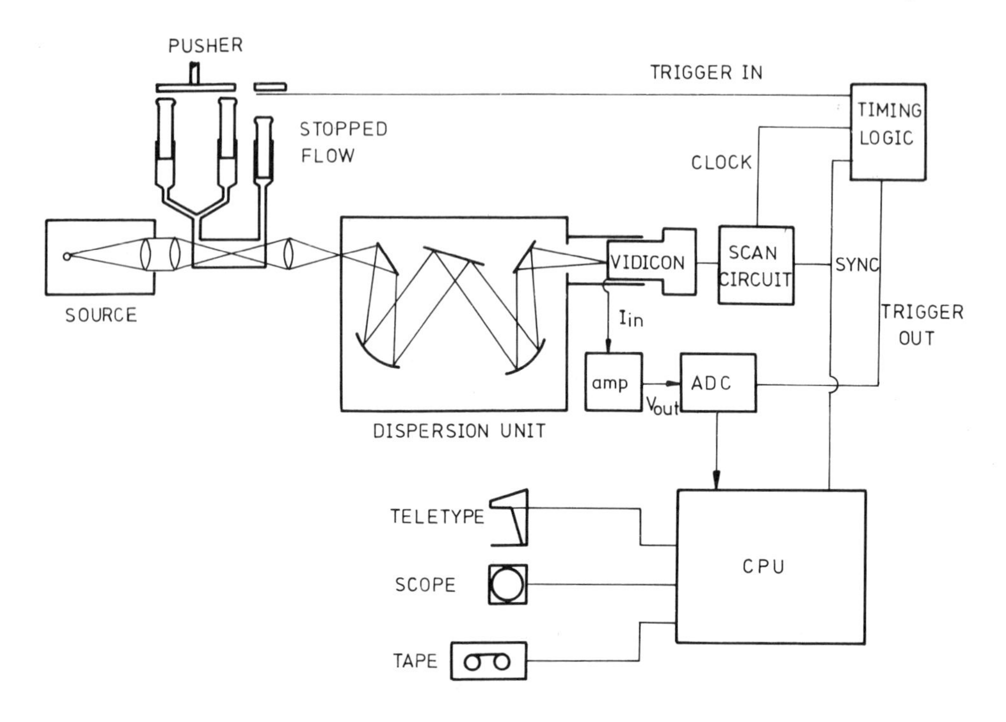

Fig. 7.24 — Block diagram of vidicon stopped-flow rapid-scanning spectrometer interfaced to a minicomputer. (Reproduced with permission, from G. M. Ridder and D. W. Margerum, *Anal. Chem.*, 1977, **49**, 2098. Copyright 1977, American Chemical Society).

Fluorescence measurements at low light levels benefit from the use of multichannel detectors such as the vidicon or diode arrays, which allow the fluorescence intensity to be integrated at the detector itself, and all the spectral elements to be monitored simultaneously. The low signal-to-noise ratio of these devices has been improved by Ingle and Ryan [94–96a], who have developed an intensified diode-array system (IDA) for kinetic fluorimetric measurements. This features a booster consisting of a microchannel plate followed by an array of 512 photodiodes. The emission monochromator and photomultiplier tube of the conventional spectrofluorimeter are replaced by a holographic diffraction-grating spectrograph which spreads the fluorescence spectrum across the active surface of the diode-array detector, as shown in Fig. 7.25. This multichannel IDA system has been used in the simultaneous determination of thiamine and riboflavin [96] in multi-vitamin tablets by recording the fluorescence signal yielded at two different wavelengths, with the aid of a computerized system.

7.6.3 Electroanalytical detectors

Electroanalytical techniques (particularly potentiometry and to a lesser extent amperometry) are the basis for the monitoring of a variety of reactions on which kinetic methods rely. In fact, electroanalytical sensing offers advantages such as the

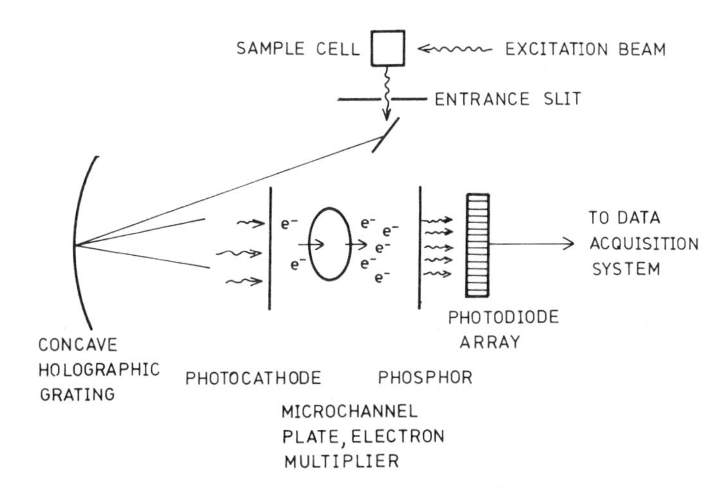

Fig. 7.25 — Operational scheme of a intensified diode-array fluorescence detector. (Reproduced from J. D. Ingle and M. A. Ryan, in *Modern Fluorescence Spectroscopy*, E. L. Wehry (ed.), 1981, p. 130, by permission. Copyright 1981, Plenum Press).

easy thermostatic control of the unknown solution and continuous recording of the electrical signal yielded (current, potential). Electrochemical detectors have been the subject of two recent books [96b,96c].

7.6.3.1 Potentiometric and amperometric detectors

Potentiometric sensing usually involves use of the variable-time method, i.e. measuring the time elapsed until the required change between two preselected potentials has taken place.

Selective electrodes are of especial use in potentiometric techniques applied to kinetic analysis as they are sufficiently sensitive to changes in concentration and convenient to use in the study of reaction mechanisms [97]. A typical example is the iodide-selective electrode, which is used to monitor the Fe(II)-induced reaction between perbromate and iodide through automatic measurements of the time required for the potential to change by 10.0 mV. The use of 1,10-phenanthroline as activator permits the determination of between 4 and 40 ng of iron (1.8×10^{-8}–$1.8 \times 10^{-7}M$), in addition to that of perbromate (5×10^{-6}–$7.5 \times 10^{-5}M$), EDTA, DTPA and EGTA on the basis of their inhibitory effect on the induced reaction [98]. A chlorate-selective electrode permits the automated determination of various glycols by the constant-concentration method [99].

The glass electrode (pH electrode) is very often used in the determination of enzymes and their substrates, as the reactions in which these species take part usually involve the release or consumption of hydrogen ions. The ammonia-selective membrane electrode has also been employed in enzymatic analysis in the determination of creatinine [100].

The oxygen electrode is normally employed in enzymatic analysis with oxygen-consuming systems (e.g. in the determination of glucose in blood). This amperometric sensor is a Clark electrode consisting of an electrolytic cell accommodating a platinum cathode and an Ag/AgCl anode in a phosphate/potassium chloride buffer. The electrode reactions involved are

$$O_2 + 4H^+ + 4e^- = 2H_2O \qquad \text{(cathode)}$$

$$Ag = Ag^+ + e^- \qquad \text{(anode)}$$

A gas-permeable polypropylene membrane separates the electrode solution from the unknown and obviates interference from proteins or other oxidants. Oxygen, in the absence of which the electrodes are polarized, gives rise to a current with an intensity proportional to the oxygen concentration. However, the application of gas-sensing electrodes to biological fluids is limited to a certain extent by blocking of the membrane pores. This shortcoming has been circumvented by the introduction of air-gap electrodes [101], a much more efficient means of isolating the surface of the indicator electrode from the sample solution. This type of electrode has been successfully used for kinetic initial-rate measurements [102].

The more recent enzyme electrodes combine the high selectivity of these biocatalysts with a straightforward electrochemical technique [103]. The extensive use of these electrodes is partly due to the degree of development achieved in enzyme immobilization. The simplest version of these electrodes consists of a membrane accommodating an artificial enzyme fixed on the transducer [100]. The substrate diffuses through the thin catalyst layer and yields an electroactive substance which is detected by the potentiometric or amperometric sensor. Potentiometric sensing is more advantageous on account of its lower cost and the ease of preparing the electrode but it is rather slow owing to the sluggish diffusion of the substrate and its product through the enzyme layer, particularly when artificial membranes of a thickness of 0.1 mm or greater are used. Potentiometric enzyme electrodes have been preferentially used in the determination of amino-acids [104], urea [101,105], glucose and penicillin.

Amperometric enzyme electrodes are commonly used with enzymatic reactions involving the consumption or release of oxygen or hydrogen peroxide. They normally consist of a platinum electrode covered with a plastic membrane onto which a gel layer of the enzyme is deposited. The current measured is proportional to the amount of electroactive substance generated on the membrane. Thus, in the determination of glucose, the current is proportional to the loss of oxygen in the immobilized glucose oxidase layer as the glucose from the bulk solution penetrates into the electrode and reacts according to:

$$\text{Glucose} + O_2 + H_2O \xrightarrow[\text{oxidase}]{\text{glucose}} \text{Gluconic acid} + H_2O_2$$

These amperometric detectors are more advantageous than their potentiometric counterparts in terms of linearity range. The current generated has been found to depend on the rates of reaction and diffusion of the enzyme [106]. A further advantage is that the detectors can be made specific by use of a double enzyme system, the first enzyme yielding a substrate that reacts with the detector reaction enzyme [106a].

The different optimum conditions required for the enzymatic reaction and the electrode reaction have led to the development of enzyme reactor electrodes, in which the immobilized enzyme (e.g. on glass beads) is placed in a small reactor through which the sample (in an appropriate buffer) is circulated. The substrate is converted into products with 100% efficiency. The effluent from the reactor is mixed with a buffer to fix the optimum pH for measurement, which is performed by a flow-through electrode. The enzyme reactor electrode, which can be used in conjunction with other instrumental techniques, has been applied to the determination of urea and amino-acids [107,108].

In Fig. 7.26 are depicted various electroanalytical devices used to monitor the reaction of glucose described above. This enzymatic reaction can be monitored directly by means of a straightforward molecular oxygen sensor (Fig. 7.26a) or, alternatively, through a second reaction detecting the release of hydrogen peroxide. Such a reaction is the formation of I_2 from iodide in the presence of molybdenum as catalyst, which can be followed potentiometrically or amperometrically. The application of the potentiometric technique in this case [109] entails selecting a potential interval of 6 mV and measuring the time needed for the system to traverse it, usually 10–100 sec (variable-time method). The potential needs to be measured with high precision (± 0.02 mV); this is facilitated by the use of a commercially available concentration comparator furnished with auxiliary relays for automatic measurements and the control of temperature and time (within ± 0.02 sec) by means of a timer actuated by the relays, which in turn respond to changes in potential. The sample solution is added to a composite reagent solution containing buffer-catalyst, iodide and enzyme in the sample cell (Fig. 7.26b), which is electrically connected to the reference solution (buffer-catalyst and iodine/iodide) by an asbestos fibre sealed in the bottom of the cell. On addition of the glucose sample, the iodide concentration is decreased and a potential difference (ΔE) is generated between the reference and sample solutions; at that moment, the timer is started and is stopped when $\Delta E = 6$ mV. The reciprocal of the time elapsed is proportional to the glucose concentration between μg/ml 5 and 500 μg/ml.

As stated above, the reaction can also be monitored amperometrically [110] by using a static and a rotary platinum electrode. The substrate is oxidized enzymatically, and the resulting hydrogen peroxide oxidizes iodide to iodine, which is then determined amperometrically with a pair of polarized platinum electrodes (Fig. 7.26c). The current between the electrodes is again measured with a concentration comparator aided by a series of relays, one of which controls the time elapsed until a preselected current is reached. Glucose can thus be determined in the range 0.5–3 mg/ml, somewhat wider than that afforded by the potentiometric technique.

Differential potentiometry and biamperometry are also commonplace in kinetic analysis. The latter involves measurement of the potential difference between two platinum electrodes (polarized by a small direct current) with a high-input impedance voltmeter (follower) and recording the variation of the signal as a function of time. This procedure is of especial use for catalysed reactions and has been chiefly used in the determination of enzymes. In addition, it requires a difference between the redox potentials of the reactants and products involved. The potential difference,

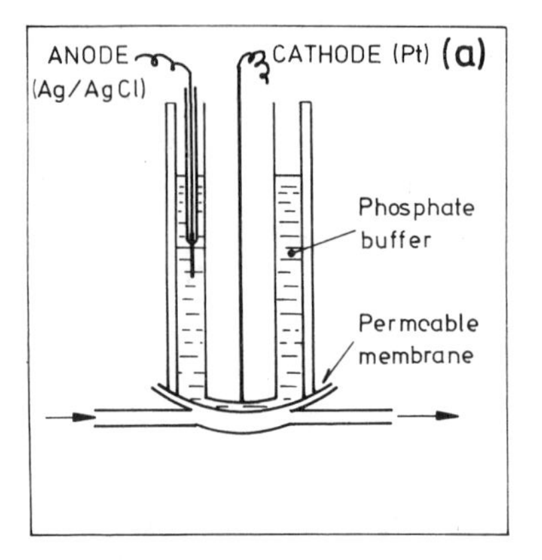

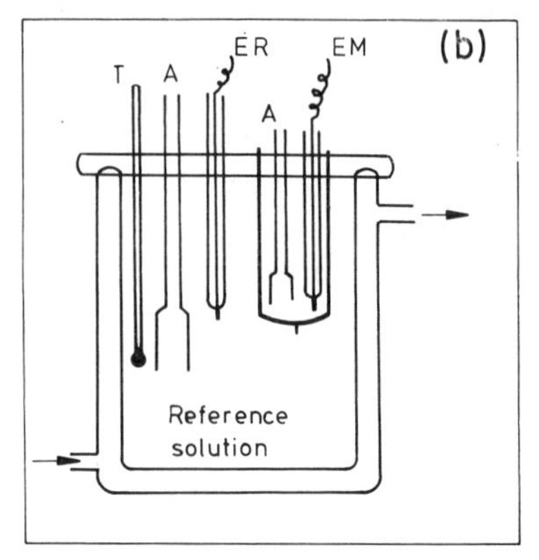

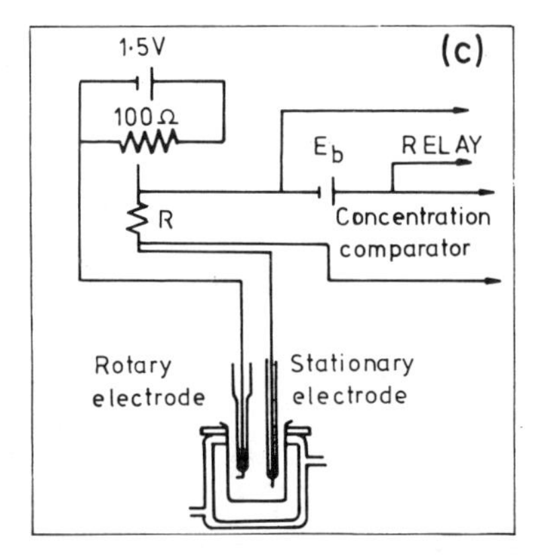

Fig. 7.26 — Electroanalytical techniques used to monitor enzymatic reactions: (a) with an oxygen electrode; (b) by potentiometry (T = thermometer, A = stirrer; ER = reference electrode; EM = sample measurement electrode); (c) by amperometry. [Reproduced with permission, from (b) [109], (c) [110]. Copyright (b) 1961, (c) 1963, American Chemical Society].

which is zero in the absence of the catalyst, varies (as does the reaction rate) as this is added to the solution, making the differential initial-rate method particularly useful.

Differential potentiometry has also been used for monitoring purposes in kinetic analysis. The most usual experimental set-up is composed of two platinum electrodes polarized by a constant current. As with biamperometric sensing, the variation of the potential difference between the two electrodes is recorded as a function of time. Organic and inorganic species that can undergo oxidation reactions are suitable substrates. If the concentration of electroactive oxidant decreases throughout the reaction, then the potential–time curve obtained shows a rising portion followed by a maximum and a falling segment, the peak width, Δt, being proportional to the reaction rate.

The differential potentiometric technique has been applied to the determination of phenols such as 2,6-dichlorophenol by bromination and of hydroxylamine by oxidation with ferricyanide [111]. These require the careful selection of the currents I_1 and I_2 used to polarize the two electrodes. The integrated rate equation of a pseudo first-order reaction between the species of interest, B, and a reagent A will be given by:

$$\ln([A]_t/[A]_0) = k[B]_0\Delta t$$

where $[A]_t$ is the concentration of A corresponding to the peak width Δt. If currents I_1 and I_2 are assumed to correspond to the voltammetric limits, then the equation $I = nFADC/\delta$ can be transformed into:

$$[B]_0 = (1/\Delta t)(1/k)\ln(I_2\delta' A/I_1\delta A')$$

where k is the rate constant. Since I_2/I_1 is a constant and δ (the thickness of the diffusion layer) is a function of the cell geometry and the stirring speed, the logarithmic term is also a constant for a given set of conditions and hence the concentration of B can be expressed as:

$$[B]_0 \approx 1/\Delta t$$

7.6.3.2 Coulometric and conductimetric detectors

These, though less usual than potentiometric or amperometric detectors, are also worth describing. Coulometrically generated reagents have been used in some enzymatic reactions [112].

Conductimetric measurement (one of the most accurate and precise of the electrochemical techniques), used in the earliest kinetic determinations of organic substances [113,114], has been revitalized by the use of bipolar pulses [115]. This technique is much more sensitive than conventional conductimetry and involves the application of two voltage pulses (of the same magnitude and length but of opposite polarity) to the cell, followed by measurement of the instantaneous current at the end of the second pulse. Other variants are possible [116,117]. A computer-assisted bipolar pulse conductivity technique has been applied in conjunction with a stopped-flow mixing system to the study of the dehydration of carbonic acid and the reaction of nitromethane with bases [118].

7.6.4 Other detection systems

This section is devoted to other interesting (though less often used) detection techniques employed with kinetic methods of analysis.

Radiochemical sensing has been employed with systems containing radioisotopes, which are monitored with the aid of radioactive tracers.

Thermometric techniques have been widely used in catalytic titrations (see Chapter 4) and in the study of organic reactions and the determination of organic substrates, either alone or in mixtures. The earliest application of thermometric sensing to kinetic analysis was reported by Papoff and Zambonin [119], who employed it for the resolution of mixtures of methyl and isopropyl acetate, based on the temperature–time curves recorded for the base-catalysed hydrolysis of both species.

The differential thermometric technique used later by De Oliveira and Meites [120] for resolution of a mixture of cyclohexanone and propanal by condensation with hydroxylamine is more advantageous as it can deal with smaller rate-constant ratios and compensates for the heat of dilution and other thermal effects with the aid of a differential device designed by Meites *et al.* [121] and the application of corrections for heat exchange. The purpose-built device [120] is a colorimeter with two identical vessels holding the reactant mixture and a reference solution, respectively. The experiment is started by adding a solution of hydroxylamine, hydrochloric acid and sodium chloride (to adjust the ionic strength) of suitable concentration, to each vessel. Two syringes dispense a volume of the analyte mixture into one of the vessels and the same volume of water into the other. The temperature difference between the vessels is then measured by means of two 100-kΩ thermistors connected to a Wheatstone bridge, the voltage response of which is recorded as a function of time. The temperature difference increases with the development of the exothermic oximation reaction up to a maximum, after which it decreases exponentially as the Newtonian heat exchange balances the temperature of the reacting mixture with that of its environment. The overall temperature change is about 0.002°C and the determination is done at 25.7 ± 0.3°C. The computations are done by a multiparametric curve-fitting computer program.

Still more recent and advantageous is direct-injection enthalpimetry [122], a technique involving the injection of an excess of reagent into the unknown solution under adiabatic conditions. The resultant heat pulse is proportional to the analyte concentration and is also measured by a sensitive thermistor through a Wheatstone bridge. This technique has been applied to the determination of various enzymes and their substrates [123] and is of especial advantage to analyses of whole blood as it requires no prior deproteination.

7.6.5 Double indication systems

It is not uncommon for kinetic analysis to use two different sensing systems for the monitoring of a reaction. This so-called 'double indication' provides richer information about the system's behaviour than does the individual use of both techniques.

This principle has been applied to kinetic catalytic analyses such as the determination of thiosulphate and sulphide, catalysts for the reaction between sodium azide and iodine [124]. The release of nitrogen results in clouding and heating of the solution. Thus, the simultaneous use of the photometric (turbidimetric) and thermometric techniques improves the accuracy and precision of the determination, whether this is done in a closed or a continuous-flow system. The signal *vs.* time curves typically obtained with a dual-channel recorder for a closed system are shown in Fig. 7.27, in which the solid line corresponds to the thermometric monitoring of the fast exothermic reaction involved. The reaction is halted once all the catalyst, which is oxidized and hence inactivated by iodine in a side-reaction, has been completely consumed. The temperature change due to the oxidation of thiosulphate by iodine can be safely assumed to be negligible. The time required for the complete oxidation of the catalyst is proportional to its concentration, which can be estimated from the signal height corresponding to the maximum temperature attained in the recording. The broken line in Fig. 7.27 corresponds to the optical detection. As can be seen, there is a maximum corresponding to the turbidity measured after the catalysed reaction has developed to completion — the greater the amount of nitrogen formed, the higher the peak maximum and the shorter the time it takes to appear. The maximum height, H_m and its appearance time, t_m, are both measures of the catalyst concentration. Their ratio has been found to provide better results than those obtained by using either alone.

Dual thermometric–biamperometric indication has also been applied to other catalysed reactions [125].

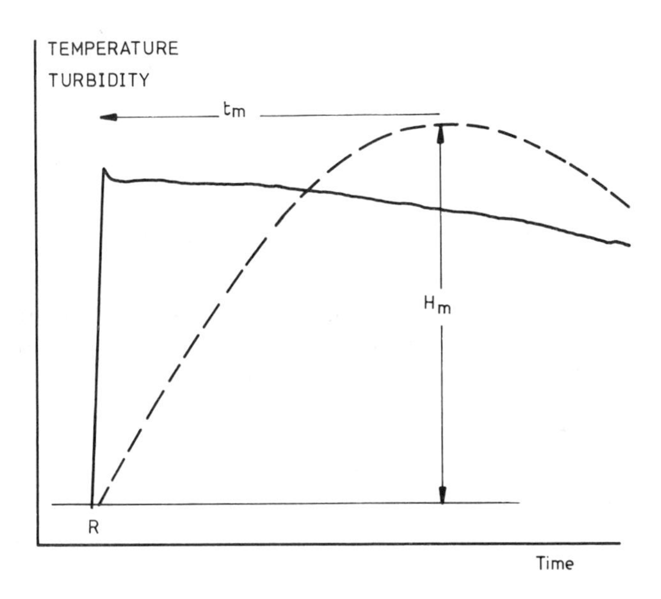

Fig. 7.27 — Plots of temperature (solid line) and turbidity (broken line) as a function of time. (Reproduced from H. Weisz, W. Meiners and G. Fritz, *Anal. Chim. Acta*, 1979, **107**, 301, by permission. Copyright 1979, Elsevier Science Publishers).

7.7 DATA ACQUISITION AND TREATMENT

This final stage, common to other analytical methodologies, is a key to obtaining acceptable results, particularly in dealing with fast reactions or differential kinetic analyses.

The manipulation of transduced signals in kinetic methods of analysis involves two sequential stages, (a) collection of the transduced signal as a function of time and (b) processing of the signal/time data pairs generated, which are transformed into reaction rates or directly into analyte concentrations.

The implementation of these final operations depends on the characteristics of the method concerned. Thus, the situation will be different according to whether the analysis is for a single species or a mixture, is based on pseudo first-order or second-order kinetics, and is performed with an open or closed system. Data collection is a critical operation in dealing with fast reactions, which call for highly responsive instruments.

Kinetic methods for determination of a single species can be classified into three broad groups according to the type of data-processing system used [126] (see Table 7.2).

Table 7.2 — Classification of kinetic methods for determination of a single species according to the data acquisition and treatment system used

Direct response methods
— One-point methods
— Two-point methods
— Multi-point methods
Delta method
Regression method
Derivative methods
Integral methods

Direct response methods, the basis for most kinetic analysers, rely on the use of one or more signal readings made at preselected time intervals to calculate the reaction rate and/or analyte concentration sought. According to the number of measurements used, these methods can be further divided into one-point, two-point and multi-point methods.

One-point methods are based on a signal reading made at a given time after the reaction has started. Insofar as they involve the determination of the reaction rate as the slope of the kinetic plot of ΔS *vs.* Δt at a given point, they are commonly referred to as 'derivative methods' [10,127,128]. However, they should not be confused with the derivative methods included in Table 7.2, which are based on signal derivative ($\Delta S/\Delta t$) *vs.* time plots. Since these methods use a single point per sample, the signal–time curves on which they are based must conform to the expected kinetic behaviour, the point used must not be on a non-linear portion of the curve, and the results are subject to errors from the blanks used.

There are two types of *two-point method*, namely fixed-time and variable-time methods (see Chapters 2 and 5 for a complete description). Two-point methods are superior to one-point methods in that they compensate for sample and reagent blanks, but may still be subject to errors resulting from non-linearity arising from lag and/or substrate depletion.

Multi-point methods can be classified into two groups: multi-point delta methods and regression methods. The delta variant (Fig. 7.28a) involves comparing discrete differences between signals taken at identical time intervals along the kinetic curve, in order to determine its degree of linearity and hence select the best possible zone for determining the reaction rate. From Fig. 7.28a it is obvious that intervals δ_1 and δ_5 are inferior as regards linearity and should not be used for calculation of the reaction rate and/or analyte concentration.

Regression methods use a mathematical algorithm to ensure the best possible fit between three or more data points on the basis of a given mathematical function, which coincides with the equation of a straight line for calculation of the reaction rate and hence should be fitted to the linear portion of the kinetic curve. Figure 7.28b illustrates the application of this type of method. The different slopes (b) obtained by applying the regression treatment to various data groups along the response curve are compared in order to determine its linear portion. As with the previous method, slopes b_1 and b_5 should be discarded for determinative purposes. This type of method is equally applicable to linear [129–131] and non-linear [42,132] responses.

Derivative methods, as their name implies, involve the electronically obtained first or second derivative of the kinetic curve, i.e. the plot of $\Delta S/\Delta t$ or $\Delta^2S/\Delta t^2$ as a function of time. Both plots (see Fig. 7.29a for the first-derivative) feature a horizontal segment defining the interval over which the signal varies linearly with

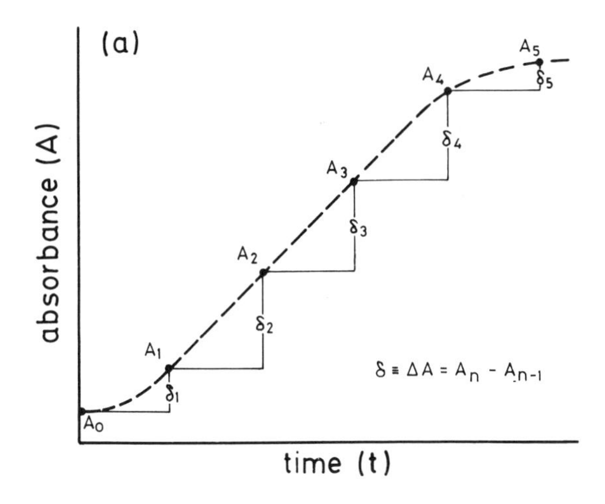

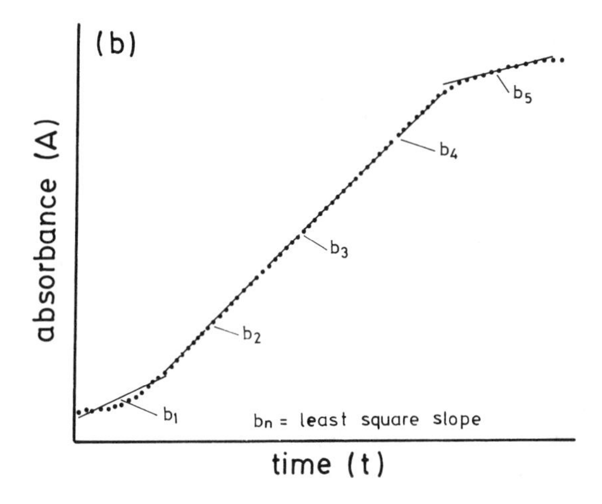

Fig. 7.28 — Conceptual representation of (a) multi-point delta method and (b) regression kinetic method. (With permission of the American Association for Clinical Chemistry).

time — the most suitable for determination of the reaction rate. These methods, which use a differentiating circuit to record the first and/or second derivative of the signal as a function of time should be distinguished from those regression methods applied to overlapping data and involving the plotting of an approximation to the derivative of the signal as a function of time, which is not done electronically.

As shown in Fig. 7.29b, integration methods are based on the integration of the kinetic curve over two different intervals and computing the difference between both areas [133], which should be proportional to the analyte concentration. The differentiation can be done either by an analogue circuit or by a digital integrator.

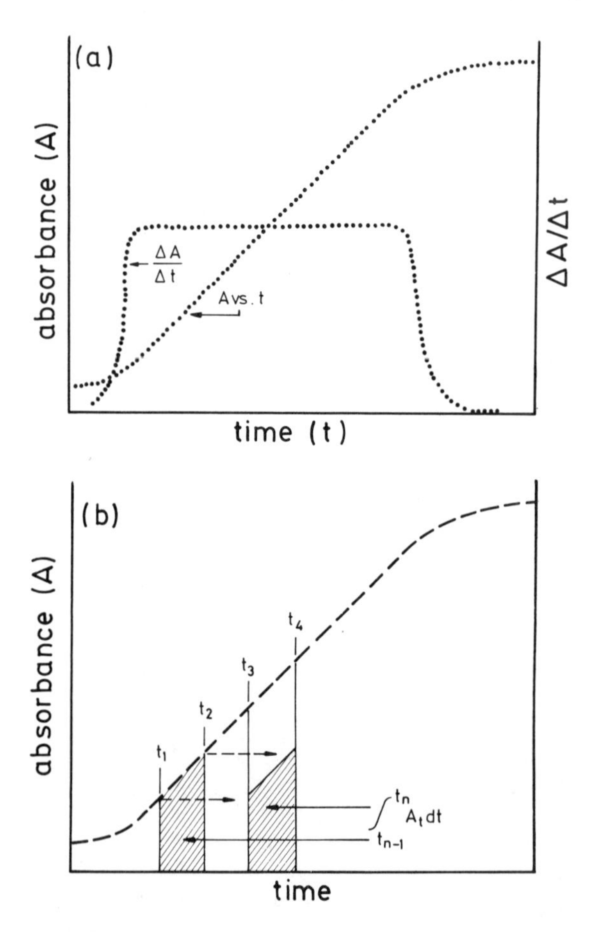

Fig. 7.29 — Conceptual representation of (a) derivative method and (b) integral method. (Reproduced with permission, from E. Cordos, S. R. Crouch and H. V. Malmstadt, *Anal. Chem.*, 1968, **40**, 1812. Copyright 1968, American Chemical Society).

As with other areas of analytical chemistry, developments in data processing in kinetic analysis have paralleled the advances in electronics and computer technology. The data collection process has gone through two basic stages: the use of analogue systems consisting of electronic circuits, and the later use of digital, minicomputer-interfaced systems which, in addition, allow the complete automation of the process and the development of kinetic procedures for fast reactions. As far as data treatment is concerned, microcomputers are invaluable in improving the accuracy and precision of measurements, and facilitating the application to kinetic data of the different estimation and simulation strategies based on the use of complex mathematical treatments relying on a large number of points.

7.7.1 Data acquisition

It should be noted that the presentation of an analogue reading on a digital display is not equivalent to the use of a digital device, as the digital display makes use of the binary code or one of its derivatives. Digital devices require a software system to store the digitized signal from the analogue-to-digital converter (ADC) as a function of time, in a computer peripheral.

7.7.1.1 Analogue data acquisition by electronic systems

Data collection by means of analogue devices is based on the use of so-called 'rate-meters' which receive the signals from the transducer and transform them into electrical signals which may be correlated with the reaction rate or the analyte concentration. Their design is normally dictated by the particular rate method used (see Table 7.2).

The electronic system commonly used for reaction-rate measurements by the one-point (derivative) method is very simple. As shown in Fig. 7.30, it consists of an operational amplifier (OA), an input capacitor (C) and a feed-back resistance (R) [134]. This differentiating circuit yields an output signal proportional to the rate of change of the input signal and provides the instantaneous value of the slope at any one point of the kinetic curve. The system has the disadvantage that low noise in the input signal is boosted to high levels in the output signal. This effect can be reduced by decreasing the differentiation response frequency or using a slope-matching system [134,135], i.e. by generating a reference kinetic curve by means of an integrator OA and continuously comparing and matching it to the unknown curve.

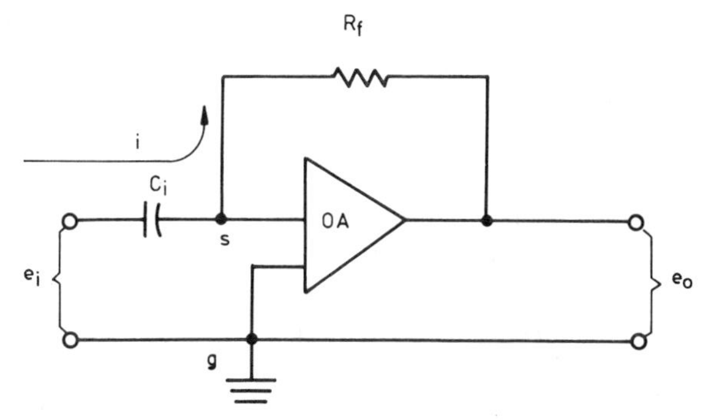

Fig. 7.30 — Classical operational amplifier differentiator. (Reproduced with permission, from H. V. Malmstadt and S. R. Crouch, *J. Chem. Educ.*, 1966, **43**, 340. Copyright 1966, American Chemical Society).

Analogue systems used with two-point methods are intended to overcome the problems stemming from the adverse influence of noise on the measurement of the reaction rate.

The most serious problem with variable-time methods is the accurate determination of the reciprocal of the time. This is overcome by using variable-time ratemeters, which are capable of accurately starting and stopping time readings on receipt of the transducer signals at two preselected levels. The earliest automatic system for measurement of $1/\Delta t$ of this kind was devised by Pardue [135]. Later, Stehl *et al.* designed a completely electronic system which, after measuring Δt, calculated its logarithm and differentiated it in order to obtain an analogue signal directly proportional to $1/\Delta t$ [17]. A similar system was described by James and Pardue [136]. There are also analogue–digital hybrid systems using time-to-voltage converters coupled to voltage/frequency converters [137] or direct time-to-period converters [138].

Electronic circuits used for fixed-time measurements should be furnished with storage devices, as they must store the transduced signal data generated over the preselected time interval. The unavailability of storage peripherals in earlier times was made up for as follows: two observation cells (t_1 and t_2) were placed at a given distance apart in a continuous-flow system, resulting in a constant time interval $t_2 - t_1$, and the signals from both cells were sent to a differentiator, the output from which was proportional to the reaction rate [139].

The systems for reaction rate measurements based on the integration method provide the least noisy signals [135]. As shown in Fig. 7.31a, the electronic circuit normally used is made up of two operational amplifiers connected to each other and to a signal modifier through a switch network. In the first integration period, the signal reaches the integrator OA2 or not, depending on the position of switches 1 and 2. If the modifier output is of the type shown in Fig. 7.31b, then line BC represents the signal applied to the integrator, which charges the capacitor C with an output voltage V_1 (Fig. 7.31c). In the second integration period, the modifier signal is first sent to the unit-gain inverter (OA1). The voltage represented by line DE in Fig. 7.31b is applied to the integrator OA2, which discharges the capacitor, thereby decreasing the integrator output to V_2 (Fig. 7.31c). The voltage difference, ΔV, is proportional to the difference in the integrated areas and hence to the slope of the curve (reaction rate). This system is capable of measuring slopes in the range from 0.45 mV/sec to 450 V/sec, so it is suitable for fast and slow reactions alike. As with the variable-time method, there are also hybrid systems based on voltage-to-frequency converters [26].

Analogue systems are used for reaction-rate measurements not only in the determination of a single species, but also in the resolution of mixtures (differential reaction-rate methods). One such system is the exponential function generator proposed by Worthington and Pardue [140]. This consists of a straightforward analogue circuit coupled to an x–y recorder in such a manner that the linear segments of the graph recorded can be related to the concentrations of the different components of the mixture (see Chapter 6). The novel system developed by Gary and Lagrange [141] is particularly useful for relatively slow reactions.

7.7.1.2 Computer-aided digital data collection

The advent of inexpensive commercial microcomputers and interfaces has revolutionized the design of instruments devoted to the measurement of chemical reactions. The analogue instruments commonly used for rate measurements are no doubt a less

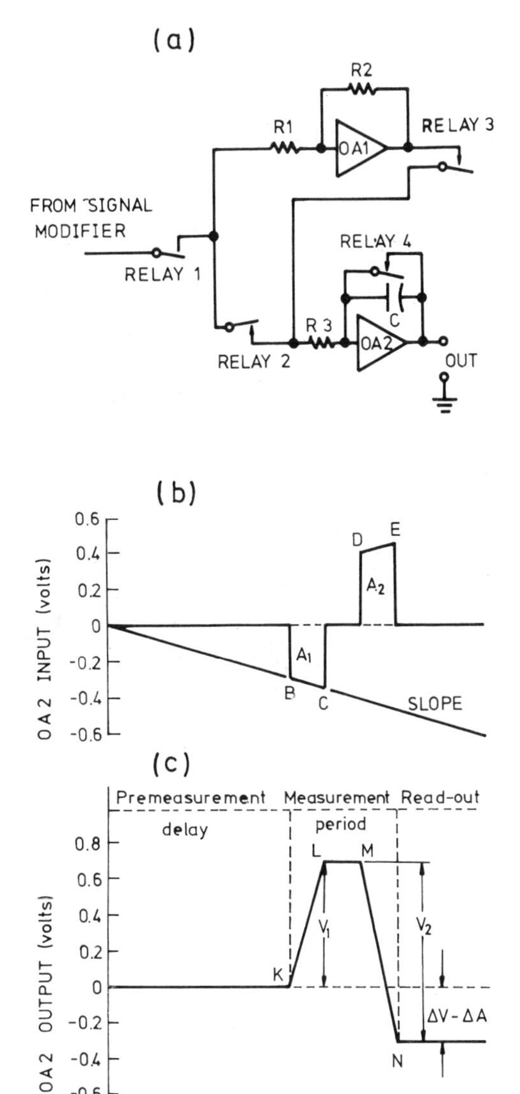

Fig. 7.31 — (a) Integration and subtraction circuit of rate-meter. (b) Oscilloscope trace of slope from ramp generator and input voltage of OA2. (c) Oscilloscope trace of output voltage of OA2. (Reproduced with permission from H. L. Pardue, C. S. Frings and C. J. Delaney, *Anal. Chem.*, 1965, **37**, 1426. Copyright 1965, American Chemical Society).

expensive alternative for specific purposes but a suitably interfaced microcomputer programmed to command the instrumental set-up does not significantly increase costs and endows the system with elegance in design and operation.

The acquisition of real-time data by an instrument governed by a computer requires both to be adequately interfaced. The particular design of the interface will be dictated by the nature of the data to be acquired and by the mini- or microcomputer's capability.

The purpose of an interface is thus to collect experimental data as delivered by the instrument and transmit them to the computer in an intelligible manner. The interface can be an item of hardware located between the computer and the instrument or a part of the instrument itself. Inasmuch as most analytical instruments handle an analogue signal proportional to the physical property measured, the data collection process is ideally carried out through the computer's I/O interface. The general scheme of such an interface is shown in Fig. 7.32. As can be seen, the object of an analogue interface is to convert a voltage or any other analogue output into a binary (digital) form that can be handled by the computer. These analogue-to-digital (A/D) converters are characterized by their wide dynamic range, high resolving power and sampling rate [142].

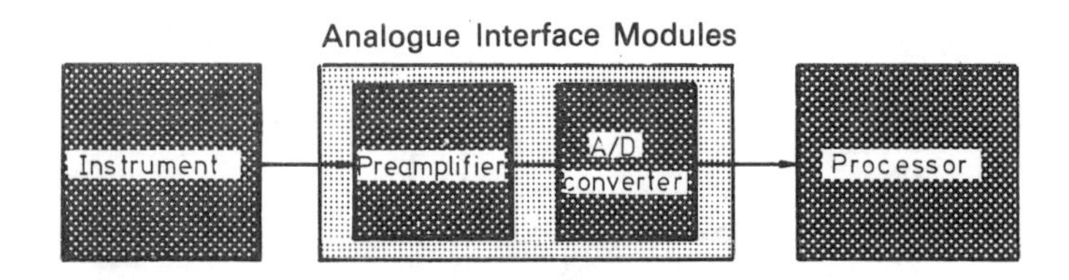

Fig. 7.32 — Analogue interface for conversion of the analogue dc signals from a primary measuring instrument into binary digits. (Reproduced with permission, from J. G. Liscouski, *Anal. Chem.*, 1982, **54**, 849A. Copyright 1982, American Chemical Society).

There are two chief types of A/D converters, namely *integrating* and *successive-approximation*. In the former, the input voltage is used to charge a capacitor, the discharge of which is measured by a high-frequency clock as a number of counts proportional to the input charging voltage. This system has a wide dynamic range and is hardly affected by electrical noise; however, the slow discharge of the capacitor results in a low data acquisition throughput (about 20–200 points per second), which is obviously insufficient to monitor some fast reactions accurately.

The successive-approximation converter transforms an analogue signal into its digital counterpart by a series of approximations, each doubling the resolution of its predecessor. This converter is superior to the integrating version as regards throughput, which can be as high as 10^5 points per second.

The resolution of an A/D converter is defined as the smallest change in the analogue signal which it is capable of discriminating. From the point of view of digital signals, the resolution can be regarded as the number of steps into which the overall analogue signal is split. Most A/D converters are 12-bit systems and hence carry out twelve successive approximations, dividing the analogue voltage into 4096 (2^{12}) steps. Thus, for a converter with an input voltage range of ± 5 V dc, the smallest detectable change in the analogue voltage will be 2.44 mV (10/4096). Integrating converters are 16-bit devices and can thus divide the input voltage range into 65536 (2^{16}) steps. They thus have a wider dynamic range than successive-approximation converters, as their full-scale input voltage is 16 times as great. A preamplifier located between the instrument and the converter is occasionally used to improve the resolution in either case (Fig. 7.32).

The total cost per channel of the successive-approximation converter is lower than that of the integration converter. In addition, the former can acquire data through 16 independent channels thanks to its multiplexer module — up to 64 channels can be handled if additional multiplexers are added. In contrast, the integrating converter can scarcely collect data from 2–4 channels per unit simultaneously.

By virtue of its overall performance, the successive-approximation A/D converter is the more commonly used, despite its associated noise problems. Its high data-acquisition speed makes it especially suitable for fast reactions investigated in stopped-flow systems.

Reaction-rate determinations based on the use of these data-acquisition systems are simple: they involve the use of one of the methods listed in Table 7.2 and the application of a suitable computation program to the data stored in the computer's RAM. Multi-point methods are the most suitable on account of the speed of computers, which can handle a large number of signal/time data in a reasonably short time.

Data acquisition, like many other operations aimed at complete instrumental automation, can now be considered routine thanks to the development and relative cheapness of computer systems. The number of commercial instruments (e.g. spectrophotometers) marketed with integral microprocessors increases day by day. Computers are supplied with programs for kinetic (generally enzymatic) determinations, that demand the minimum of analytical knowledge from the user — to the point that the application of an equilibrium or a kinetic method requires only choosing the appropriate software, leaving all the rest to the computer. The only problem is that the manufacturers never divulge the algorithms used, so the user is unaware of any consequent hidden sources of error [142a,142b].

Nevertheless, the advent of computer-controlled data-acquisition systems is not recent. In fact, the early 1970s saw the development of hardware and software devoted to improving the accuracy, precision, resolution and sampling rate of the stopped-flow technique applied to fast reactions. In Fig. 7.33 are shown the basic components of a data-acquisition system for rapid kinetics, developed by Holtzman [143]. The machine code executive program used permits the collection of data at varying time intervals during a given run. Moreover, the computer is fast enough to calculate results simultaneously with data acquisition. In an interesting paper, Alcocer *et al.* [44] describe a microcomputer-based high-speed data-acquisition system for a stopped-flow spectrophotometer, as well as the microcomputer interfaces developed for use with the stopped-flow methodology in order to avoid the slowness of manual procedures such as those involving the use of an oscilloscope and recording the resulting curves on film.

7.7.2 Data treatment

Manual procedures for treatment of kinetic data are becoming a thing of the past. The signal/time curve can be readily treated by geometric or mathematical methods to abstract valuable information. The one-point rate method can be implemented by

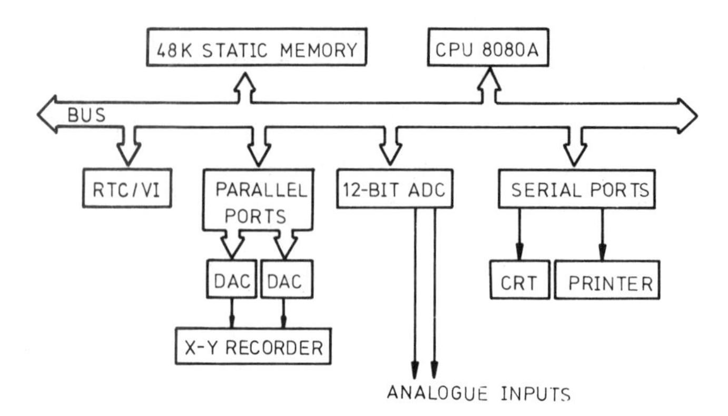

Fig. 7.33—Basic components of a data-acquisition system for rapid kinetics. (Reproduced with permission from J. L. Holtzman, *Anal. Chem.*, 1980, **52**, 989. Copyright 1980, American Chemical Society).

the mirror procedure [10] if a non-linear curve is handled. Though fixed- or variable-time methods pose no problem, integral methods are rather more time-consuming.

The manual determination of the initial rate is subject to major personal errors, in addition to others arising from the timing of the operation of the oscilloscope, recorder or printer and that of the transducer. The use of automatic data manipulation and display systems increases the precision of kinetic measurements by a few orders of magnitude.

Recent advances in instrumentation, data acquisition and treatment have placed kinetic methods of analysis in a prominent place by virtue of their simplicity, rapidity and precision [3,10], all of which are comparable to those of many equilibrium methods. This is undoubtedly the result of the use of microcomputers, which have the following assets: the ability to treat, average out and smooth a large number of data, from either first- or second-order kinetics; a linearity-checking capability, and automatic compensation for blanks (e.g. in uncatalysed reactions).

The methods described so far for the treatment of kinetic data are largely based on mathematical procedures intended to determine the accurate value of the reaction rate and/or other related parameters. This strategy has also been applied to the resolution of mixtures by differential kinetics.

One of the most popular methods for determination of pseudo first-order rate constants is that developed by Guggenheim [144], which is useful for decay curves conforming to the expression $D(t) = K\exp(-t/\tau)$, where $D(t)$ is the measured property, $K = [A]_0$ and $1/\tau = k$. This function can be linearized to $Y = A_0 + A_1 t$ by taking $D(t)$–$D(t + \Delta t)$ pairs.

The terms of this linear function are defined as:

$$Y = \ln\,[D(t) - D(t + \Delta t)]$$

$$A_0 = \ln\,\{k[1 - \exp(-\Delta t/\tau)]\}$$

$$A_1 = 1/\tau$$

As the plot of Y *vs.* t is a straight line, τ and K can be calculated from its slope and intercept, respectively. In general, the problem lies in the selection of Δt, which should be at least 2–3τ unless the data-acquisition operation is carried out in a time shorter than this difference.

This method is simpler to implement than non-linear least-squares methods. However, it has some disadvantages, namely the problem of selection of Δt and its unsuitability for decay curves, which are normally affected by uncontrollable noise. These curves conform to the general expression $D(t) = K\exp(-t/\tau) + B$, where B is the baseline level. This last parameter, easy to calculate for stable noiseless baselines, can be subtracted from the curve to obtain a new curve that can be handled by the Guggenheim method, but this is not a very common situation, as the baseline can be affected by irreproducibility arising from the blank (luminescence techniques), oscillations from the instrument [145] or too short a data-acquisition time (so that the baseline is not reached) [146,147].

This problem has been tackled by Bacon and Demas [148] with their phase-plane method, which allows direct calculation of k, B and τ, is insensitive to noise and calls for no parameter-fitting by the operator.

Another method for calculation of first-order rate constants has been developed by Mieling and Pardue [149]. These authors propose a mathematical model for fitting of the kinetic curve, based on a simplified Taylor series expansion of the expression

$$A_t = A_\infty - (A_\infty - A_0)\exp(-kt)$$

The model affords the simultaneous determination of A_0, A_∞ and k from a series of experimental data taken during three or four reaction half-lives at the beginning of the reaction. The results predicted by this method are usually in good agreement with the absorbances measured. A multiple-regression program fits the experimental data to a first-order model.

A recently introduced method for estimation of rate constants of systems of varying complexity is based on the statistical moment theory [150]. This was widely used in the 1950s by chemical engineers to evaluate the phenomena of mass transfer through a tube [151]. Lately, it has been applied to the study of chromatographic elution [152,153], though rarely to kinetic studies [154,155].

The method involves calculating the area enclosed under the kinetic curve and the first-moment curve, and requires no explicit solutions, integrated kinetic equations or fitting. Knowledge of the relative concentrations of the species involved is only occasionally needed. The method is suitable for irreversible first-order and consecutive first-order kinetics, as well as for consecutive first-order kinetics with a reversible step and parallel first-order kinetics. Thus, with simple irreversible first-order reactions, the rate constant is given by

$$k = \int_0^\infty C\mathrm{d}t \Big/ \int_0^\infty tC\mathrm{d}t$$

where C is the analyte concentration.

The method has been tested both with simulated and experimental data and can be applied to drug transfer, sequential drug metabolism and other decay processes.

Higher-order reactions have also received attention in this context. According to Toby *et al.* [156], the rate law of a complex chemical reaction (e.g. a chain-reaction) between two species A and B can be written as:

$$-\mathrm{d}[\mathrm{A}]/\mathrm{d}t = k_1[\mathrm{A}][\mathrm{B}] + k_\mathrm{m}[\mathrm{A}]^m[\mathrm{B}]^p \quad (7.1)$$

As the kinetic study is usually performed with one of the reactants in excess, Eq. (7.1) can be simplified to:

$$-\mathrm{d}[\mathrm{A}]/\mathrm{d}t = k_1'[\mathrm{A}] + \mathrm{k}_\mathrm{m}'[\mathrm{A}]^m \quad (7.2)$$

where k_1' and k_m' are pseudo rate constants since they include the concentration of B.

Equation (7.2) is normally solved either by the graphical method involving the plot of $(-\mathrm{d}[\mathrm{A}]/\mathrm{d}t)/[\mathrm{A}]$ *vs.* $[\mathrm{A}]^{m-1}$, the slope and intercept of which allow the calculation of k_m' *and* k_1', respectively [157], or by a procedure based on the iterative minimization of the difference between what is actually observed and what is predicted [158].

Toby *et al.* [156] also proposed an integrated alternative to Eq. (7.2):

$$(1/n)\ln\{[\mathrm{A}]^n/(1 + k[\mathrm{A}]^n)\} = -k_1't + (1/n)\ln\{[\mathrm{A}]_0^n/(1 + K[\mathrm{A}]_0^n\}$$

where $K = k_\mathrm{m}'/k_1'$, $n = m - 1$ and $[\mathrm{A}]_0$ is the concentration of A at zero time.

The 'one-pass' method developed more recently by Bertrand *et al.* [159] is simpler and is valid for any value of m. Neither iterative nor repetitive, it can be readily implemented with the aid of a computer or even a programmable pocket calculator. It is based on data taken at regular time intervals and uses the same program irrespective of the reaction order (first, m or mixed).

Several methods have been developed for the particular case of $m = 2$ in Eq. (7.2). Thus, Shank and Dorfman [160] have reported a graphical procedure based on an extension of the Guggenheim method, and Kelter and Carr [161] have developed a new algorithm for determination of k_1' and k_2', featuring advantages such as great simplicity, rapidity and accuracy, as well as the possibility of discarding data not conforming to the kinetic model considered.

Computers, as stated above, play a major role in data treatment in differential kinetic analysis [5]. The graphical methods described in Chapter 6 generally use a small number of data and rely on the ease with which the behaviour of the more slowly reacting component can be monitored once the other component has reacted. However, multicomponent mixture analysis is better implemented by considering the overall response of the product as a function of time. Thus binary and ternary mixtures of alkaline-earth metals have been resolved with considerably increased accuracy and precision by the linear least-squares method [132]. Ridder and Margerum [162] have applied the variable data-acquisition speed, data-centring, correlation matrix and linear regression methods to the resolution of mixtures of zinc, cadmium, mercury and copper in the range 10^{-6}–$10^{-5}M$ by means of a minicomputer coupled to a stopped-flow system. Pinkle and Mark [163] have proposed the use of an analogue computer for the resolution of the simultaneous equations involved in the proportional-equation method.

REFERENCES

[1] H. V. Malmstadt, D. L. Krottinger and M. S. McCracken, in *Topics in Automatic Chemical Analysis*, J. K. Foreman and P. B. Stockwell (eds.), Chapter 4, Horwood, Chichester, 1979.
[2] H. V. Malmstadt, E. A. Cordos and C. J. Delaney, *Anal. Chem.*, 1972, **44**, 26A.
[3] H. V. Malmstadt, C. J. Delaney and E. A. Cordos, *Anal. Chem.*, 1972, **44**, 79A.
[4] R. M. Reich, *Anal. Chem.*, 1971, **43**, No. 12, 85A.
[5] S. R. Crouch, in *Computers in Chemistry and Instrumentation*, J. S. Mattson, H. B. Mark and H. C. McDonald (eds.), Vol. 6, Dekker, New York, 1972.
[6] T. E. Weichselbaum, W. H. Plumpe, Jr. and H. B. Mark, Jr., *Anal. Chem.*, 1969, **41**, No. 3, 103A.
[7] K. B. Yatsimirskii, *Kinetic Methods of Analysis*, Pergamon Press, Oxford, 1966.
[8] H. B. Mark and G. A. Reichnitz, *Kinetics in Analytical Chemistry*, Wiley–Interscience, New York, 1968.
[9] M. Kopanica and V. Stará, in *Wilson and Wilson's Comprehensive Analytical Chemistry*, G. Svehla (ed.), Vol. XVIII, Elsevier, Amsterdam, 1983.
[10] H. V. Malmstadt, C. J. Delaney and E. A. Cordos, *CRC Crit. Rev. Anal. Chem.*, 1972, **2**, 559.
[11] H. B. Mark, Jr., *Talanta*, 1972, **19**, 717.
[12] A. Moreno, M. Silva, D. Pérez-Bendito and M. Valcárcel, *Talanta*, 1983, **30**, 107.
[13] J. Růžička and E. H. Hansen, *Flow Injection Analysis*, 2nd Ed., Wiley, New York, 1988.
[14] M. Valcárcel and M. D. Luque de Castro, *Flow Injection Analysis: Principles and Applications*, Horwood, Chichester, 1987.
[15] H. Hartridge and F. J. W. Roughton, *Proc. Roy. Soc.*, 1923, **A104**, 376.
[16] B. Chance, *J. Franklin Inst.*, 1940, **229**, 455.
[17] R. H. Stehl, D. W. Margerum and J. J. Latterell, *Anal. Chem.*, 1967, **39**, 1346.
[17a] H. Gerischer and W. Heim, *Z. Phys. Chem. (Frankfurt)*, 1965, **46**, 345.
[18] G. D. Owens and D. W. Margerum, *Anal. Chem.*, 1980, **52**, 91A.
[19] G. D. Owens, R. W. Taylor, T. Y. Ridley and D. W. Margerum, *Anal. Chem.*, 1980, **52**, 130.
[20] S. A. Jacobs, M. T. Nemeth, G. W. Kramer, T. Y. Ridley and D. W. Margerum, *Anal. Chem.*, 1984, **56**, 1058.
[21] B. Chance, *J. Franklin Inst.*, 1940, **229**, 613, 737.
[22] B. Chance, in *Techniques of Organic Chemistry*, Vol. 8, Part 2, A. Weissberger (ed.), Wiley–Interscience, New York, 1963.
[23] K. Hiromi, in *Methods of Biochemical Analysis*, 1980, **26**, 137.
[24] F. J. Holler, S. R. Crouch and C. G. Enke, *Anal. Chem.*, 1976, **48**, 1429.
[25] P. M. Beckwith and S. R. Crouch, *Anal. Chem.*, 1972, **44**, 221.
[26] J. D. Ingle, Jr. and S. R. Crouch, *Anal. Chem.*, 1970, **42**, 1055.
[27] M. A. Koupparis, K. M. Walczak and H. V. Malmstadt, *J. Autom. Chem.*, 1980, **2**, 66.
[28] H. V. Malmstadt, K. M. Walczak and M. A. Koupparis, *Int. Lab.*, 1981, **11**, No. 1, 32.
[29] M. A. Koupparis, E. P. Diamandis and H. V. Malmstadt, *Clin. Chim. Acta*, 1985, **149**, 225.
[30] Q. H. Gibson and L. Milnes, *Biochem. J.*, 1964, **91**, 161.
[31] S. Ainsworth and L. Milnes, *Int. J. Biochem.*, 1977, **8**, 835.
[32] P. Papoff, B. Morelli and D. Guidarini, *Italian Patent* No. 2374A/73.
[33] P. Papoff, B. Morelli and L. Lampugnani, *Gazz. Chim. Ital.*, 1974, **81**, 104.
[34] B. Morelli, *J. Chem. Educ.*, 1976, **53**, 119.
[35] D. Sanderson, J. A. Bittikofer and H. L. Pardue, *Anal. Chem.*, 1972, **44**, 1934.
[36] G. E. Mieling, R. W. Taylor, L. G. Hargis, J. English and H. L. Pardue, *Anal. Chem.*, 1976, **48**, 1686.
[37] S. Stieg and T. Nieman, *Anal. Chem.*, 1980, **52**, 796.

[38] R. M. Reich, in *Instrumentation in Analytical Chemistry*, A. J. Senzel (ed.), ACS, Washington DC, 1973.
[39] R. Q. Thompson and S. R. Crouch, *Anal. Chim. Acta*, 1982, **144**, 155.
[40] P. K. Chattopadhyay and J. F. Coetzee, *Anal. Chem.*, 1972, **44**, 2117.
[41] D. Hanahan and D. S. Auld, *Anal. Biochem.*, 1980, **108**, 86.
[42] B. G. Willis, J. A. Bittikofer, H. L. Pardue and D. W. Margerum, *Anal. Chem.*, 1970, **42**, 1340.
[43] J. L. Holtzman, *Anal. Chem.*, 1980, **52**, 989.
[44] J. A. Alcocer, D. J. Livingston, G. F. Russell, C. F. Shoemaker and W. D. Brown, *J. Autom. Chem.*, 1983, **5**, 83.
[45] J. C. Thompsen and H. A. Mottola, *Anal. Instrum.*, 1984, **13**, 89.
[46] I. A. G. Roos, L. P. G. Wakelin, J. Hakkennes and J. Coles, *Anal. Biochem.*, 1985, **146**, 287.
[47] I. R. Bonnell and J. D. Defreese, *Anal. Chem.*, 1980, **52**, 139.
[48] P. Bertels, M. K. S. Mak and C. H. Langford, *Can. J. Spectrosc.*, 1981, **26**, 234.
[49] S. Lever and J. Crooks, *Int. Lab.*, 1985, **7** No. 10, 106.
[50] A. Loriguillo, M. Silva and D. Pérez-Bendito, *Anal. Chim. Acta*, 1987, **199**, 29.
[51] J. E. Erman and G. G. Hammes, *Rev. Sci. Instrum.*, 1966, **37**, 746.
[52] H. L. Pardue, *Rec. Chem. Progr.*, 1966, **27**, 151.
[53] H. V. Malmstadt and E. H. Piemmeier, *Anal. Chem.*, 1965, **37**, 34.
[54] R. E. Karcher and H. L. Pardue, *Clin. Chem.*, 1971, **17**, 214.
[55] R. E. Adams, S. R. Betso and P. W. Carr, *Anal. Chem.*, 1976, **48**, 1989.
[56] H. Weisz and K. Rothmaier, *Anal. Chim. Acta*, 1975, **75**, 119.
[57] S. Pantel and H. Weisz, *Anal. Chim. Acta*, 1974, **70**, 391.
[58] S. Pantel and H. Weisz, *Anal. Chim. Acta*, 1977, **89**, 47.
[59] S. Pantel and H. Weisz, *Anal. Chim. Acta*, 1975, **74**, 275.
[60] H. Weisz and H. Ludwig, *Anal. Chim. Acta*, 1972, **60**, 385.
[61] G. G. Hammes (ed.), *Investigation of Rates and Mechanisms of Reactions*, 3rd Ed., Part II, *Investigation of Elementary Reaction Steps in Solution and Very Fast Reactions*, Wiley–Interscience, New York, 1974.
[62] A. F. Yapel and R. Lumry, *Methods Biochem. Anal.*, 1971, **20**, 169.
[63] C. F. Bernasconi, *Relaxation Kinetics*, Academic Press, New York, 1976.
[64] H. Strehlow and W. Knoche, *Fundamentals of Chemical Relaxation*, Verlag Chemie, New York, 1977.
[65] H. Strehlow and S. Kalaricakal, *Ber. Bunsenges. Phys. Chem.*, 1966, **70**, 139.
[66] D. H. Turner, G. W. Flynn, N. Sutin and J. V. Beitz, *J. Am. Chem. Soc.*, 1972, **94**, 1554.
[67] H. Strehlow and M. Becker, *Z. Elektrochem.*, 1959, **63**, 457.
[68] A. Jost, *Ber. Bunsenges. Phys. Chem.*, 1966, **70**, 1057.
[69] L. Onsager, *J. Chem. Phys.*, 1934, **2**, 599.
[70] E. F. Caldin, *Fast Reactions in Solution*, Wiley, New York, 1971.
[71] K. K. Smith, K. J. Kaufman, D. Huppert and M. Gutman, *Chem. Phys. Lett.*, 1979, **64**, 522.
[72] J. H. Clark, S. L. Shapiro, A. J. Campillo and K. A. Winn, *J. Am. Chem. Soc.*, 1979, **101**, 746.
[73] M. Gutman, D. Huppert and E. Pines, *J. Am. Chem. Soc.*, 1981, **103**, 3709.
[74] J. N. Bradley, *Fast Reactions*, Clarendon Press, Oxford, 1975.
[75] A. K. Chibisov, *Usp. Khim.*, 1970, **10**, 1886.
[76] M. J. Bronskill and J. W. Hunt, *J. Phys. Chem.*, 1968, **72**, 3762.
[77] H. A. Schwarz, *J. Phys. Chem.*, 1962, **66**, 255.
[78] H. Weisz and G. Ludwig, *Anal. Chim. Acta*, 1971, **55**, 303.
[79] H. L. Pardue and P. A. Rodríguez, *Anal. Chem.*, 1967, **39**, 901.
[80] H. L. Pardue and S. N. Deming, *Anal. Chem.*, 1969, **41**, 986.
[81] J. D. Ingle, Jr. and S. R. Crouch, *Anal. Chem.*, 1972, **44**, 785.
[82] J. D. Ingle, Jr. and S. R. Crouch, *Anal. Chem.*, 1972, **44**, 1709.
[83] R. E. Santini, M. J. Milano and H. L. Pardue, *Anal. Chem.*, 1973, **45**, 915A.
[84] J. D. Ingle, Jr. and M. A. Ryan, in *Modern Fluorescence Spectroscopy*, E. L. Wehry (ed.), Vol. 3, Chapter 3, Plenum Press, New York, 1981.
[85] M. Valcárcel and F. Grases, *Talanta*, 1983, **30**, 139.
[86] R. L. Wilson and J. D. Ingle, Jr., *Anal. Chem.*, 1977, **49**, 1060.
[87] A. K. Chibisov, *Russ. J. Anal. Chem.*, 1983, **38**, 842.
[88] H. L. Pardue, in *Topics in Automatic Chemical Analysis*, J. K. Foreman and P. B. Stockwell (eds.), Chapter 6, Horwood, Chichester, 1979.
[89] T. A. Nieman and C. G. Enke, *Anal. Chem.*, 1976, **48**, 619.
[90] M. J. Milano and H. L. Pardue, *Clin. Chem.*, 1975, **21**, 211.
[90a] H. L. Pardue, in *Topics in Automatic Chemical Analysis*, J. K. Foreman and P. B. Stockwell

(eds.), Horwood, Chichester, 1979, p. 186.
[91] T. A. M. Carrick and F. W. McLafferty, *J. Chem. Educ.*, 1984, **61**, 463.
[92] G. M. Ridder and D. W. Margerum, *Anal. Chem.*, 1977, **49**, 2098.
[93] D. W. Johnson, J. G. Gallis and and G. D. Christian, *Anal. Chem.*, 1977, **49**, 747A.
[94] M. A. Ryan and J. D. Ingle, Jr., *Pittsburgh Conference on Analytical Chemistry and Applied Spectroscopy*, Cleveland, Ohio, 1979.
[95] M. A. Ryan, R. J. Miller and J. D. Ingle, Jr., *Anal. Chem.*, 1978, **50**, 1772.
[96] M. A. Ryan and J. D. Ingle, Jr., *Pittsburgh Conference on Analytical Chemistry and Applied Spectroscopy*, Atlantic City, New Jersey, 1980.
[96a] J. D. Ingle and M. A. Ryan, in *Modern Fluorescence Spectroscopy*, E. L. Wehry (ed.), Plenum Press, New York, 1981, p. 130.
[96b] K. Štulik and V. Pacáková, *Electroanalytical Measurements in Flowing Liquids*, Horwood, Chichester, 1987.
[96c] V. Linek, V. Vacek, J. Sinkule and P. Beneš, *Measurement of Oxygen by Membrane-Covered Probes*, Horwood, Chichester, 1988.
[97] G. A. Rechnitz and L. Zui-Feng, *Anal. Chem.*, 1967, **39**, 1406.
[98] L. A. Lazarou and T. P. Hadjiioannou, *Anal. Chem.*, 1979, **51**, 790.
[99] G. H. Efstathiou and T. P. Hadjiioannou, *Anal. Chem.*, 1975, **47**, 864.
[100] H. Thompson and G. A. Rechnitz, *Anal. Chem.*, 1974, **46**, 246.
[101] T. Anfalt, A. Graneli and D. Jagner, *Anal. Lett.*, 1973, **6**, 969.
[102] N. R. Larsen, E. H. Hansen and G. G. G. Guilbault, *Anal. Chim. Acta*, 1975, **9**, 79.
[103] L. Clark and A. Lyons, *Ann. N. Y. Acad. Sci.*, 1962, **29**, 102.
[104] G. G. Guilbault and E. Hrabankova, *Anal. Chim. Acta*, 1971, **56**, 285.
[105] G. G. Guilbault, G. Nagy and S. S. Kuan, *Anal. Chim. Acta*, 1973, **67**, 195.
[106] L. D. Mell and J. T. Maloy, *Anal. Chem.*, 1975, **47**, 299.
[106a] G. E. Baiulescu and V. V. Coşofreţ, *Applications of Ion Selective Membrane Electrodes in Organic Analysis*, Horwood, Chichester, 1977, p. 177.
[107] G. Johansson and L. Ogren, *Anal. Chim. Acta*, 1976, **84**, 23.
[108] G. Johansson, K. Edstrom and L. Ogren, *Anal. Chim. Acta*, 1976, **85**, 55.
[109] H. V. Malmstadt and H. L. Pardue, *Anal. Chem.*, 1961, **33**, 1040.
[110] H. L. Pardue, *Anal. Chem.*, 1963, **35**, 1240.
[111] J. B. Sand and C. O. Huber, *Anal. Chem.*, 1970, **42**, 238.
[112] R. K. Simon, G. D. Christian and W. C. Purdy, *Clin. Chem.*, 1968, **14**, 463.
[113] J. F. Norris and V. W. Ware, *J. Am. Chem. Soc.*, 1939, **61**, 1418.
[114] L. J. Papa, H. J. Patterson, H. B. Mark and C. N. Reilley, *Anal. Chem.*, 1963, **35**, 1889.
[115] D. E. Johnson and C. G. Enke, *Anal. Chem.*, 1970, **42**, 329.
[116] P. H. Daum and D. F. Nelson, *Anal. Chem.*, 1973, **45**, 463.
[117] S. G. Ballard, *Rev. Sci. Instrum.*, 1976, **47**, 1157.
[118] K. J. Caserta, F. J. Holler, S. R. Crouch and C. G. Enke, *Anal. Chem.*, 1978, **50**, 1534.
[119] P. Papoff and P. G. Zambonin, *Talanta*, 1967, **14**, 581.
[120] W. A. De Oliviera and L. Meites, *Anal. Chim. Acta*, 1974, **70**, 383.
[121] T. Meites, L. Meites and J. N. Jaitly, *J. Phys. Chem.*, 1969, **73**, 3801.
[122] J. Jordan, J. K. Grime, D. M. Waugh, C. D. Miller, H. M. Cullis and D. Lohrs, *Anal. Chem.*, 1976, **48**, 427A.
[123] R. E. Adams, S. R. Betso and P. W. Carr, *Anal. Chem.*, 1976, **48**, 1989.
[124] H. Weisz, W. Meiners and G. Fritz, *Anal. Chim. Acta*, 1979, **107**, 301.
[125] H. Weisz and W. Meiners, *Anal. Chim. Acta*, 1977, **90**, 71.
[126] H. L. Pardue, *Clin. Chem.*, 1977, **23**, 2189.
[127] H. L. Pardue in *Advances in Analytical Chemistry and Instrumentation*, Vol. 7, C. N. Reilley and F. W. McLafferty (eds.), Interscience, New York, 1969.
[128] J. D. Ingle, Jr. and S. R. Crouch, *Anal. Chem.*, 1971, **43**, 697.
[129] H. L. Pardue and M. M. Miller, *Clin. Chem.*, 1972, **18**, 928.
[130] M. T. Kelley and J. M. Jansen, *Clin. Chem.*, 1971, **17**, 701.
[131] N. G. Anderson, *Science*, 1969, **166**, 317.
[132] B. G. Willis, W. H. Woodruff, J. R. Frysinger, D. W. Margerum and H. L. Pardue, *Anal. Chem.*, 1970, **42**, 1350.
[133] E. Cordos, S. R. Crouch and H. V. Malmstadt, *Anal. Chem.*, 1968, **40**, 1812.
[134] H. V. Malmstadt and S. R. Crouch, *J. Chem. Educ.*, 1966, **43**, 340.
[135] H. L. Pardue, C. S. Frings and C. J. Delaney, *Anal. Chem.*, 1965, **37**, 1426.
[136] G. E. James and H. L. Pardue, *Anal. Chem.*, 1968, **40**, 796.
[137] S. R. Crouch, *Anal. Chem.*, 1969, **41**, 880.
[138] R. A. Parker, H. L. Pardue and B. G. Willis, *Anal. Chem.*, 1970, **42**, 56.
[139] W. J. Blaedel and G. P. Hicks, *Anal. Chem.*, 1962, **34**, 388.
[140] J. B. Worthington and H. L. Pardue, *Anal. Chem.*, 1972, **44**, 767.

[141] A. Gary and P. Lagrange, *Bull. Soc. Chim. France*, 1974, 1219.
[142] J. G. Liscouski, *Anal. Chem.*, 1982, **54**, 849A.
[142a] J. F. Tyson, *Analyst*, 1984, **109**, 313.
[142b] G. H. Morrison, *Anal. Chem.*, 1983, **55**, 1.
[143] J. L. Holtzman, *Anal. Chem.*, 1980, **52**, 989.
[144] E. A. Guggenheim, *Phil. Mag.*, 1926, **2**, 538.
[145] L. J. Cline Love and M. Skrilec, *Anal. Chem.*, 1981, **53**, 2103.
[146] S. L. Friess, E. S. Lewis and A. Weissberger, *Techniques of Organic Chemistry*, Vol. VIII, Part II, Wiley, New York, 1963.
[147] K. J. Laidler, *Chemical Kinetics*, 2nd Ed., Chapter I. McGraw-Hill, New York, 1965.
[148] J. R. Bacon and J. N. Demas, *Anal. Chem.*, 1983, **55**, 653.
[149] G. E. Mieling and H. L. Pardue, *Anal. Chem.*, 1978, **50**, 1611.
[150] K. K. Chan, M. B. Bolger and K. S. Pang, *Anal. Chem.*, 1985, **57**, 2145.
[151] E. Th. van der Laan, *Chem. Eng. Sci.*, 1957, **7**, 187.
[152] D. A. McQuarrie, *J. Chem. Phys.*, 1963, **38**, 437.
[153] K. Yamaoka and T. Nakagawa, *J. Chromatog.*, 1974, **92**, 213.
[154] W. Perl and P. Samuel, *Circ. Res.*, 1959, **25**, 191.
[155] J. Oppenheimer, H. Schwartz and M. I. Surks, *J. Clin. Endocrinol. Metab.*, 1975, **41**, 319.
[156] B. Toby, F. S. Toby and S. Toby, *Int. J. Chem. Kinet.*, 1978, **10**, 417.
[157] H. Linschitz and K. Sarkanen, *J. Am. Chem. Soc.*, 1958, **80**, 4826.
[158] R. Fletcher and M. J. D. Powell, *Comput. J.*, 1963, **6**, 163.
[159] R. Bertrand, J. E. Dubois and J. Toullec, *Anal. Chem.*, 1981, **53**, 219.
[160] N. E. Shank and L. M. Dorfman, *J. Chem. Phys.*, 1970, **52**, 4441.
[161] P. B. Kelter and J. D. Carr, *Anal. Chem.*, 1979, **51**, 1828.
[162] G. R. Ridder and D. W. Margerum, *Anal. Chem.*, 1977, **49**, 2090.
[163] D. Pinkle and H. B. Mark, *Talanta*, 1965, **12**, 491.

8

Sensitivity, selectivity, accuracy and precision

8.1 SENSITIVITY

The term 'sensitivity' is still one of the most controversial in the context of analytical chemistry. Although it is clearly different from the detection limit, it is used as a synonym for this in many papers and textbooks. The controversy is found not only in the literature, but also in practical work; in fact, the detection limit is calculated differently by different authors, many of whom disregard the model proposed by IUPAC in 1978 [1] and endorsed by the ACS Subcommittee of Environmental Analytical Chemistry two years later [2]. Unfortunately, the analytical community has not adhered unanimously to the IUPAC recommendations, so that comparisons between the performance of different methods in terms of their detection limits have still to be made with some reservations [3]. Consistent comparisons can thus only be made if detection limits are calculated according to the same criteria (such as those in the IUPAC guidelines).

8.1.1 Definitions

The concept of sensitivity is currently identified with the slope of the calibration graph, i.e. with IUPAC's *analytical sensitivity* [1]. However, sensitivity measurements should not be limited to calculation of the slope in question, since the calibration curve graph will be affected by some degree of dispersion, resulting in inaccurate determinations. According to Mandel [4], the sensitivity should be calculated as the ratio between the slope and its standard error of fit. Accordingly, the sensitivity will increase with increasing slopes and decreasing standard deviations. This is thus an excellent criterion by which to compare analytical methods and techniques insofar as the higher the sensitivity, the narrower the confidence intervals and the greater the precision.

The ability to determine a trace element or molecule in a chemical or biological matrix by a given analytical method is often expressed in terms of the detection limit. According to the IUPAC definition, "the detection limit is a number, expressed in units of concentration (or amount) that describes the lowest concentration level (or amount) of the element that an analyst can determine to be statistically different from an analytical blank."

Despite its apparent comprehensiveness, this definition leaves some doubts with the reader as to how the expression 'statistically different' should be interpreted. Thus, the detection limit calculated for a given element can differ by a few orders of magnitude depending on the particular statistical approach used.

According to IUPAC, a detection limit expressed in terms of a concentration C_L corresponds to an analytical signal X_L given by:

$$X_L = \overline{X}_B + kS_B \qquad (8.1)$$

where $\overline{X}_B$ is the mean blank response for n_B observations (generally $\geqslant 20$) defined by

$$\overline{X}_B = \frac{1}{n_B} \sum_{j=1}^{n_B} X_{B_j}$$

and S_B is the standard deviation of the blank, expressed as

$$S_B^2 = \frac{1}{(n_B - 1)} \sum_{j=1}^{n_B} (X_{B_j} - \overline{X}_B)^2$$

and k is a numerical factor selected according to the desired confidence level.

The concentration C_L is related to X_L through

$$C_L = (X_L - \overline{X}_B)/m \qquad (8.2)$$

where m is the analytical sensitivity (IUPAC version). Substitution of Eq. (8.1) into (8.2) yields:

$$C_L = kS_B/m \qquad (8.3)$$

The C_L value is a good measure of the detection limit provided m has a definite value and the intercept of the calibration graph is zero or virtually zero.

For $k = 3$, Eq. (8.3) allows for a confidence level of 99.86%, so $X_L \geqslant (\overline{X}_B + 3S_B)$ for measurements based on the error in the blank signal and conforming to a normal distribution. It should be noted that if X_B does not conform to a normal distribution, then the probability of X_L being equal to or greater than $(\overline{X}_B + 3S_B)$ will be only 89% according to Tschebyscheff's inequality [5]. Hence, no k values below 3 should be used in calculating the detection limit.

To obtain a detection limit reflecting more faithfully the value of m in Eq. (8.3), this should be expressed as a function of its confidence range, i.e. as $m \pm t_\alpha S_m$, where

S_m is the standard deviation of the slope and t_α the Student's t value corresponding to a desired confidence level, α, and the number of degrees of freedom of the system, ν. Under these conditions, Eq. (8.3) can be rewritten as:

$$C_L = kS_B/(m \pm t_\alpha S_m) \tag{8.4}$$

It is therefore the value of $t_\alpha S_m$ which dictates the expression to be used to calculate the detection limit. Thus, if this term is zero or much less than m, C_L can be calculated from Eq. (8.3); otherwise, appropriate correction should be made and the limit calculated from Eq. (8.4).

It is also of interest to take into account the error made in measuring the intercept of the calibration curve. As a rule, this is neglected or assumed to be zero, as measurements are normally made against a blank. However, the intercept turns out to be non-zero in most regression analyses. Bearing this in mind, IUPAC has proposed an alternative expression for calculation of the detection limit, namely:

$$C_L = k(S_B^2 \pm S_i^2)^{1/2}/m \tag{8.5}$$

where S_i is the standard deviation of the intercept. This expression is equivalent to Eq. (8.3) for $S_i = 0$ when no significant errors are made in measuring the slope.

Finally, some mention should be made of the so-called *quantification limit* [2], namely the analyte concentration corresponding to an analytical signal X_Q given by:

$$X_Q = \overline{X}_B + kS_B$$

This expression should not be applied for $k < 10$, and results in a quantification limit

$$C_Q = 10S_B/m \tag{8.6}$$

which can be transformed into two expressions similar to Eqs. (8.4) and (8.5) by the same considerations as above.

In short, there are three basic analytical regions connected with the determination of a substance (Fig. 8.1), namely: (a) that in which the analyte cannot be reliably detected (i.e. at concentrations below C_L); (b) that lying between C_L and C_Q, the detection region; (c) the quantification region, corresponding to concentrations $\geqslant C_Q$.

8.1.2 Sensitivity and detection limit in kinetic methods

It is widely known that the high sensitivity and low detection limits afforded by kinetic methods are the result of using catalysed reactions; hence literature reports dealing with the determination of these two parameters are largely concerned with this type of reaction.

The identification of sensitivity with the slope of the calibration curve was introduced in kinetic methodology by Mottola in 1975 [6]. Other definitions reported in the literature have more to do with the detection limit proper than with sensitivity.

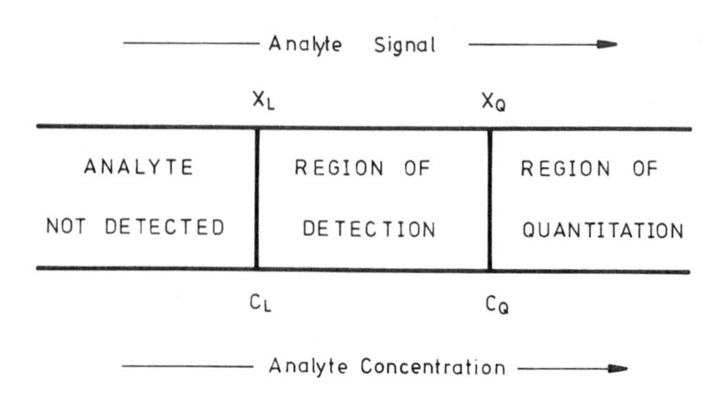

Fig. 8.1 — Regions of an analytical determination.

In his monograph, Yatsimirskii [7] deals briefly with the minimum concentration of a substance that can be determined by a kinetic catalytic method. The treatment used therein is related to the detection limit rather than to sensitivity. Despite the terminological confusion, the book gives a clear exposition of the great potential of kinetic methods compared with their equilibrium counterparts.

Yatsimirskii's treatment is limited to initial-rate measurements and relies on the rate law corresponding to the reaction $A + B \xrightarrow{C} P$,

$$d[P]/dt = k\pi_c[C]_0 = k'[C]_0 \tag{8.7}$$

where π_c is a factor comprising the reactant concentrations (i.e. $[A]_0[B]_0$) and where the contribution of the uncatalysed reaction has been assumed to be zero.

According to the expression above, the catalyst concentration will be given by

$$[C]_0 = (d[P]/dt)/k\pi_c \tag{8.8}$$

or in incremental form:

$$[C]_0 = (\Delta[P]/\Delta t)/k\pi_c \tag{8.9}$$

The denominator of this equation represents the number of cycles in which the catalyst takes part over a time interval Δt. From Eq. (8.9) it follows that the minimum catalyst concentration that can be determined depends on the minimum $\Delta[P]$ that can be measured by the particular instrumental technique used. The longer the observation time, Δt, the higher the reactant concentrations (π_c) and the larger the rate constant, k, the lower the determinable catalyst concentration.

This minimum detectable catalyst concentration is known as the *detection power* [8, 9]. Substituting Beer's law, $\Delta D = \varepsilon l \Delta[P]$, into Eq. (8.9) gives

$$[C]_{0,\min} = \Delta D_{\min}/t_{\max}\pi_{c,\max}k_{\max}\varepsilon_{\max}l_{\max} \tag{8.10}$$

i.e. the detection power of a catalytic spectrophotometric determination.

For a molar absorptivity of 10^5 l.mole^{-1}.cm^{-1}, a light-path of 5 cm, a minimum average absorbance difference of 0.05, a maximum analysis time of 10 min, a maximum rate constant of 10^8 min^{-1} and initial reactant concentrations of $1M$, the minimum detectable catalyst concentration will be:

$$[\mathrm{C}]_{0,\mathrm{min}} = 0.05/10 \times 1 \times 10^8 \times 10^5 \times 5 = 10^{-16}M$$

Such a low limit has never been reached in practice; in fact, the most sensitive catalytic methods allow measurement of concentrations of the order of 10^{-12} g/ml [10, 11], so the detection limit exceeds the theoretical detection power by a few orders of magnitude.

There are various reasons for the discrepancy between the theoretical and practical detection limit, the most significant of which is the so-called 'background effect' resulting from the uncatalysed reaction. If the contribution of this to the process is taken into account, the overall rate will be given by:

$$\upsilon = \upsilon_c + \upsilon_0$$

or by

$$\mathrm{d}[\mathrm{P}]/\mathrm{d}t = k\pi_c[\mathrm{C}]_0 + k'\pi_c$$

Hence, the rate of the catalysed reaction will be given by $\upsilon_c = \upsilon - \upsilon_0$. Obviously, the smaller υ_0, the greater the difference in this expression and the lower $[\mathrm{C}]_{0,\mathrm{min}}$ will be. The contribution of the uncatalysed reaction though the background effect can be expressed by means of a factor, α, which will vary between 0 (absence of uncatalysed reaction) and 1 (absence of catalysed reaction). Such a contribution can be incorporated into Eq. (8.10), which becomes:

$$[\mathrm{C}]_{0,\mathrm{min}} = (\Delta[\mathrm{P}]_{\mathrm{min}}/\Delta t_{\mathrm{max}} k\pi_c) + (\alpha\upsilon_0/k\pi_c)$$

If $\upsilon_0 = k'\pi_c$,

$$[\mathrm{C}]_{0,\mathrm{min}} = (\Delta[\mathrm{P}]_{\mathrm{min}}/\Delta t_{\mathrm{max}} k\pi_c) + \alpha(k'/k) \tag{8.11}$$

The k'/k ratio can be expressed as:

$$k'/k = \exp[-(E_0 - E_c)/RT]$$

where E_0 and E_0 are the activation energies of the catalysed and uncatalysed reactions, respectively.

According to Eq. (8.11), the lowest detection limit will be obtained for the greatest difference between the rate constants (or activation energies) of the catalysed and uncatalysed reactions, i.e. when k'/k is negligible and the second term in Eq. (8.11) can safely be neglected.

The minimum catalyst concentration or detection limit can also be calculated by other methods described in the literature. Thus, Liteanu *et al.* [12] have estimated the photometric detection limit for the catalytic determination of vanadium by the bromate–Bordeaux R reaction to be of the order of 10^{-8} g/l. by applying the Neyman–Pearson statistical signal detection criterion and making a statistical analysis of the noise oscillations. Proskuryakova *et al.* [13] have reviewed the mathematical expressions used to calculate the minimum amount, n_{min}, of a given chemical species that can be determined by an absorptiometric method and recommend the equation $n_{min} = (D - D_0)/m = 0.05/m$, where D and D_0 are the absorbances of the sample and blank, respectively, and m is the slope of the calibration curve. This procedure is applicable both to equilibrium photometric determinations and to kinetic determinations based on the fixed-time method.

In an interesting review published in 1975, Mottola [6] reports other useful expressions for calculation of both the sensitivity and the detection limit of catalytic initial-rate, fixed time and variable-time kinetic methods. Through the expressions

$$-\Delta[A]/\Delta t = [A]_0(k' + k[C]_0)$$

he identifies the intercept ($k'[A]_0$) of the plot of $-\Delta[A]/\Delta t$ *vs.* $[C]_0$, with the detection limit. The same concept is applied to the fixed-time and variable-time methods. In addition, he uses the term *limit of guarantee of purity*, which is intended to answer the question of the minimum purity that can be guaranteed if a substance that might be present cannot be detected because the experimental value is below the detection limit [14]. Its evaluation is based on the mean and standard deviation of the blank values.

The currently accepted definitions of sensitivity and detection limit having been given and the evolution of these two concepts within the context of kinetic methods of analysis discussed, it is worth outlining the different expressions developed for their calculation in the various available kinetic techniques.

Table 8.1 lists the expressions defining the sensitivity and detection limit for

Table 8.1 — Sensitivity and detection limit in kinetic determination methods

Method	Calibration graph	Sensitivity	Detection limit
Initial-rate	$\Delta[P]/\Delta t = k\pi_c[C]_0$	$k\pi_c$	$3S_B/k\pi_c$
Fixed-time	$\Delta[P] = k\pi_c\Delta t[C]_0$	$k\pi_c\Delta t$	$3S_B/k\pi_c\Delta t$
Variable-time	$1/\Delta t = k\pi_c[C]_0/\Delta[P]$	$k\pi_c/\Delta[P]$	$3S_B\Delta[P]/k\pi_c$

initial-rate, fixed-time and variable-time methods applied to catalysed reactions. Similar expressions could be obtained in integral form or for modified catalytic or uncatalytic reactions. As can be seen, for simplicity no provision has been made for the contribution of the uncatalysed reaction (background effect). If such a contribution is appreciable, the expressions corresponding to the detection limit should be modified by introducing a term proportional to the ratio of the rate constants of the uncatalysed and catalysed reactions in Eq. (8.11). The expression for calculation of the sensitivity need not be altered, though.

The calculation of the detection limit according to the expressions given in Table 8.1 requires taking account of the blank, from which S_B can be determined. The blank signal corresponds to the uncatalysed reaction in the case of catalytic determinations and to the catalysed (unmodified) reaction in the case of modified catalytic reactions. As the blank undergoes no change in uncatalysed reactions, S_B will be zero in rate terms. In such a case it pays to use the criterion established by Wolf and Stewart [15] for flow systems, based on taking S_B as the standard deviation of a given sample for 30 determinations and calculating the detection limit from this. This procedure is useful both for fixed-time and variable-time methods, although in the former, the oscillations in the $\Delta[P]$ measurements corresponding to the blank will arise from instrumental noise and hence S_B can be identified with the standard deviation of the noise.

The expressions for calculation of the sensitivity and detection limit listed in Table 8.1 allow the following conclusions to be drawn: (a) as a rule, the sensitivity increases and the detection limit decreases with increasing rate constants and reactant concentrations (increasing π_c); (b) the two parameters vary in the same way as above with Δt in the fixed-time method, so that it is preferable to use this in integral rather than in differential form; (c) $\Delta[P]$ has a critical effect on these parameters in the variable-time method and should be kept as small as possible — hence it is preferable in this case to use the differential form.

8.1.3 Sensitivity and detection limit in catalysed reactions

Despite the intrinsic difficulty involved (the lack of homogeneity in the calculation) it is worth comparing the performance of the different kinetic methods based on catalysed reactions in terms of their detection limits. Figure 8.2 presents the method

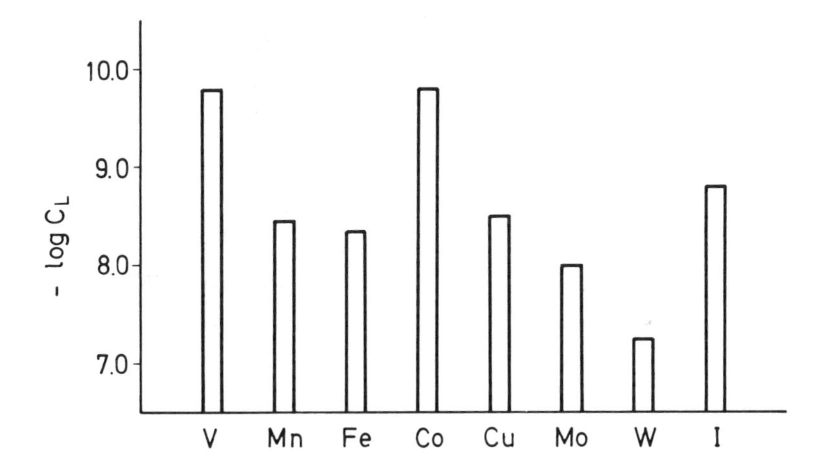

Fig. 8.2 — Comparison between the detection limits afforded by the different kinetic catalytic methods in the determination of the commonest ions.

used for determination of common ions on the basis of their catalytic effect. As can be seen, the lowest detection limits achieved so far (about $10^{-10}M$) correspond to the kinetic determinations of cobalt and vanadium and are smaller than those afforded by the use of chemiluminescent indicator reactions. Such low detection levels are sometimes achieved with the aid of activators [e.g. Fe(III) and Cu(II) in the case of vanadium] [16].

The low detection limits afforded by kinetic methods have some disadvantages; in fact, they demand especial care in preparing, storing and transferring the solutions of low concentration used, in order to avoid errors arising from loss or contamination. Thus, great care should be taken in washing laboratory ware (particularly glassware, which can give trace metal contamination at concentrations below $10^{-7}M$ [17]). In addition, the trace metal content of analytical reagents, mineral acids or even the water used to prepare the standards and solutions, imposes practical restraints on the detection limit. It is therefore surprising to find papers reporting detection limits below the expected contamination levels in the water used. Such is the case with the determination of copper at the $10^{-8}M$ level, which calls for high-purity water that cannot be obtained even after a triple distillation, since use of a borosilicate glass still results in copper contents of about $10^{-8}M$ in the distilled water [17]. This shortcoming can be circumvented by using ion-exchange resins, but though these lower the heavy metal content, they may also have the undesirable effects of introducing complexing species and reductants interfering with some redox systems [18]. A combination of both techniques, distillation and ion-exchange, is usually the best way to obtain water of acceptable purity.

On the other hand, storage in polyethylene flasks may result in contamination from plasticizers which absorb ultraviolet light and can be potential reductants. However, this consideration is not universally valid; thus, plastic is preferred to glass in the determination of zinc at the ng level, as the presence of organic species at low concentrations poses no problem in this case.

Another major problem in dealing with very dilute solutions is their tendency to suffer loss of metals by adsorption on the container walls. This is usually avoided by using polytetrafluoroethylene (PTFE) bottles and keeping the solutions at low pH in a refrigerator. If all these steps are not sufficient to preserve the solution stability, the addition of masking agents that do not interfere with the subsequent kinetic catalytic determination should be the definitive solution. The stability of the very dilute manganese solutions used in the determination of this metal in food samples has been studied by using tartrate as a complexing agent [19]. Figure 8.3 shows the results obtained by using glass and PTFE containers and different pH values and concentrations of tartrate (from zero upwards). A 100 ng/ml solution of manganese to which tartrate has been added to give a concentration of 10 μg/ml remains stable for over 50 hr when kept at pH 3.7 in a PTFE bottle. No comparable stability can be obtained in the absence of the complexing agent.

8.1.4 Ways to improve the sensitivity and detection limit

This section deals exclusively with catalysed reactions on account of the great sensitivity and low detection limits of the methods based on them.

As stated above, the detection limit is highly influenced by the degree of development of the uncatalysed reaction (background effect); its variations can be

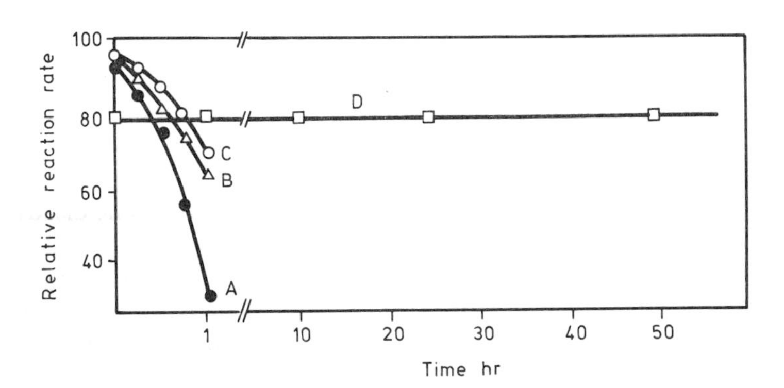

Fig. 8.3 — Effect of preparation conditions on the stability of dilute manganese(II) solutions. A, prepared at pH 0.6 or 4.6 and stored in glass or PTFE bottles; B, prepared at pH 0.6 with 1 μg/ml tartrate and stored in PTFE bottles; C, prepared at pH 3.7 with 1 μg/ml tartrate and stored in PTFE bottles; D, prepared at pH 3.7 with 10 μg/ml tartrate and stored in PTFE bottles. (Reproduced from S. Rubio, A. Gómez-Hens and M. Valcárcel, *Analyst*, 1984, **109**, 717, by permission. Copyright 1984, Royal Society of Chemistry).

minimized by taking due care in the washing and manipulation of laboratory ware. This entails avoiding differences in the matrix (i.e. between standards and samples) as regards pH, saline composition, ionic strength, impurities introduced by the water or the reagents, and even the presence of apparently insignificant amounts of solid particles in the reaction vessel.

Nevertheless, this section is concerned with ways of increasing the sensitivity or lowering the detection limit of methods based on catalysed reactions.

8.1.4.1 Optimization of the procedure

This calls for accurate knowledge of the catalysed reaction involved. Once this has been established, the experimental conditions best suited to the determination can readily be found by graphical procedures [20]. The knowledge of the mechanism facilitates obtaining the overall reaction rate for stationary concentrations of the catalyst, as well as determination of the time-dependence of the system, which can thus be compared with that found experimentally.

This is the way Yatsimirskii *et al.* [21] proceeded in the catalytic determination of iridium and ruthenium with the reaction between Mn(III) and Hg(I) as indicator. The mechanism of the catalysed reaction for iridium is as follows:

$$\mathrm{Ir(III)} + \mathrm{Mn(III)} \underset{k_{-1}}{\overset{k_1}{\rightleftharpoons}} \mathrm{Mn(III)} \ldots \mathrm{Ir(III)}$$

$$\mathrm{Mn(III)} \ldots \mathrm{Ir(III)} \xrightarrow{k_2} \mathrm{Mn(II)} + \mathrm{Ir(IV)}$$

$$\mathrm{Ir(IV)} + \mathrm{Hg(I)} \xrightarrow{k_3} \mathrm{Ir(III)} + \mathrm{Hg(II)}$$

This system is schematically shown in Fig. 8.4. According to Meson and Zimmermann [22], the overall reaction rate of this process is given by:

$$\upsilon = C_{total}\Sigma\upsilon_i D_i/\Sigma D_i$$

where C_{total} is the overall catalyst concentration, υ_i denotes the rate of formation of the different products and D_i the determinants derived for the various nodal points of the graph.

After simplification and taking into account the low Hg(I) concentration used, the reaction rate can be expressed as:

$$\upsilon_{Ir} = k_1k_3C_{total}[Mn(III)][Hg(I)]/(k_3[Hg(I)] + k_1[Mn(III)]) \quad (8.12)$$

In the case of ruthenium, and taking into account that $k_2 = k_3[Hg(I)]$, the corresponding expression is:

$$\upsilon_{Ru} = k_1k_3C_{total}[Mn(III)][Hg(I)]/(k_2 + k_1[Mn(III)] \quad (8.13)$$

By applying Eqs. (8.12) and (8.13), these authors have established the most favourable conditions for determination of both catalysts, which they determined at concentrations below 10^{-2} and 10^{-3} μg/ml for Ir and Ru, respectively.

Another major parameter to optimize in the procedure is the temperature. According to the Arrhenius equation, the reaction rate increases with increasing temperature. However, this trend is only important for the catalysed reactions, since the rates of uncatalysed reactions are so low that the increase caused by a temperature rise has no practical influence on the results. Unfortunately, too high a temperature may give rise to practical problems, so most kinetic catalytic determinations are performed at not more than 40–50°C. The optimization of other parameters such as the reactant concentrations (π_c, Section 8.1.2), ionic strength or dielectric constant (Chapter 1), can also have a beneficial effect on the reaction rate.

Unlike selectivity (see Section 8.2.4.2), there are few references to the use of separation techniques to improve the sensitivty and detection limit of methods based on catalysed reactions. Kawashima *et al.* [23] have developed a kinetic fixed-time method for the determination of selenium, based on its catalytic effect on the oxidation of *p*-hydrazinobenzenedisulphonic acid to the *p*-diazobenzenediazonium ion, which is subsequently converted into a yellow azo-dye by coupling with *m*-phenylenediamine. If the azo-dye is extracted into an organic solvent such as methyl isobutyl ketone, or amyl or isoamyl alcohol, the sensitivity (expressed by these authors as the effective molar absorptivity) is doubled from 1.2×10^6 l.mole^{-1}.cm^{-1} in an aqueous medium to 2.4×10^6 l.mole^{-1}.cm^{-1} in the organic solution. Another example worthy of note is the joint use of kinetic catalytic methods and gas chromatography. Thus, Dieztler *et al.* [24] suggest the liquid–liquid extraction of the catalysed reaction product at a preselected time and the injection of an

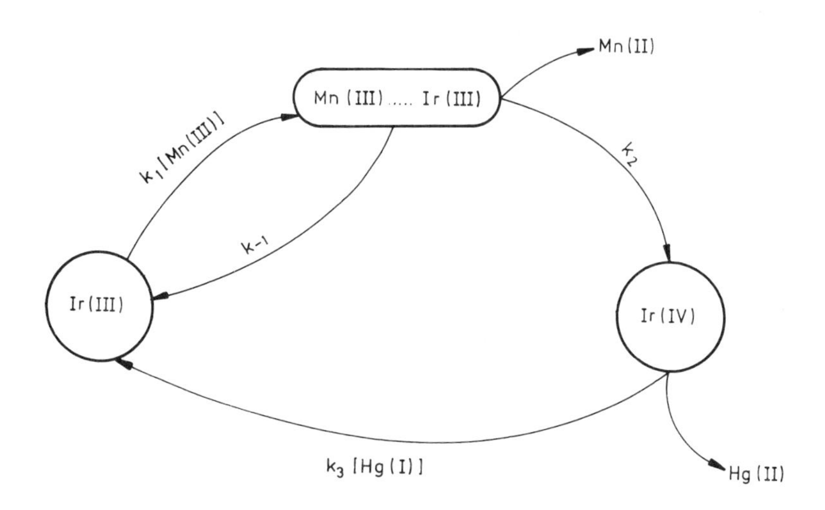

Fig. 8.4 — Mechanism of the iridium-catalysed redox reaction between Hg(I) and Mn(II).

aliquot of the extract into a gas chromatograph. The practical procedure is described in detail in Section 9.2.4.

8.1.4.2 Activation of catalysed reactions

The best way to increase the rate of a catalysed reaction is no doubt through the use of activators (see Chapter 3).

The activating effect is closely related to the ability of metal ions to form complexes. In addition to increased sensitivity and lower detection limits (by two to four orders of magnitude), the use of activators results in increased selectivity stemming from the influence of the catalyst on the reactants (see Section 8.2.4.1).

By way of example, Fig. 8.5 illustrates the activating effect of Zn^{2+} on the manganese-catalysed oxidation of 2-hydroxybenzaldehyde thiosemicarbazone by hydrogen peroxide [25]. As can be seen, the sensitivity is increased (the slope of lines B and C is greater than that of A) and the detection limit is lowered (line C *vs*. A) as a result of the presence of the activator. Thus, manganese can be determined at concentrations up to 10 ng/ml in the presence of 250 ng/ml zinc. A similar effect has been found in the kinetic determination of indium(III) based on its activating effect on a copper-catalysed reaction [26].

8.2 SELECTIVITY

According to IUPAC [27], selectivity is a quality indicative of the degree or extent to which other substances interfere with the determination of an analyte by a given procedure. Specificity, the ultimate degree of selectivity, is the absence of any interferences.

An interferent is usually defined as a species causing a systematic error in

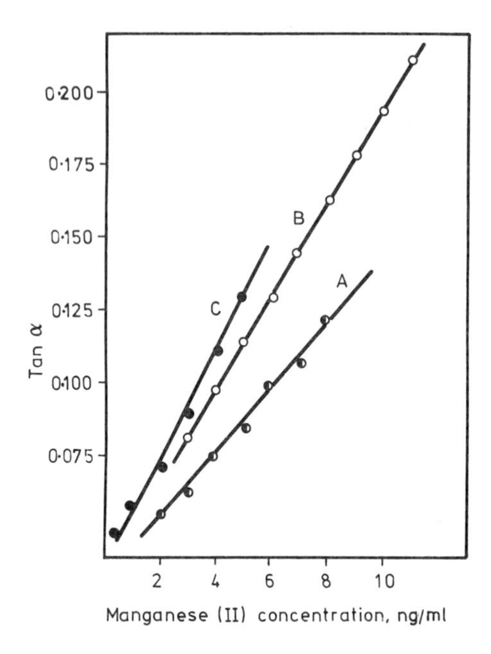

Fig. 8.5 — Dependence of the rate of the oxidation of 2-hydroxybenzaldehyde thiosemicarbazone by H_2O_2 on the Mn(II) concentration. A, without zinc; B and C, in the presence of 250 ng/ml and 1 μg/ml zinc(II), respectively. (Reproduced from A. Moreno, M. Silva, D. Pérez-Bendito and M. Valcárcel, *Analyst*, 1983, **108**, 85, by permission. Copyright 1983, Royal Society of Chemistry).

determination of the analyte, the magnitude of which should be established in terms of the standard deviation for a single determination.

Analytical interferences can be classified as additive or multiplicative according to whether the interfering signal is yielded by the same mechanism as the analytical signal.

A distinction should also be made between *active interferents* (i.e. those species directly interacting with the analyte, one of the ingredients of the indicator reaction or even the solvent) and *inert interferents* (those contributing positively or negatively to the analytical signal without interacting with the analyte or any of the substances involved in the reaction system).

As demonstrated above, sensitivity is one of the most outstanding features of kinetic methods based on catalysed reactions. Unfortunately, these methods are not as selective as they are sensitive, though in many cases they are better than their equilibrium counterpart in this respect.

8.2.1 Selectivity of kinetic methods versus equilibrium methods

The relative absence of parasitic signals and the possibility of implementing kinetic discrimination make kinetic methods more selective than equilibrium methods.

An essential feature of kinetic methods is the differential rather than the absolute nature of the measurements on which they are based (signal increments measured at preselected time intervals). This results in no interference from the signals yielded by other components of the samples as long as these do not interact with any of the species involved in the system. Thus, photometric analysis of wines may experience serious problems arising from the sample colour and requiring the use of separation techniques. In measurement of the rate of reaction between the analyte and reagent, such problems do not arise. Such is the case with the kinetic photometric determination of iron in wine by the stopped-flow technique, based on formation of the Fe(III)–thiocyanate complex [28]. Unlike its conventional counterpart, the kinetic method can be applied to red wines with excellent results.

The greater selectivity of kinetic methods can be attributed to the dynamics of the process involved. The short sample–reagent mixing time and the fact that measurements are started immediately after mixing may minimize the effect of interferents affecting equilibrium methods much more adversely. This is because of the long contact time needed to reach equilibrium, during which interferents (active interferents) may react with one or several of the system components, including the analyte.

On comparing unsegmented-flow methods such as FIA, also based on measurements made under non-equilibrium conditions, with manual kinetic methods described for the kinetic catalytic determination of a given species, it is evident that the former are subject to a larger number of interferences. Such is the case when copper, as a catalyst for the oxidation of an organic compound, is determined by FIA [29] and by the conventional technique based on reaction-rate measurements [30]. Since both the residence time and that at which the initial rate is measured (30 sec) are the same for the two methods, the greater selectivity of the conventional kinetic method can only be the result of taking differential rather than absolute fixed-time measurements (peak height corresponding to residence time).

8.2.2 Kinetic discrimination

As pointed out above, the selectivity of kinetic methods is partly the result of the so-called kinetic discrimination, based on the different rates at which the analyte and the other sample components interact with the reagent.

The determination of iron(III) by formation of a complex with pyridoxal thiosemicarbazone (λ_{max} = 425 nm) is subject to fewer interferences when carried out by a kinetic method [31] than by an equilibrium method [32]. The kinetic method relies on the fact that the complex formation is slowed down by the presence of acetate. Table 8.2 compares the selectivity of both methods. As can be seen, the

Table 8.2 — Comparison between the selectivities afforded by equilibrium and kinetic methods for determination of iron

Foreign ion	Tolerated foreign ion-to-Fe ratio: Equilibrium method	Tolerated foreign ion-to-Fe ratio: Reaction-rate method	Selectivity factor
Zn	0.4	10	25
Cd^-	8	24	3
Pb	120	>1000	>8
CN^-	1	5	5

kinetic method is much more tolerant to foreign species than is the equilibrium method.

The short time elapsing between mixing of the reactants and the first measurement occasionally results in decreased influence of foreign species (side-reactions), as shown by comparing the kinetic and equilibrium methods. Accordingly, the selectivity of a chemical system used in different kinetic methods will also increase as the measurement and mixing time become closer. This effect is illustrated below by three examples.

Induction-period measurements (Landolt effect) are more selective than initial-rate measurements in the kinetic–fluorimetric determination of Fe(III) on the basis of its promoting effect on the oxidation of pyridoxal 2-pyridylhydrazone by hydrogen peroxide [33], as shown in Table 8.3. The induction-period measurements are

Table 8.3 — Selectivity levels afforded by different kinetic methodologies

Analyte	Foreign species	Tolerated foreign ion-to-analyte ratio		Selectivity factor
		Initial rate	*Induction period*	
Iron	Ni	3	25	8.4
	I^-	1	3	3.0
	Be, Ba, Cr, Pb, tartrate	50–75	100	≤2.0
		Fixed time	*Initial rate*	
Histidine	3-Methyl-histidine	1	10	10.0
	cysteine, Mg	1	5	5.0
	glycine	5	10	2.0
		Without activator	*With activator*	
Manganese	Pb	50	1500	30.0
	Co	20	300	15.0
	Ca, Ce, Cr, iodate	400	3000	7.5
	Mg, Sn	200	800	4.0
	Ag	50	150	3.0
	Ni, Cd, Hg, S^{2-}	200–500	300–750	1.5

necessarily made after a shorter time interval than the initial-rate measurements are.

The fixed-time method is subject to a greater number of more serious interferences than the initial-rate method since, although the former is applied to the initial straight portion of the curve, the latter involves measurements made soon after mixing. This is shown for example, by comparing the results obtained in the kinetic–fluorimetric determination of histidine, which accelerates the copper-catalysed oxidation of 2-amino-1,1,3-tricyano-1-propene by hydrogen peroxide, by an initial-rate and a fixed-time method [34], the former of which is, according to Table 8.3, the more selective.

The use of activators in the demonstration of catalysts is also a source of increased sensitivity (and selectivity) as shown above. When the initial-rate method is applied to the determination of a catalyst with and without an activator, the slope of the kinetic curve is found to be steeper and its measurement zone closer to the mixing zone, when the activator is used. Such is the case with the determination of Mn(II) as a catalyst for the oxidation of 2-hydroxybenzaldehyde thiosemicarbazone by hydrogen peroxide to yield a highly fluorescent product. As can be seen from Table 8.3, the tolerance ratio (and hence selectivity) obtained by the initial-rate method is greater in the presence [25] than in the absence [35] of an activator such as Zn(II).

The application of the stopped-flow technique can also provide increased selectivity as a result of bringing the measurement times nearer to the mixing time. Table 8.4 gives the results of a study on the influence of foreign ions on the

Table 8.4 — Comparison between the selectivity levels afforded by conventional and stopped-flow methods for determination of copper

	Tolerated foreign ion-to-Cu ratio		
Foreign species	Conventional method	Stopped-flow method	Selectivity factor
Hg	1	5	5
Pt	5	10	2
Mn	10	25	2.5
Cr, As, Pb	25	100	4
Oxalate, pyrophosphate	25	100	4

determination of copper as a catalyst for the hydrogen peroxide oxidation of imidazole by the conventional kinetic and stopped-flow techniques [36].

Differential reaction-rate methods are the most significant exponents of kinetic discrimination as they allow the simultaneous determination of two or more analytes in a mixture (see Chapter 6) and are thus endowed with extreme selectivity. On the other hand, synergistic effects on the reaction rate result in decreased selectivity and make some differential methods impractical. Such effects can also be turned to advantage, however, as described in Chapter 6.

8.2.3 Selectivity of catalytic methods

Methods based on catalysed reactions are characterized by their extremely low detection limits, and are thus far superior to many conventional analytical methods in this respect. However, their poorer selectivity has prevented them from becoming commonplace in routine analysis. In fact, few indicator reactions are catalysed by only one species (e.g. the dimerization of quinolinaldehyde [37], catalysed exclusively by cyanide).

Enzyme-catalysed reactions are an exception to this rule. Their increasing use (either in solution or immobilized) for analyses of complex matrices is a result of the high selectivity (often verging on specificity) of the reactions in which they are involved.

The selectivity of non-enzymatic catalysed reactions and methods of improving it have been dealt with at length in the literature [38–40].

The catalytic properties of a chemical species depend on its size, structure, charge and co-ordination sphere; hence chemically related species may have similar catalytic effects and render some determinations difficult or unfeasible. Thus, the key step in the oxidation of organic or inorganic substrates by hydrogen peroxide in an acid medium is the formation of a complex between the metal catalyst and the peroxide, involving the cleavage of the O–O bond of the latter. Metals of a similar nature will have a similar effect on peroxide decomposition and hence on the overall catalysed reaction, so a selective determination will only be possible after a suitable modification of the procedure (see Table 8.5) [39].

Table 8.5 — Indicator reactions with hydrogen peroxide as oxidant

Reaction	Catalysts
$H_2O_2 + I^-$	Ti, Zr, Hf, Th, V, Nb, Ta, Cr, Mo, W, Fe(III), $Cr_2O_7^{2-}$, PO_4^{3-}
$H_2O_2 + S_2O_3^{2-}$	Ti, Zr, Hf, V, Nb, Ta, Mo, W
$H_2O_2 + e^-$	Hf, V, Mo, W, Fe(III)
$H_2O_2 + C_6H_5OH$, $C_6H_5NH_2$	Cr, Mn, Fe, Co, Ru, Os, Cu
$H_2O_2 + H_2(NHCS)_2$	Mo, W
H_2O_2 + org. dyes	Fe, Cu, Cr, Mn, Co, Ni

Interferences are much more difficult to predict in practice. Thus, V(V) catalyses the oxidation of Bordeaux R [41], and this reaction can be interfered with positively or negatively by a given species, depending on its concentration; this makes quantitative handling of interferences rather complicated. Kaiser [42] has so far been the only author to develop a mathematical expression for selectivity and specificity. According to this author, selectivity is the capability to determine individually all the components of a sample, and therefore refers to the system as a whole.

The selectivity or specificity of a catalytic determination depends on that of the particular catalysed reaction and monitoring technique used. The rate of an indicator reaction $A + B \rightarrow P$ catalysed by various species ($C_1, C_2 \ldots C_n$) can be expressed as [39]:

$$\upsilon_{tot} = d[P]/dt = k_1[A]^p[B]^q[C_1] + k_2[A]^p B]^q[C_2] + \ldots + k_n[A]^p[B]^q[C_n]$$

If the overall reaction is carried out under n different sets of experimental conditions, the resultant rates will be:

$$\upsilon_1 = k_{11}\pi_{11}[C_1] + k_{12}\pi_{12}[C_2] + \ldots + k_{1n}\pi_{1n}[C_n]$$

$$\upsilon_2 = k_{21}\pi_{21}[C_1] + k_{22}\pi_{22}[C_2] + \ldots + k_{2n}\pi_{2n}[C_{2n}]$$

$$\ldots$$

$$v_n = k_{n1}\pi_{n1}[C_1] + k_{n2}\pi_{n2}[C_2] + \ldots + k_{nn}\pi_{nn}[C_n]$$

where $v_1 \ldots v_n$ denote the total catalysed reaction rates and $\pi_{nn} = [A]^p_{ik}[B]^q_{ik}$, with $i, k = 1, 2 \ldots n$.

A catalytic determination will be selective if the catalytic efficiency of each of the individual catalysts varies significantly with the experimental conditions ($k_n > k_{ik}$) or if the catalysts present react according to different mechanisms ($p_n, q_n > p_{ik}, q_{ik}$). Unfortunately, little is known about the particular mechanisms followed by different catalysts in an indicator reaction, so only their effects can be determined with some degree of accuracy.

Quantification of selectivity and specificity requires the formation of a matrix of partial selectivities. By solving the equations [40]

$$Z = \min_{i=1\ldots n} \left| \frac{T_{ii}}{T_{ik} - T_{ii}} - 1 \right| ; \qquad \Psi = \frac{T_{aa}}{T_{kk} - T_{aa}} - 1$$

a numerical estimation of the selectivity, Z, and specificity, Ψ, is obtained. As can be seen, the selectivity increases as all the elements in the matrix (not only those on the main diagonal) approach zero. Specificity requires only one of the elements, namely T_{aa}, to be non-zero. This mathematical treatment yields good results provided the catalysts involved have no influence on one another.

The selectivity of reactions catalysed by chemically dissimilar species is usually high. Such is the case with the determination of molybdenum in the presence of alkali and alkaline-earth metals, cobalt, copper, iron and manganese [43], or that of metals of the platinum family in the presence of copper, iron and related metals [44]. Conversely, the greater the similarity between the analyte and the interferent(s), the lower the selectivity of the analytical determination. Thus, very little selectivity is to be expected in the determination of Ti, Zr, Hf, Fe, Nb, Ta, Mo or W (all of which are catalysts for the oxidation of different organic substances by H_2O_2) unless the matrix contains none of these metals other than the analyte (see Table 8.5).

Vanadium, ruthenium, rhenium and osmium are frequently catalysts for reactions involving halates, and so are Ag, Fe and V for persulphate, and Os, Ru and I^- for arsenite. Ligand-exchange reactions also exhibit non-specific catalytic effects at times [45, 46].

Most of the catalytic determinations reported in the literature are optimized for sensitivity, selectivity, the key to their routine analytical applicability, being frequently disregarded.

Although selectivity can be improved in several ways, it is often unavoidable to resort to separation techniques to remove interferences.

8.2.4 Ways to improve selectivity

The selectivity of a determination can be improved in two ways, for both catalysed and uncatalysed reactions (and in many cases for equilibrium methods for trace analysis), namely by optimizing the procedure and by using separation techniques.

8.2.4.1 Optimization of the procedure

The optimization of the procedure can be approached from an instrumental or a chemical point of view. Only chemical optimization will be dealt with here, on account of its greater possibilities.

One requisite for the optimization of a chemical system is knowledge of the reaction mechanism involved. More often than not, lack of this knowledge compels the experimenter to find the most suitable reaction conditions empirically by: (a) altering the catalyst–substrate interaction; (b) using masking agents and activators (either individually or in conjunction); (c) varying the reaction conditions; (d) irradiating the system.

Catalyst–substrate interaction. The first step of a catalysed reaction is often the formation of a complex between the metal catalyst and the substrate. The catalyst's co-ordination sphere, the nature of the bond and the structure of the substrate will therefore be of great significance to the reaction.

When the catalyst forms inert complexes with the substrate (as is the case with metals of the platinum family), the co-ordination sphere of the former influences the complexation reaction. Thus, the reaction of *p*-benzoquinone with the Mn(III)–pyrophosphate complex, has been found to be catalysed by Ru and Pd [47]. The chloro-complexes of these two metals have the same catalytic effect, whereas a sulphate medium enhances the activity of ruthenium and inhibits that of palladium. Similarly, the reaction between luminol and periodate is catalysed by rhodium only in sulphate medium, but the catalytic action of iridium takes place even in chloride media [48].

Organic substrates readily lend themselves to modification of their interaction with the catalyst, thereby providing a means of regulating the selectivity. Thus, studies on the catalytic oxidation of substituted phenols by H_2O_2 have revealed the catalytic activity of copper to be especially high for *para*-derivatives, and *ortho*-derivatives to be particularly susceptible to the action of nickel and cobalt [49], as shown in Table 8.6 [39].

Table 8.6 — Substituted phenols as substrates in metal-catalysed oxidations with hydrogen peroxide

Substrate	Position of OH-groups	Detectable concentration (μg/ml)		
		Cu(II)	Ni(II)	Co(II)
Quinol	*p*-	0.005	Inactive	0.02
Resorcinol	*m*-	0.01	Inactive	0.002
Pyrocatechol	*o*-	0.05	0.1	0.0002
Pyrogallol	*o*-	0.05	0.1	0.002
Tiron	*o*-	—	0.01	0.00002

Reproduced from H. Müller, *CRC Crit. Rev. Anal. Chem.*, 1982, **13**, 313, by permission. Copyright 1982, CRC Press.

Masking and activation. Masking agents, the commonest of which include EDTA, CN^-, F^-, citrate and sulphosalicylic acid (see Table 8.7) [40], inactivate all potential catalysts for a reaction except the analyte.

The indicator reaction is occasionally the source of the masking agent. Thus, the reaction between iodide and hydrogen peroxide is catalysed by Nb and Ta. However,

Table 8.7 — Masking in catalytic analysis

Species of interest	Indicator reaction	Masking agent
Os	lucigenin + H_2O_2	EDTA (Co, Ni, etc.)
Zn, Cd, Pb	$Ni(trien)^{2+}$ + $Cu(EDTA)^{2-}$	CN^- (Fe, Co, Cu)
Cu	$I_2 + N_3^-$ + thioammeline	N_3^- (metals)
Ta	$I^- + H_2O_2$	H_2O_2 (Nb)
Co	$I_2 + N_3^-$ + thiocarbamate	F^- (Fe, Al)
Mo, W	*o*-aminophenol + H_2O_2	$C_2O_4^{2-}$ (Mo)
	$I^- + H_2O_2$	citric acid (W)
Ir, Ru	Hg(I) + Ce(IV)	HNO_3 (Ru)
Co, Ni	diphenylcarbazone + H_2O_2 + *o*-hydroxybiphenyl	2-nitroso-1-naphthol (Co)
Cr	Bromopyrogallol Red + H_2O_2	EDTA (Fe, Co, Ni)

Reproduced from G. Werner, *Quim. Anal.* (*Barcelona*), 1983, **2,** (Extra), 68, by permission. Copyright 1983, Sociedad Española de Química Analítica.

the addition of peroxide to the sample before the other reagents masks niobium and makes the determination of tantalum interference-free. The indicator reaction between iodide and sodium azide is also quite selective, as the latter compound is a natural masking agent for many interfering materials. This masking ability is taken advantage of in the determination of copper based on its inhibitory effect on this reaction, which is virtually specific when catalysed by *m*-thioammeline [50].

Some ligands can act both as masking agents and as activators. Bontchev was the first to propose the exploitation of this dual nature to regulate the selectivity of a catalysed reaction such as that between *p*-phenetidine and chlorate, which is catalysed by V(V), Fe(III) and Cu(II). Citric acid is an activator for vanadium in this reaction and, in addition, masks ferric and cupric ions [51]. Similar properties are exhibited by other activators such as 2,2′-bipyridyl, ethylenediamine and sulphosalicylic acid [52, 53].

A suitable selection of the masking agents and activators to be used allows the individual determination of Ru and Os in the presence of each other [54]. This involves monitoring the absorbance of the azo dye formed from nitrite, sulphanilic acid and excess of 1-naphthylamine in the oxidation of the amine by nitrate. Ruthenium is determined in the presence of 1,10-phenanthroline, which masks osmium, which in turn is determined in the presence of 8-hydroxyquinoline, which besides activating the catalysis by osmium, inhibits the catalytic effect of ruthenium if a sufficiently low nitrate concentration is used. This system has a selectivity of $Z = 9$, as calculated by Kaiser's expressions.

As shown in Chapter 4, the use of masking agents, whether ligands or metal ions, can improve the inherently low selectivity of catalytic titrations.

Reaction conditions. The nature of the catalyst–substrate interaction can be modified to obtain the highest selectivity possible, by introduction of suitable changes in the pH, concentration of reactants, composition of the buffer, and temperature.

Changes in the hydrogen-ion concentration may influence the formation of the catalyst–substrate or catalyst–activator complex. Thus, copper catalyses the oxidation of phenol and arylamines in alkaline media. Acid media considerably decrease the stability of Cu–phenol complexes, so the catalytic effect of the metal is exerted exclusively on arylamines. This effect can be exploited in the determination of iron in the presence of excess of copper, based on the monitoring of the Fe-catalysed oxidation of phenols, which is not affected by the presence of copper in acid medium [55].

As most catalysts are active only over a narrow pH range, a suitable selection of the pH of the medium can be an effective means of improving the selectivity of a catalytic determination. Thus, Zr and Hf, both of which catalyse the reaction between iodide and hydrogen peroxide, can be determined in the presence of each other simply by use of two different pH values, namely 1.1 for Zr and 2.0 for Hf, which mark the optimum acidity for the formation of their catalytically active hydroxo-complexes, the species effectively involved in the process [56].

The selectivity of a catalysed reaction is sometimes dependent on the concentration of the reactants if the reaction rate has a different kinetic dependence on the catalysts present. Hence the need to elucidate the mechanism and kinetics of the process and establish the participation of all the species in the system if their concentrations are to be optimized.

The comprehensive study of the kinetics of the Ce(IV)–As(III) system by Worthington and Pardue [57] reveals that the rate of the osmium-catalysed reaction does not depend on the Ce(IV) concentration, while that of the ruthenium-catalysed reaction is independent of the As(III) concentration:

$$\upsilon_1 = 4 \times 10^{10}[\mathrm{Ru}][\mathrm{Ce(IV)}]^{2.5}/(1 + 2.1 \times 10^3[\mathrm{Ce(IV)}]^{1.5})$$

$$\upsilon_2 = 8 \times 10^2[\mathrm{Os}][\mathrm{As(III)}](4.3 \times 10^{-3} + [\mathrm{As(III)}])$$

By use of two different As(III)/Ce(IV) ratios (e.g. with a constant arsenite concentration of $10^{-3}M$ and cerium concentrations of $5 \times 10^{-3}M$ and $5 \times 10^{-4}M$), the two catalysts can be determined at the ng level in the same sample on the basis of the following equations:

$$\upsilon_1 = 240[\mathrm{RuO_4}] + 90.4[\mathrm{OsO_4}] \tag{8.14}$$

$$\upsilon_2 = 2.69[\mathrm{RuO_4}] + 78.0[\mathrm{OsO_4}] \tag{8.15}$$

The selectivity achieved is $Z = 1.65$.

Another interesting study has been made of the reaction between Tiron and H_2O_2, catalysed by Co, Mn, Ni and Cu, in order to elucidate the reaction mechanism [58] and identify the most suitable conditions for determination of the first two metals [59]. These conditions were optimized by the simplex and response-surface methods. The shape of the plot of the reaction rate as a function of both pH and the concentration of the two reactants shows the different catalytic behaviour of Co, Mn and Ni. The increase in activity of cobalt with increasing Tiron and H_2O_2 concentration is the result of the formation of a catalytically active ternary complex. However, exceedingly high concentrations of Tiron and the peroxide result in the formation of a less active Co(Tiron)$_2$ complex and the destruction of the indicator, respectively.

The influence of temperature on the activity of a series of catalysts can also be used to increase selectivity, provided that they are sufficiently differently affected by this parameter. Wolff and Schwing [60] have shown that the oxidation of iodide by bromate is catalysed by Mo, W and Cr (in oxidation state VI) to an extent dependent on temperature. However, since the temperature-dependence of all the catalysts is quite similar, any attempt at determination of these metals in binary mixtures will result in poor precision.

The use of a buffer of suitable composition can also result in increased selectivity. Thus, the oxidation of Alizarin S by H_2O_2 in a sodium hydrogen carbonate buffer is poorly selective as it is catalysed simultaneously by cobalt and manganese [61]. However, the reaction becomes stable to the latter in an ammonium carbonate buffer, where cobalt forms stable, catalytically inactive ammine complexes [62]. This allows the determination of manganese in the presence of excess of cobalt, either by catalytic titration [61] or by photometric monitoring of the indicator reaction (in the presence of a hundredfold ratio of cobalt) [37].

Photoactivation. Werner has studied the effect of irradiation on catalysed reactions and has shown the influence of light on their rate and selectivity [40]. Interferences are considerably lessened by irradiation of samples, as demonstrated in the direct photoactivation of a reactant [e.g. the copper(II)-catalysed decomposition of H_2O_2], indirect photosensitization [e.g. the manganese(II)-catalysed oxidation of sulphite by oxygen] or catalyst photoactivation (e.g. oxidation of Methyl Orange by bromate, catalysed by the iron–oxalate complex).

8.2.4.2 Use of separation techniques

When neither the use of a masking agent nor modification of the catalyst–substrate interaction has the desired effect on selectivity, this can still be enhanced by isolating the analyte from other interfering catalysts.

Many of the separation techniques used in trace analysis are adequate for this purpose. A good account of them is given by Minczewski *et al.* [62a]. Thus, liquid–liquid extraction and (to a lesser extent) ion-exchange and chromatographic techniques are commonplace in kinetic analysis. Table 8.8 summarizes the application of these techniques to typical catalytic determinations. Paper and thin-layer chromatography are rarely used, because of the adverse influence of the matrix on catalytic reactions; liquid and gas chromatography, on the other hand, are frequently used, particularly where chemiluminescent reactions are involved (see Chapter 9). Liquid–liquid extraction involves the isolation of the species of interest in an organic solvent as a prior step to its determination in the extract, or dissolved in a mixed solvent (e.g. water–acetone–chloroform). On account of their resemblance to spectrophotometric extractive methods, these are called 'extractive catalytic' by Otto *et al.* [38], who have contributed a great deal to their development. The application of the extraction technique to a catalysed reaction entails two prerequisites: (a) the indicator reaction should be feasible in an organic medium or a mixed solvent, wherein the catalyst activity must be preserved; (b) the extraction system should isolate the metal of interest from all potential interferents.

Table 8.8 — Combination of catalytic methods with separation procedures

Species	Indicator reaction	Separation system or technique	Matrix
Ni	diphenylcarbazone + Tiron + H_2O_2	I: benzildioxime/$CHCl_3$	La, Y and P oxides
Co	Alizarin S + H_2O_2	I: 2-nitroso-1-naphthol	P
Au	Hg(I) + Ce(IV) + ClO_4	I: HCl/ethyl acetate	Ores
V	phenylhydrazine-*p*-sulphonic acid + ClO_4^-	I: oxine/$CHCl_3$	Blood, urine
Cr	*o*-dianisidine + H_2O_2	I: HCl/MIBK	$AlCl_3$, blood
Mn	H_2O_2 decomp., 1,10-phenanthro-line as activator	I: PAN/$CHCl_3$	Alkali-metal halides
Mo	Fe(III) + $SnCl_2$	II: Dowex 1− ×8	Sea-water
Se	Methylene Blue + S^{2-}	II: Dowcx 50W− ×2	Water
Cr	*o*-dianisidine + H_2O_2	II: Dowex 50W− ×8	Industrial dust
Mn	Malachite Green + IO_4^-	II: Amberlite	Ta, Nb
F^-	perborate + I^- [Zr(IV) as inhibitor]	Microdiffusion	Biological materials
Ru	benzidine + H_2O_2	Paper chromatography	Water
IR	Mn(II) + BrO^-	Thin-layer chromatography	Water
Ru	*o*-dianisidine + IO_4^-	Distilled as RuO_4	Ores
Cu	$Fe(SCN)_3 + S_2O_3^{2-}$	Copreciptn. on Hg_2S	Me(II, III) salts
CO_3^{2-}	Accel. of formation of Cr(III)–XO complex	Directional crystln. of aq. salt soln. of eutectic composition	CsI
	—		

I: liquid–liquid extraction, II: ion-exchange, PAN: 1-(2-pyridylazo)-2-naphthol, XO: Xylenol Orange. (Reproduced from [39] with permission of the copyright holders, CRC Press).

The first of the two requisites is difficult to meet, as the catalytic activity depends in most cases on the formation of a complex between the catalyst and one of the components of the indicator reaction. Consequently, the conditions for extraction of the catalyst must be such that the catalytically active complex is formed after the extraction, by a ligand-exchange or dissociation reaction. The extraction technique is thus inadvisable whenever very stable complexes are involved, as these undergo very slow exchange reactions or none at all.

Mixed solvents are suitable for some catalytic reactions and result in determinations of roughly the same order of precision as those in aqueous media; however, the solubility of common extractants in non-polar media is rather limited and occasionally results in phase separation.

The procedure devised by Otto *et al.* [38] has been used by these authors for the determination of silver [63], copper [64, 65], molybdenum [66], iron [67] and chromium [68]. Table 8.9 gathers some more recent examples reported in the literature [69].

8.3 ACCURACY AND PRECISION

8.3.1 Accuracy of kinetic methods

Accuracy is a broad concept related to the closeness of a measurement to the actual value of the property measured. It is also related to the bias of a measurement method. Strictly speaking, accuracy cannot be determined unless the real value of the measured property is known. However, limits of the error can be estimated by

Table 8.9 — Recent examples of extraction catalytic methods

Analyte	Extraction system	Indicator reaction
Fe	1,10-phen/$CHCl_3$	*p*-phenetidine + H_2O_2
Cu	neocuproine/$CHCl_3$	*p*-phenetidine + H_2O_2
Ti	pyrocatechol + dioctylamine/butanol	*o*-phenylenediamine + H_2O_2
V	pyrocatechol + octyldimethylamine/butanol	*o*-phenylenediamine + bromate
Nb	benzoinoxime/$CHCl_3$	*o*-aminophenol + H_2O_2

considering all the potential sources of systematic errors and their maximum possible effect on the measurement, though this has the implicit risk of inadvertently omitting a major source of error in the estimation (but may also lead to overestimation of the error).

Whenever the real value of the measured property is unknown, the experimenter usually resorts to comparing the results with those provided by one or several other analytical techniques, or by analysis of reference materials.

The accuracy of kinetic methods depends to a great extent on the degree of adherence of the process to the proposed mechanism (i.e. on the absence of side-reactions involving the reactants or products). Obviously, the accuracy is also dependent on the reliability of the analytical technique used to measure the changes in the concentration of a given component as a function of time, and on that of the experimental conditions (pH, temperature, ionic strength), which should be kept rigorously constant.

8.3.1.1 Evaluation of methods

The choice between the fixed-time and variable-time methods is normally dictated by the reaction time, the nature of the species of interest and the type of monitoring technique used (optical, electroanalytical). When the analytical signal is linearly related to concentration, the fixed-time method is in theory superior for pseudo first-order reactions and for the determination of substrates involved in enzymatic reactions, while the variable-time method is better suited to the determination of enzymatic or catalytic activity in general.

On the other hand, if the analytical signal varies non-linearly with concentration, the variable-time method is to be preferred. In any case, the error introduced by the non-linear relationship should be checked against that found by applying the method to a pseudo first-order reaction. Both techniques are complementary and suited to different types of reactions and transducers.

Many analytical methods rely on pseudo first-order reaction kinetics and on a linear relationship between the measured signal and the monitored concentration. Ingle and Crouch [70] have evaluated the accuracy of the methods, starting with the simplest possible irreversible reaction, namely:

$$A \xrightarrow{k} P$$

The integrated rate equation for the reactant, relating its initial concentration to measurable parameters, is of the form (see Chapter 3):

$$[\mathrm{A}]_0 = -\Delta[\mathrm{A}]/\exp(-kt_1)[1-\exp(-k\Delta t)] \tag{8.16}$$

Differential methods are usually based on pseudo zero-order kinetics, i.e. on a linear relationship between the concentration and time. The validity of this assumption can be checked by expanding $\exp(-k\Delta t)$ and substituting the resultant Maclaurin series into Eq. (8.16):

$$[\mathrm{A}]_0 = -\Delta[\mathrm{A}]/\exp(-kt_1)\{k\Delta t - [(k\Delta t)^2/2!] + \ldots\} \tag{8.17}$$

If $k\Delta t$ is small enough, this equation can be simplified to:

$$[\mathrm{A}]_0 = -\Delta[\mathrm{A}]/k\Delta t \exp(-kt_1) \tag{8.18}$$

The concentration change that can be tolerated without detracting from accuracy can be obtained by introducing into Eq. (8.19)

$$\ln([\mathrm{A}]_2/[\mathrm{A}]_1) = -k\Delta t \tag{8.19}$$

(where $[\mathrm{A}]_2$ is the concentration corresponding to the end of the measuring period, t_2, and $[\mathrm{A}]_1$ that measured at time t_1) the $k\Delta t$ value obtained from Eq. (8.17).

Table 8.10 lists the errors resulting from ignoring the higher order $k\Delta t$ terms in

Table 8.10 — Accuracy of the initial-rate approximation for first and pseudo first-order reactions

$k\Delta t$	Error (%)	$[\mathrm{A}]_2/[\mathrm{A}]_1$	Relative change in $[\mathrm{A}]_1$ (%)
0.002	0.10	0.9980	0.20
0.005	0.25	0.9950	0.50
0.010	0.50	0.9900	1.00
0.020	1.00	0.9802	1.98
0.050	2.52	0.9512	4.88

(Reproduced with permission, from J. D. Ingle, Jr. and S. R. Crouch, *Anal. Chem.*, 1971, **43**, 697. Copyright 1971, American Chemical Society.)

calculating the acceptable concentration change, and shows that the initial-rate approximation is valid only for a small fraction of the reaction development if a high accuracy is required (the tolerable change in concentration for an error of 1% in the initial rate is less than 2%). This is a major limitation to variable-time methods, though not to the fixed-time method when applied to first-order and pseudo first-order reactions.

Variable-time methods entail measuring Δt while keeping $\Delta[A]$ constant — Eq. (8.16). Since both Δt and t_1 are dependent on $[A]_0$, the relationship between this initial concentration and the reciprocal of Δt will always be non-linear. If $\Delta[A]$ represents only a small fraction of the reaction, Eq. (8.18) will be roughly valid and $[A]_0$ will be proportional to $1/\Delta t$. However, $\Delta[A]$ should be sufficiently large in practice to ensure measurable changes in the analytical signal. In addition, the smallest possible error will be obtained if the concentration of A measured at time t_1 is as close to its initial value, $[A]_0$, as the mixing time and induction period permit, so that the first exponential term in the denominator of Eq. (8.16) approaches unity.

To illustrate the potential error of a variable method even for small concentration changes, Ingle and Crouch [70] used a pseudo first-order reaction with a rate constant of 10^{-3} sec^{-1} and assayed initial concentrations in the range 1–10mM. If the reaction is monitored through the appearance of the product, P, it would be reasonable to choose two concentration levels such as $[P]_1 = 0.01$mM and $[P]_2 = 0.02$mM. Lower concentrations of A would require at least 2% of the reactant to have reacted by the end of the measurement interval. The time elapsed between $[P]_1$ and $[P]_2$ can be calculated from:

$$[P]_1 = [A]_0[1 - \exp(-kt_1)] \tag{8.20}$$

$$[P]_2 = [A]_0[1 - \exp(-kt_2)] \tag{8.21}$$

The results from use of these equations are shown in Table 8.11 for concentrations of 1, 5 and 10mM; the 5mM solution was used as the standard for calculating the other concentrations. As can be seen, the error obtained for a concentration range ten times as wide is 1.21% (sampling or measurement errors neglected). Also, the error exceeds the 0.5% indicated in Table 8.10 for a concentration change of 1% — this is a result of Table 8.10 considering errors due exclusively to Δt and ignoring the effect of t_1.

Table 8.11 — Accuracy of the variable-time approximation

$[A]_0$ taken (mM)	Δt (sec)	$1/\Delta t$ (sec^{-1})	$[A]_0$ calc. (mM)	Error (%)
1	10.1524	0.09850	0.9879	−1.21
5	2.0060	0.49850	—	—
10	1.0015	0.99850	10.015	+0.15

(Reproduced with permission, from J. D. Ingle, Jr. and S. R. Crouch, *Anal. Chem.*, 1971, **43**, 967. Copyright 1971, American Chemical Society.)

The fixed-time method involves keeping t_1 and Δt constant throughout the measurements, so that the measured property, $\Delta[A]$, and the initial concentration are linearly related [Eq. (8.16)] whether or not the kinetic curve is linear over the measurement interval. This is a characteristic feature of first-order reactions. The relative change in concentration over a fixed time interval remains constant and the absolute change is proportional to the initial concentration throughout. By using the procedure described above for the fixed-time method, it is easy to see that no error is made in the calculation of $[A]_0$ [70]. As measurements are not limited to the initial stages of the reaction, the fixed-time method has a broader dynamic range than the variable-time method does, when applied to pseudo first-order reactions.

An experimental comparison between the fixed- and variable-time methods has been reported by Ingle and Crouch [70] on the basis of the determination of phosphate by photometric monitoring of the formation of phosphomolybdenum blue (pseudo first-order kinetics). The data collected from both methods were obtained manually from the recorded absorbance–time curves. The time intervals (1 and 2 min) of the fixed-time method and the absorbance intervals (0.005 and 0.010) of the variable-time method were chosen so that the extent of reaction was similar in both methods, about 5% in the longer time interval of the fixed-time method, and approximately 7% in the course of an absolute change of 0.01 for the lower phosphorus concentration (2 ppm) used in the variable-time method. Table 8.12

Table 8.12 — Comparison of phosphate determinations by fixed and variable-time methods

P taken (ppm)	Fixed-time method				Variable-time method			
	Found[a]	Error (%)	Found[b]	Error (%)	Found[c]	Error (%)	Found[d]	Error (%)
5	5.12	+2.4	4.95	−1.0	5.10	+2.0	5.37	+7.4
10	10.26	+2.6	10.07	+0.7	9.90	−1.0	10.78	+7.8

Concentrations found (expressed in ppm) based on average of 5 determinations referred to a 2 ppm standard.

[a] t_1 = min, Δt = 1 min, [b] t_1 = 1 min, Δt = 2 min, [c] $A_1 = 0.005$, $\Delta A = 0.005$, [d] $A_1 = 0.005$, $\Delta A = 0.010$.
(Reproduced with permission, from J. D. Ingle, Jr. and S. R. Crouch, *Anal. Chem.*, 1971, **43**, 697.

gives the data obtained in the determination of 5 and 10 ppm P, based on calibration with a 2 ppm standard. The data treated by the fixed-time method show no significant differences (Student's t, 95% confidence level) from the expected values calculated from the standard for either interval. Nor are significant errors obtained for the shorter interval of the variable-time method. However, appreciable errors (Student's t for a confidence level of 99%) are obtained for the longer interval (0.01). This is a result of using Eq. (8.18), a poorer approximation than Eq. (8.16) for the 2 ppm standard, which yields positive errors in the measurement of larger phosphate concentrations. The data also indicate that the accuracy and precision of the results obtained by the fixed-time method on a pseudo first-order reaction increase with increasing length of the time intervals used. Too long intervals should, however, be avoided in practice.

Catalysed reactions. Ingle and Crouch have also shown the difficulty involved in developing a general treatment for catalysed reactions, which arises owing to the large variety of possible mechanisms encountered [70]. The rate law of these reactions is usually of the form

$$d[P]/dt = F[(X_1), (X)_2, \ldots (X)_i][C]_0 \tag{8.22}$$

where F is a function of the rate constant and the concentration of the different reactants and $[C]_0$ is the initial catalyst concentration. Equation (8.22) can be rearranged and expressed in integral form as:

$$\int_{[P]_1}^{[P]_2} \frac{d[P]}{F[(X_1), (X_2) \ldots (X_i)]} = \int_{t_1}^{t_2} [C]_0 dt \tag{8.23}$$

If $G(P) = \int d[P]/F(X_i)$, then Eq. (8.23) is reduced to:

$$[C]_0 = \{G([P]_2) - G([P]_1)\}/\Delta t \tag{8.24}$$

Equation (8.24) will be of use provided that the rate law is not very complicated. For the variable-time method, $[C]_0$ will be directly proportional to $1/t$ since the numerator of Eq. (8.24) remains constant from run to run. This endows the variable-time method with wider dynamic ranges on account of the unavoidable adherence to pseudo zero-order kinetics. Thus, the variable-time methodology allows for measurements over a broad dynamic catalyst concentration range.

Differential reaction-rate methods. Although the chief sources of error and the various factors influencing the accuracy of differential methods were described in detail in Chapter 6, it is worth making a few general observations.

The greatest possible accuracy is provided by the proportional-equation method, the determining factor in this case being the time t_2. Extrapolation methods are in general less accurate (for a similar accuracy in the measurement technique used) as they rely on graphical extrapolation. On the other hand, the single-point method is as accurate as the proportional-equation method [71, 72].

All these methods are suitable (at least in theory) for the resolution of multicomponent mixtures. In practice, however, only the graphical extrapolation and proportional-equation methods are adequate for this purpose. In addition, they call for greater differences between the rate constants of the different components than are needed in the resolution of binary mixtures. The proportional-equation method, the more suitable [73], allows the resolution of mixtures of three or four components with acceptable errors as long as their rate constants are sufficiently different.

8.3.1.2 Influence of the detection system

Ingle and Crouch have studied the influence of the method used to monitor the reaction, on the signal–concentration relationship [70]. In spectrophotometric detection, the photocurrent, which is proportional to the transmittance, is not linearly related to the concentration. The use of logarithmic amplification to obtain a signal proportional to the absorbance introduces no significant complications. The non-linearity of the signal–concentration relationship results in no additional error in the variable-time method, as measurements are made between two fixed transmittance levels, so that the change in absolute concentration remains constant from run to run. The use of a signal proportional to transmittance in the fixed-time method may lead to error unless the measuring interval is limited to very small changes in $\%T$, so that the relative changes in transmittance are proportional to the changes in relative absorbance.

The relative change in $\%T$ tolerated by the fixed-time method can be calculated from Beer's law [70]. The change in the analyte concentration, $\Delta C = C_2 - C_1$, can be expressed in terms of transmittance as:

$$\Delta C = (-1/\varepsilon l) \log (T_2/T_1) = (-1/2.303\varepsilon l) \ln (T_2/T_1) \tag{8.25}$$

This equation can be expressed in terms of the change in relative transmittance, $\Delta T/T_1$. Substitution for T_2 and expansion of the logarithmic terms yields

$$\Delta C = (-1/2.303\varepsilon l) \ln \{(\Delta T/T_1) - [(\Delta T^2/T_1^2)/2] + \ldots\} \tag{8.26}$$

As the changes in relative transmittance for the fixed-time method must be rather small, Eq. (8.26) can be reduced to:

$$\Delta C = (-1/2.3\varepsilon l)\,(\Delta T/T_1) \tag{8.27}$$

This approximate expression results in errors of less than 1% relative to its exact counterpart, Eq. (8.26), for changes smaller than 2% in the relative transmittance [70]. This results in a somewhat decreased dynamic range, which can be broadened by logarithmic amplification.

The signals provided by the fluorimetric and amperometric techniques are both linear as long as sufficiently dilute solutions are used, so the selection of a particular measurement method is dictated by the nature of the reaction monitored and the kinetic behaviour of the species of interest.

The non-linearity of the signal *vs.* concentration plots obtained by the potentiometric technique may result in complications. Pardue has studied in depth the different possible types of curve and found the variable-time method to be superior to its fixed-time counterpart whenever the signal varies non-linearly as a function of time [74]. However, some catalytic reactions yield composite response curves linearly related to time, thus allowing any type of measurement method to be used.

8.3.2 Precision in kinetic methods

The advanced instrumentation and data-processing methods available today make kinetic methods comparable to their equilibrium counterparts as regards rapidity, precision and simplicity, but the kinetic methods are more strongly dependent on the experimental conditions (pH, temperature, ionic strength, etc.), so small differences in the experimental data obtained from kinetic measurements may give large variations in the reaction rates calculated. The methods based on first-order reactions are somewhat more precise, as they allow the corresponding rate constants to be calculated without the need for prior knowledge of the concentrations of the species involved in the reaction. This allows the precision of a method to be increased substantially by fitting the experimental data obtained to a first-order model with the aid of a linear regression program [75].

The precision of reaction-rate measurements depends essentially on three factors: (a) the noise and drift of the monitored signal; (b) the reproducibility of the physicochemical variables involved (i.e. the constancy of the reaction conditions); (c) the magnitude of the errors made by data-acquisition systems in time measurements. This last source of error can be safely ignored thanks to the brilliant performance of modern instrumentation in this respect. Thus, relative standard deviations (rsd) of about 0.1% have been obtained [76] by checking the rate-calculating system with reproducible, electronically generated slopes simulating the response of the monitored reaction. The rsd values obtained for real measurements close to the detection limit typically range between 0.3 and 3.3%.

The rsd of a rate measurement, S_υ/υ, is given by:

$$S_\upsilon/\upsilon = [(S_N/\upsilon)^2 + (S_{RC}/\upsilon)^2]^{1/2} \tag{8.28}$$

where υ is the reaction rate or the change in detector signal over a given time interval, S_υ is the overall standard deviation of υ, and S_N^2 and S_{RC}^2 are the variances of the monitored signal and the reaction conditions, respectively.

For differential measurements (i.e. at the beginning of the reaction), S_N will be virtually independent of the reaction rate, as the absolute magnitude of the noise remains constant throughout the measurement period. On the other hand, the fraction S_{RC}/υ is normally constant and roughly equal to 1%, so the irreproducibility due to the reaction conditions will be a fraction of the reaction rate and independent of it. In summary, S_N/υ will be the significant term for low reaction rates or low analyte concentrations (i.e. in the vicinity of the detection limit); conversely, it will lose significance as the reaction rate increases, to the point of becoming negligible relative to S_{RC}/υ at sufficiently high rates. Thus, noise will be the limiting factor of the precision of measurements close to the detection limit (or up to 50 C_L), above which the irreproducibility in the reaction conditions will become increasingly significant. Obviously, these considerations will be valid only as long as the blank is not significant relative to the monitored reaction signal.

The precision of rate measurements can be controlled and improved through a better knowledge of the factors affecting S_N and S_{RC}, which are commented on below.

8.3.2.1 Influence and evaluation of noise

As stated above, noise limits the precision obtainable from kinetic measurements whenever small reaction rates are to be determined (differential methods) or measurements are made in the vicinity of the detection limit.

According to Ingle *et al.* [77], there are two general procedures available for evaluation of the effect of noise on the precision of kinetic methods. One is based on theoretical calculations of the signal-to-noise ratio; the other relies on the use of an experimental device to compare the relative precision obtained from repetitive runs for a simulated rate with that found in real kinetic runs for the same rate in order to determine what fraction of the overall run-to-run imprecision in the measurement of the actual rate is due to noise.

Theory of the signal-to-noise ratio. One of the major limitations of rate measurements compared to equilibrium measurements is that the signal-to-noise (S/N) ratio of the former is inherently lower as a result of use of only a small portion of the available signal. Inasmuch as kinetic measurements are made with dynamic signals, the noise-equivalent bandwidth of the measurement system cannot be reduced indefinitely to increase the S/N ratio without introducing some distortions. Since the S/N ratio is a measure of the instrumental precision with which the measurements are made, knowledge of the factors influencing it will be of aid both in evaluating the precision and in optimizing the experimental conditions.

Ingle and Crouch have developed different expressions for evaluation of the S/N ratio, corresponding to the different kinetic methods, by use of an automatic spectrophotometric sensing system [78].

Of the various possible ways to define the S/N ratio, the commonest involves considering the signal, S, as the ratio $\Delta M/\Delta t$ (where M is the output of the sensing system), and the rms (root mean square) relative standard deviation of the measured signal, S_S. According to the mathematical theory of error propagation, the relative standard deviation of the signal will be given by:

$$S_S = \frac{\Delta M}{\Delta t}\left[\frac{S^2_{\Delta M}}{(\Delta M)^2} + \frac{S^2_{\Delta t}}{(\Delta t)^2}\right]^{1/2}$$

Hence, the S/N ratio, equal to the ratio of S to S_S, will be:

$$(\mathrm{S/N})_{v} = \frac{S}{S_s} = \left[\frac{S^2_{\Delta M}}{(\Delta M)^2} + \frac{S^2_{\Delta t}}{(\Delta t)^2}\right]^{-1/2} \tag{8.29}$$

In Eq. (8.29) $\Delta M = M_2 - M_1$ and $\Delta t = t_2, - t_1$, where M_1 and M_2 are the signals measured at times t_1 and t_2, respectively. Since $S_{\Delta M}$ is $< \Delta M$ and $S_{\Delta t}$ is $< \Delta t$, ΔM and Δt are independent of each other in Eq. (8.29), which is useful for both the fixed- and the variable-time methods.

The fixed-time method, for example, assumes Δt to be constant in Eq. (8.29). Since the reproducibility of the time base is much better than that of the sensing system, the variance of the time increment, $S^2_{\Delta t}$, can be neglected and Eq. (8.29) reduced to:

$$(\mathrm{S/N})_{\mathrm{ft}} = \Delta M/S_{\Delta M} = \Delta M/(S^2_{\mathrm{M1}} + S^2_{\mathrm{M2}})^{1/2}$$

where S^2_{M1} and S^2_{M2} are the variances of M_1 and M_2 respectively. An additional consideration is that the measurements made at t_1 and t_2 are referred to the same initial time, t_0, which is taken as the start of the reaction.

The variable-time method assumes the constancy of ΔM in Eq. (8.29) and the imprecision in the measurements to be due to the noise in the detector output signal.

The error introduced through $S^2_{\Delta M}$ can be safely neglected on account of the precision of modern sensors; hence the S/N ratio can be expressed as:

$$(\mathrm{S/N})_{vt} = (\Delta t/S_{\Delta t}) = \Delta t/(S^2_{t1} + S^2_{t2})^{1/2}$$

where S^2_{t1} and S^2_{t2} are the variances of the determination of times t_1 and t_2 corresponding to signals M_1 and M_2, respectively. In this case, the determination of the S/N ratio requires prior knowledge of the variances of t_1 and t_2, the calculation of which involves complicated probability density functions. However, Ingle and Crouch [78] have developed a special treatment for evaluation of the S/N ratio of fixed-time spectrophotometric measurements, the imprecision of which arises from (a) the photocurrent shot noise and (b) the source flicker noise.

The precision of reaction-rate measurements of fast reactions is chiefly affected by the photocurrent shot noise and can be increased by using automatic stopped-flow systems allowing a large number of measurements to be made over a short interval of time [79]. The source flicker noise is a major limiting factor in the precision achievable in the measurement of the rate for slow reactions, though not necessarily the predominant one. The situation in this case can also be improved by making repetitive measurements, though this can be rather tedious and the use of a stopped-flow system may result in problems stemming from diffusion phenomena. The only safe way to increase precision in this case seems to be stabilization of the source.

Instrumental comparison method. This method for evaluation of the noise of rate measurements comprises four basic stages.

(a) Electronic generation, by a simple operational amplifier (OA), of a ramp simulating the actual change in signal yielded by the monitored reaction.

(b) Evaluation of the steady-state noise from the signal yielded by a blank or any other suitable solution that can give rise to an absolute signal and a noise level equivalent to the average noise produced by the software or the measuring circuit in an actual kinetic run.

(c) Addition of the calculated noise to the electronically generated ramp.

(d) Comparison of the relative precision obtained in repetitive runs by the simulated system with that found in real kinetic runs, in order to find what fraction of the run-to-run imprecision in the actual rate measurement is due to noise.

In this regard, Ingle *et al.* [77] have shown that the precision of reaction-rate measurements is proportional to the ratio between the peak-to-peak noise (N_{pp}) and the change in the monitored signal over a preselected time interval:

$$S_{\mathrm{N}}/v = K(N_{\mathrm{pp}}/\Delta S) \tag{8.30}$$

where K is a proportionality constant that is dependent on the nature of the noise and the type of monitoring system used (photometric, fluorimetric). These authors have also demonstrated that the precision achieved is independent of the method used to calculate the rate (fixed-time, variable-time, initial-rate) since the K values obtained are virtually identical.

The simplicity of Eq. (8.30) permits K to be used as a diagnostic tool for routine rate measurements. The ΔS value is selected beforehand in the variable-time method and measured directly in the fixed-time method, and N_{pp} can be readily determined from a single recording or with the aid of suitable software. By substitution of these parameters into Eq. (8.30), the required S_N/v ratio can be determined in a single kinetic run. Thus, the rsd can be compared with the actual run-to-run precision of real measurements (S_v/v) made at the same rate, and the fractions of the overall rsd corresponding to S_N and S_{RC} can be readily calculated from Eq. (8.28).

The best S_{RC}/v ratio obtained for a measurement time of 10 sec with the fluorimetric sensing system used by Ingle and Crouch [78] is 0.5%. If $N_{pp}/\Delta S$ is <1%, then S_N/v is <0.2% and the irreproducibility of the reaction conditions will be the limiting factor. However, for small rates, $N_{pp}/\Delta S$ is >10% and S_N/v >2%, i.e. the noise is the limiting factor in this case. This reasoning is obviously applicable to any other sensing system (e.g. spectrophotometric), with the sole exception of the greater K value involved.

8.3.2.2 Influence of the experimental conditions

Their popularity notwithstanding, kinetic methods of analysis are subject to a number of experimental errors and much experimental uncertainty unknown in equilibrium methods. The largest errors usually result from changes in the first-order or pseudo first-order rate constants, arising from variations in the experimental conditions. Such changes can take place in the same sample or from sample to sample.

The chief experimental variables affecting the precision of reaction-rate measurements, namely pH, temperature, reactant concentrations, ionic strength and mixing conditions, have been studied in depth by several authors [80–86]. According to Carr [81], it is the pH and the temperature which affect the precision of kinetic methods to the greatest extent. In fact, the relative uncertainty in the measurement of the rate constant arising from the uncertainty in the temperature can be readily calculated from a differential form of the Arrhenius equation, after application of the theory of error propagation, namely:

$$\sigma_k/k = |E_a|\sigma_T/RT^2$$

where E_a is the activation energy, R is the universal gas constant, T is the absolute temperature, and σ_T its uncertainty. This equation indicates that an uncertainty of 1 K in the temperature or 1 kcal/mole in the activation energy can cause an uncertainty of 5.6% in the measurement of k at 300 K.

A similar estimation can be made for the pH, the uncertainty of which can be calculated from another expression proposed by Carr [81]:

$$\sigma_k/k \approx 2.303\sigma\text{pH}$$

according to which an uncertainty of only 0.01 in the pH results in about ±2.3% uncertainty in the rate constant.

Carr has also demonstrated that the precision in the determination of an analyte depends significantly on the extent of development of the reaction concerned. The relative uncertainty in the initial concentration of one of the components of a reaction with first-order or pseudo first-order kinetics, of the type $A+B \rightarrow P$, is given by:

$$\sigma_{[A]^*}/[A]_0 = (1-x)[\ln(1-x)]\sigma_k/kx \quad (8.31)$$

where $\sigma_{[A]^*}$ denotes the uncertainty in the measurement of $[A]^*$ (the concentration to be calculated), $[A]_0$ the initial analyte concentration, x the fractional extent of reaction, and σ_k the uncertainty in the measurement of the first-order rate constant k.

From Eq. (8.31) it follows that the uncertainty in the measurement of the concentration of A is proportional to the relative uncertainty in the measurement of the rate constant at any x -value. In Fig. 8.6 is shown the variation of the relative

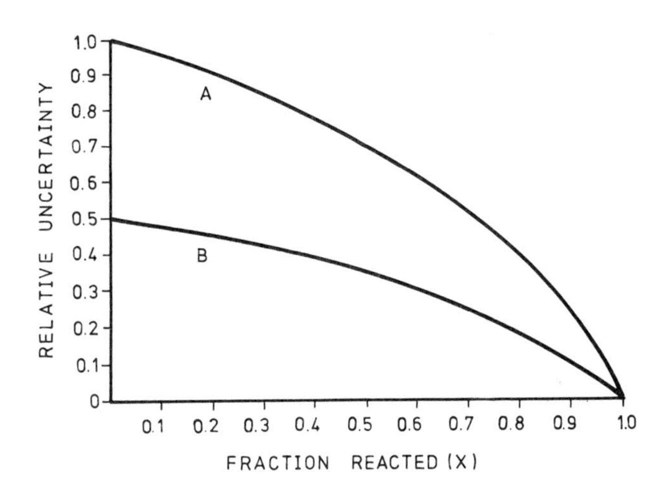

Fig. 8.6 — Plot of per cent relative uncertainty in estimated initial concentration, $[A]^*$, as a function of the extent of reaction for first-order kinetics. (a) $\sigma_k/k = 0.01$ and (b) $\sigma_k/k = 0.005$. (Reproduced with permission, from P. W. Carr, *Anal. Chem.*, 1978, **50**, 1602. Copyright 1978, American Chemical Society.)

uncertainty in the measured concentration of species A, as a function of the extent of reaction. As can be seen, the relative precision improves monotonically with increasing x. The graph has two distinct zones: (a) at small x-values, the relative uncertainty in [A] tends to σ_k/k and makes the uncertainty in the measurement of the rate constant the limiting factor; (b) when the reaction nears completion ($x \rightarrow 1$), the uncertainty in the measurement of k causes no appreciable variation in the estimated value $[A]^*$.

Since the estimated precision increases with the duration (and hence extent) of the reaction, the plot of the precision per unit time as a function of the extent of reaction should have a maximum. On the other hand, based on the fact that the precision increases with decreasing relative uncertainty, Carr has created the so-called 'precision index', which is defined as the reciprocal of $\sigma_{[A]^*}/[A]_0$ per unit extent of reaction rather than time, as this simplifies the computations. Such an index, a measure of the analytical efficiency, is given by:

$$[A]_0/\sigma_{[A]^*}x = k/(1-x)[\ln(1-x)]\sigma_k$$

Figure 8.7 shows the variation of this efficiency index as a function of the extent of reaction. The minimum of the curve is the result of two opposing factors. On the one hand, the analytical precision increases very slowly at the beginning of the reaction (see Fig. 8.6), but this effect is offset by the more marked increase in x and t. By the end of the reaction, though, there is a net increase in precision aided by the fact that the uncertainty approaches zero (Fig. 8.6). It should be noted that the plot in Fig. 8.7 does not predict a minimum-precision point, but rather a point of minimal precision for the time expended on the assay.

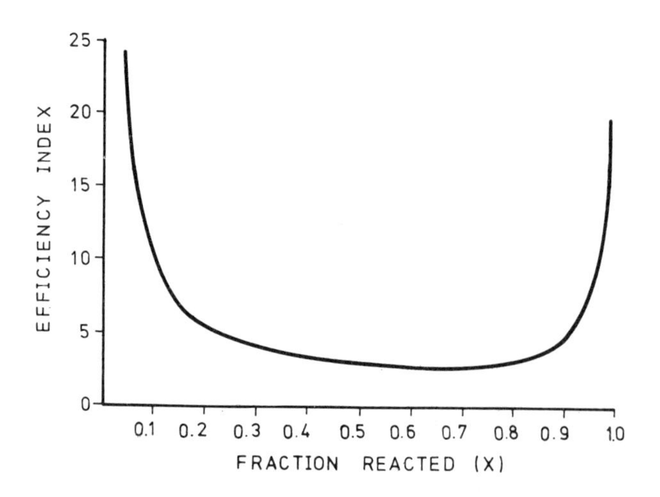

Fig. 8.7—Variation of the relative precision (per cent) as a function of the extent of reaction for first-order kinetics. $\sigma_k/k = 0.01$. (Reproduced with permission, from P. W. Carr, *Anal. Chem.*, 1978, **50**, 1602. Copyright 1978, American Chemical Society.)

Consistent with the reasoning above, it may be concluded that under identical working conditions, equilibrium methods are invariably more precise than kinetic methods based on the same reaction. However, kinetic methods can be more accurate than their equilibrium counterparts if side-reactions are involved. A longer assay time is usually the price to be paid for increased precision. Thus, an increase in precision by a factor of 3.5 calls for an x-value of 0.9 instead of 0.1, which in turn corresponds to a reaction time 23 times as long.

8.3.3 Critical comparison of methods of determination

As stated above, the errors usually encountered in the application of kinetic methods of analysis normally arise from variations in the experimental conditions (between-run variations in the rate constant) or from instrumental noise.

Errors resulting from between-run variations in the rate constant have been given much attention in the last few years. In their outstanding work on this subject, Landis *et al.* [83] have succeeded in optimizing the time of rate measurements ($t = 1/k = \tau$), thereby minimizing these variations, and propose a linear-regression procedure to compensate for the changes in the kinetic curve [75]. Recently, Wentzell and Crouch [87] developed a 'two-rate method' overcoming the dependence of the rate constant on between-run variations, by measuring the rate at two different reaction times. Thus, for a first-order or pseudo first-order reaction of the type A + B → P, the rate at a given time t is given by:

$$v_t = (\mathrm{d[P]/d}t)_t = k[\mathrm{A}]_0 \exp(-kt) \quad (8.32)$$

where $[\mathrm{A}]_0$ is the initial concentration of A and k is the first-order rate constant. The ratio between two rate measurements made at times t_1 and t_2 will be:

$$v_{t1}/v_{t2} = \exp[k(t_2 - t_1)] = \exp(k\Delta t)$$

Hence

$$k = [\ln(v_{t1}/v_{t2})]/\Delta t \quad (8.33)$$

Substitution of k into Eq. (8.32) yields:

$$[\mathrm{A}]_0 = \frac{v_{t1}\Delta t}{\ln(v_{t1}/v_{t2})}(v_{t1}/v_{t2})^{t1}/\Delta t = \mathrm{TRP}(t_1, t_2)$$

where $\mathrm{TRP}(t_1, t_2)$ stands for two-rate parameter.

By way of example, Table 8.13 testifies to the accuracy of this expression through

Table 8.13 — Effect of pH variations on the initial-rate method (IRM) and two-rate method (TRM)

$[Ca^{2+}]$ taken (μM)	pH[a]	k[b] (sec^{-1})	$[Ca^{2+}]$ found (μM)		Error (%)	
			IRM	TRM	IRM	TRM
1.00	6.0	0.305	1.118	1.044	12	4.4
1.00	5.8	0.476	1.681	0.962	68	−3.8
1.00	6.2	0.191	0.602	0.988	−40	−1.2

[a] Standards run at pH 6.0, [b] Determined from non-linear least-squares fit of data. (Reproduced with permission from P. D. Wentzell and S. R. Crouch, *Anal. Chem.*, 1986, **58**, 2851. Copyright 1986, American Chemical Society.)

the results obtained in a study of the effect of pH on the determination of calcium by a metal-complex exchange reaction developed by Pausch and Margerum [88]. As can be seen, the two-rate method is virtually insensitive to variations in the rate constant due to changes in pH. However, the initial rate method (IRM), with which it is compared, is significantly affected by such variations. These authors have developed a model for the propagation of random errors in this new methodology.

Figure 8.8 summarizes the results of a theoretical study on the influence of between-run variations on the rate constant in different kinetic methods, made by Wentzell and Crouch [89]. As can be seen, the variable-time and initial-rate methods are strongly dependent on the rate constant, since small variations in this result in significant errors in the determination of the analyte. The fixed-time method is less influenced by the reaction rate, though this dependence is clearly affected by the measurement time chosen [84]. Finally, the two-rate method (at least in theory) is completely independent of these variations.

To be able to determine the influence of errors resulting from instrumental noise, the researcher should minimize potential variations in the rate constant, by strict control of the experimental conditions — particularly the pH and temperature. Table 8.14 summarizes the results of a study by Wentzell and Crouch [89] on the precision of various kinetic methods at a constant reaction rate, made with the aid of the chemical system [88] for determination of calcium. As can be seen, the precision is similar for all the methods (except for the initial-rate method, which has a somewhat lower rsd). On the other hand, it is worth noting that the precision of both the fixed-time and the two-rate method improves with increasing measurement-time interval and approaches that of the equilibrium method under limiting conditions. The integral version of the fixed-time method is therefore more precise than the differential one. The variable-time method, not included in Table 8.14, is (according to Ingle and Crouch [70]) the least precise of all when applied to first-order reactions.

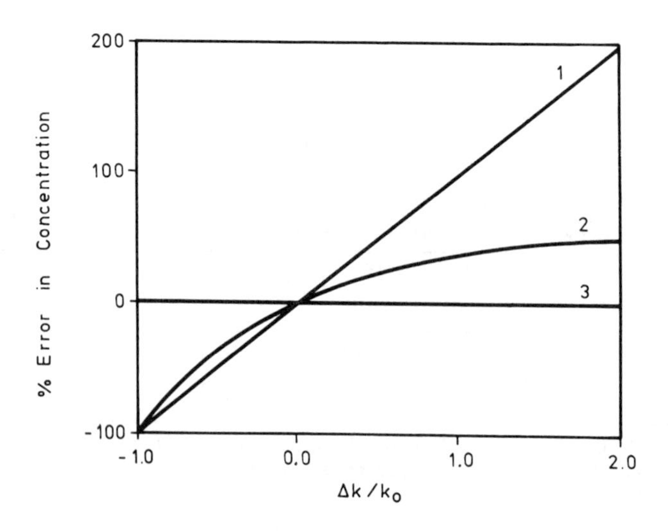

Fig. 8.8 — Theoretical errors expected in the concentration determined by various reaction-rate methods as a function of the relative deviation of the rate constant from its nominal value. (1) Variable-time and initial-rate methods: (2) fixed-time method and (3) two-rate method. (Adapted with permission from P. D. Wentzell and S. R. Crouch, *Anal. Chem.*, 1986, **58**, 2855. Copyright 1986, American Chemical Society.)

Table 8.14 — Precision of various reaction-rate methods under conditions of an invariant rate constant (adapted from [89])

Method	Parameters	rsd (%)
Initial-rate	$t = 0.5\tau$	0.57
Fixed-time	$t_1 = 0, t_2 = \tau$	1.2
Fixed-time	$t_1 = 0, t_2 = 2\tau$	0.83
Two-rate	$t_1 = 0.5\tau, t_2 = \tau$	1.1
Two-rate	$t_1 = 0.5\tau, t_2 = 1.76\tau$	0.86

REFERENCES

[1] *Spectrochim. Acta*, 1978, **33B**, 242.
[2] *Anal. Chem.*, 1980, **55**, 2242.
[3] G. L. Long and J. D. Winefordner, *Anal. Chem.*, 1983, **55**, 712A.
[4] J. Mandel, *The Statistical Analysis of Experimental Data*, Wiley–Interscience, New York, 1964.
[5] H. Kaiser, *Anal. Chem.*, 1970, **42**, 26A.
[6] H. A. Mottola, *CRC Crit. Rev. Anal. Chem.*, 1975, **4**, 229.
[7] K. B. Yatsimirskii, *Kinetic Methods of Analysis*, Pergamon Press, Oxford, 1966.
[8] H. Kaiser and H. Specker, *Z. Anal. Chem.*, 1956, **149**, 46.
[9] H. Specker, *Angew. Chem.*, 1968, **80**, 297.
[10] J. Bognár and O. Jellinek, *Mag. Kem. Foly*, 1961, **67**, 147.
[11] I. I. Alekseeva, A. P. Rysev, N. M. Sinicyn, L. P. Zitenko and A. Yaksinskii, *Zh. Analit. Khim.*, 1974, **29**, 1859.
[12] C. Liteanu, S. Sasu and D. Costache, *Rev. Chim. (Bucharest)*, 1973, **24**, 747.
[13] G. F. Proskuryakova, Yu. N. Dunaeva and B. S. Prints, *Agrokhimiya*, 1974, No. 4, 149.
[14] N. V. Nalimov, *The Application of Mathematical Statistics to Chemical Analysis*, Pergamon Press, Oxford, 1963.
[15] W. R. Wolf and K. K. Stewart, *Anal. Chem.*, 1979, **51**, 1201.
[16] K. Hirayama and N. Unohara, *Bunseki Kagaku*, 1980, **29**, 733.
[17] E. B. Sandell, *Colorimetric Determination of Traces of Metals*, 3rd Ed., Interscience, New York, 1959.
[18] G. L. Ellis and H. A. Mottola, *Anal. Chem.*, 1972, **44**, 2037.
[19] S. Rubio, A. Gómez-Hens and M. Valcárcel, *Analyst*, 1984, **109**, 717.
[20] K. B. Yatsimirskii, *Teoret. Eksp. Khim.*, 1972, **8**, 17.
[21] L. P. Tikhonova, K. B. Yatsimirskii and G. V. Kudinova, *Zh. Analit. Khim.*, 1974, **29**, 595.
[22] S. Meson and G. Zimmermann, *Elektrische Schaltungen, Signale, Systeme*, Inostrannoy Literatury, Moskow, 1964.
[23] T. Kawashima, S. Kai and S. Takashima, *Anal. Chim. Acta*, 1977, **89**, 65.
[24] M. A. Ditzler and W. F. Gutknecht, *Anal. Chem.*, 1980, **52**, 614.
[25] A. Moreno, M. Silva, D. Pérez-Bendito and M. Valcárcel, *Analyst*, 1983, **108**, 85.
[26] J. L. Ferrer-Herranz and D. Pérez-Bendito, *Quim. Anal.* 1984, **3**, 40.
[27] G. den Boef and A. Hulanicki, *Pure Appl. Chem.*, 1983, **55**, 169.
[28] A. Loriguillo, M. Silva and D. Pérez-Bendito, *Anal. Chim. Acta*, 1987, **199**, 29.
[29] F. Lázaro, M. D. Luque de Castro and M. Valcárcel, *Analyst*, 1984, **109**, 333.
[30] F. Grases, F. García-Sanchez and M. Valcárcel, *Anal. Chim. Acta*, 1981, **125**, 21.
[31] L. Ballesteros and D. Pérez-Bendito, *Analyst*, 1983, **108**, 443.
[32] D. Pérez-Bendito and M. Valcárcel, *Afinidad*, 1980, **336**, 123.
[33] S. Rubio, A. Gómez-Hens and M. Valcárcel, *Anal. Chem.*, 1984, **56**, 1417.
[34] M. C. Gutiérrez, A. Gomez-Hens and M. Valcárcel, *Anal. Chim. Acta*, 1986, **185**, 83.
[35] A. Moreno, M. Silva, D. Pérez-Bendito and M. Valcárcel, *Talanta*, 1983, **30**, 107.
[36] M. C. Gutiérrez, A. Gómez-Hens and D. Pérez-Bendito, *Z. Anal. Chem.*, 1987, **328**, 120.
[37] G. Werner, M. Hanrieder and H. Müller, *Proc. 8th Symp. Microchem. Techniques Graz*, Austria, 1980, 456.
[38] M. Otto, H. Müller and G. Werner, *Talanta*, 1978, **25**, 123.
[39] H. Müller, *CRC Crit. Rev. Anal. Chem.*, 1982, **13**, 313.
[40] G. Werner, *Quím. Anal. (Barcelona)*, 1983, **2** (Extra), 68.
[41] C. W. Fuller and J. M. Ottaway, *Analyst*, 1970, **95**, 41.
[42] H. Kaiser, *Z. Anal. Chem.*, 1972, **260**, 252.

[43] K. B. Yatsimirskii and A. P. Filippov, *Zh. Analit. Khim.*, 1965, **20**, No. 8, 815.†
[44] V. I. Shlenskaya, V. P. Khvostrova and G. I. Kadyrova, *Zh. Analit. Khim.*, 1973, **28**, 779.
[45] P. R. Bontchev, *Komplexobrasovanye i kataliticheskaya aktinovsst*, Mir, Moscow, 1975.
[46] A. Žmikić, D. Curtila, P. Pavlović, I. Murati, W. Reynolds and S. Ašperger, *J. Chem. Soc. Dalton Trans.*, 1973, 1284.
[47] H. Müller, L. P. Tikhonova, K. B. Yatsimirskii and S. N. Borkovets, *Zh. Analit. Khim.*, 1973, **28**, 2012.
[48] N. M. Lukovskaya and N. F. Kushchesvskaya, *Ukr. Khim. Zh.*, 1976, **42**, 87.
[49] I. F. Dolmanova, G. A. Zolotova, N. M. Ushakova, T. N. Chernyasvkaya and V. N. Peshkova, *Uspekhi Analitischeskoi Khimii Nauka*, Moscow, 1974.
[50] Z. Kurzawa and M. Zietkiewicz, *Chem. Anal. Warsaw*, 1976, **21**, 13.
[51] P. R. Bontchev, *Mikrochim. Acta*, 1964, 79.
[52] P. R. Bontchev, *Talanta*, 1970, **17**, 499.
[53] P. R. Bontchev, A. A. Alexiev and B. Dimitrova, *Talanta*, 1969, **16**, 597.
[54] H. Müller and G. Werner, *Z. Anal. Chem.*, 1974, **14**, 159.
[55] E. A. Bozhevolnov, S. U. Kreingol'd, R. P. Lastovskii and V. V. Sidorenko, *Dokl. Akad. Nauk. SSSR*, 1963, **153**, 97.
[56] K. B. Yatsimirskii and L. P. Raizman, *Zh. Analit. Khim.*, 1963, **18**, 29.
[57] L. B. Worthington and H. L. Pardue, *Anal. Chem.*, 1970, **42**, 1157.
[58] M. Otto and G. Werner, *Anal. Chim. Acta*, 1983, **147**, 255.
[59] M. Otto, J. Rentsch and G. Werner, *Anal. Chim. Acta*, 1983, **147**, 267.
[60] C. M. Wolff and J. P. Schwing, *Bull. Soc. Chim. France*, 1976, 679.
[61] H. Weisz and T. Janjic, *Z. Anal. Chem.*, 1967, **227**, 1.
[62] T. Janjic, G. A. Milovanovic and M. B. Celap, *Anal. Chem.*, 1970, **42**, 27.
[62a] J. Minczewski, J. Chwastowska and R. Dybczyński, *Separation and Preconcentration Methods in Inorganic Trace Analysis*, Horwood, Chichester, 1982.
[63] H. Müller, H. Schuring and G. Werner, *Talanta*, 1974, **21**, 581.
[64] M. Otto, P. R. Bontchev and H. Müller, *Mikrochim. Acta*, 1970 **I**, 193.
[65] M. Otto, H. Müller and G. Werner, *Talanta*, 1979, **26**, 781.
[66] M. Otto and H. Müller, *Talanta*, 1977, **24**, 15.
[67] M. Otto and H. Müller, *Anal. Chim. Acta*, 1977, **90**, 159.
[68] H. Müller, J. Mattusch and G. Werner, *Mikrochim. Acta*, 1980 **II**, 349.
[69] S. U. Kreingol'd, E. D. Shigina and E. V. Loginova, *Zh. Analit. Khim.*, 1983, **38**, 1397.
[70] J. D. Ingle, Jr. and S. R. Crouch, *Anal. Chem.*, 1971, **43**, 697.
[71] F. Willeboordse and F. E. Critchfield, *Anal. Chem.*, 1964, **36**, 2270.
[72] F. Willeboordse and R. L. Meeker, *Anal. Chem.*, 1966, **38**, 854.
[73] K. B. Yatsimirskii, A. G. Khachatryan and L. I. Budarain, *Dokl. Akad. Nauk. SSSR*, 1973, **211**, 1139.
[74] H. L. Pardue, *Advan. Anal. Chem. Instrum.*, 1969, **7**, 141.
[75] G. E. Mieling and H. L. Pardue, *Anal. Chem.*, 1978, **50**, 1611.
[76] R. L. Wilson and J. D. Ingle, Jr., *Anal. Chim. Acta*, 1976, **83**, 203.
[77] J. D. Ingle, Jr., M. J. White and D. Salin, *Anal. Chem.*, 1982, **54**, 56.
[78] J. D. Ingle, Jr. and S. R. Crouch, *Anal. Chem.*, 1973, **45**, 333.
[79] P. M. Beckwith and S. R. Crouch, *Anal. Chem.*, 1972, **44**, 221.
[80] T. E. Hewitt and H. L. Pardue, *Clin. Chem.*, 1975, **21**, 199.
[81] P. W. Carr, *Anal. Chem.*, 1978, **50**, 1602.
[82] T. E. Hewitt and H. L. Pardue, *Clin. Chem.*, 1973, **19**, 1128.
[83] J. B. Landis, M. Rebec and H. L. Pardue, *Anal. Chem.*, 1977, **49**, 785.
[84] J. E. Davis and B. Renoe, *Anal. Chem.*, 1979, **51**, 526.
[85] J. G. Atwood and J. L. DiCesare, *Clin. Chem.*, 1975, **21**, 1263.
[86] J. E. Davis and J. Pevnick, *Anal. Chem.*, 1979, **51**, 529.
[87] P. D. Wentzell and S. R. Crouch, *Anal. Chem.*, 1986, **58**, 2851.
[88] J. B. Pausch and D. W. Margerum, *Anal. Chem.*, 1969, **41**, 226.
[89] P. D. Wentzell and S. R. Crouch, *Anal. Chem.*, 1986, **58**, 2855.

† There was duplication of the page numbers for the July and August issues (and some with September, as well).

9

Analysis of real samples

9.1 INTRODUCTION

The literature on kinetic methods of analysis emphasizes the significant advantages offered by these methods in the analysis of real samples. This is because of the high sensitivity of catalytic and chemiluminescence methods, and the possibility of resolving mixtures of closely related species by differential reaction-rate methods. This potential has not been fully exploited so far, and with a few exceptions, the kinetic methods proposed are not usually incorporated in reference works such as 'Standard Methods for the Examination of Water and Wastewater', 'Official Methods of Analysis of the Association of Official Analytical Chemists', 'American Society for Testing and Materials Handbooks', etc. An outstanding exception is the group of kinetic methods based on enzymatic reactions, of great relevance to clinical and pharmaceutical chemistry.

In the authors' opinion, this situation is not a result of any shortcomings of the methods or their cost, but is mainly due to three factors, namely: (a) the lack of precision inherent in time measurements; (b) the need for accurate temperature control, and (c) the much longer times needed for complete analysis.

These apparently adverse factors have been greatly mitigated thanks to recent technological developments, and kinetic methods are now serious contenders for use in many practical situations. The delay in their acceptance is perhaps logical, however, since the handbooks in question still include titrimetric and gravimetric determinations which, though usually the ultimate reference procedures, are regarded by many as superseded by instrumental and chromatographic techniques, which are gradually being incorporated as standard procedures.

9.1.1 Effect of sensitivity

Undoubtedly, the advantages of kinetic methods over their equilibrium counterparts as regards their scope of application lie basically in the high sensitivity of catalysed and chemiluminescent reactions. Thus, kinetic methods allow the determination of traces of species in minute amounts of sample, which is crucial to the analysis of biological samples and forensic or certain geological materials. For example, the iodide-catalysed Sandell–Kolthoff reaction allows the determination of iodine-containing proteins in sera in sample volumes of only about 10 μl. Another advantage of use of small amounts of sample is their readier dissolution — particularly in the case of minerals and alloys, but against this must be set the greater risk of error due to inhomogeneity of the sample. Further, the use of sensitive catalytic methods avoids preconcentration stages and hence saves considerable time. Flame atomic-absorption spectrometry (AAS) has been used on occasion to check the results obtained by kinetic methods. The application of AAS to water analysis requires a preconcentration step completely unnecessary in the case of kinetic procedures.

Even those kinetic methods based on uncatalysed reactions, affording sensitivity similar to that of thermodynamic methods, offer some advantages. The usual determinations of organic compounds of biological interest are chiefly based on oxidation, hydrolysis or bromination reactions, which normally attain completion within 30–60 min. The application of kinetic procedures (e.g. initial-rate measurements) for these reactions results in considerably shortened analysis times. In addition, if the reactions develop rapidly enough, the use of the stopped-flow technique offers the added advantage of resolution of mixtures.

9.1.2 Effect of selectivity

The influence of selectivity on the practical aspects of the kinetic determinations can be classified according to whether the selectivity is high, medium or low.

Highly selective kinetic methods provide the so-called 'direct' determinations, which require no sample matrix removal. A typical example is that of manganese(II) in wines and brandies, based on its catalytic effect on a redox reaction [1]. Though this method is not very tolerant to iron, the presence of tartrate, citrate and pyrophosphate (with which this metal forms stable complexes) in the sample eliminates its interference.

The potential interference of two or perhaps three species in the determination of another by a kinetic method of medium sensitivity is sometimes more of an advantage than a disadvantage, since the application of a suitable differential reaction-rate method allows determination of all these species. Thus, Coetzee *et al.* [2] analysed atmospheric air for NO and NO_2 on the basis of their formation of a nitrosyl complex of Fe(II). The resolution of metal mixtures is also an area of growing interest to analytical chemistry, particularly in the speciation of environmental samples. The work of Truesdale and Smith [3] on the speciation of iodine species (iodate and iodide) by use of the Ce(IV)/As(III) system and a Technicon I AutoAnalyzer is worth mention in this connection.

The low selectivity of some kinetic methods is not always an obstacle in their application to real samples. Thus, the metal species of interest in waters or biological materials are usually present at concentration levels similar to those of potential interferents, so the selectivity of the kinetic determination is generally sufficient for prior separation steps to be unnecessary.

When the sample matrix poses serious interference problems, these are dealt with by removal of the matrix or isolation of the analyte prior to the determination.

One of the techniques most commonly used for separation of the sample matrix is liquid–liquid extraction. The determination of phosphate and silicate in the chemical ferric chloride requires the prior extraction of the Fe(III) present into methyl isobutyl ketone [4]. The separation technique used to isolate the analyte from the matrix must be suited to the type of sample dealt with. Thus, the lead content of galena is separated by precipitation as $PbSO_4$ prior to the catalytic determination of any copper present [5]. The determination of manganese and chromium in SiO_2 is only possible after volatilization of the silica with hydrofluoric acid [6,7].

Chromatographic techniques are often used when the analyte is to be separated from the sample. Liquid–liquid extraction or volatilization techniques are also applied for this purpose; an interesting example of application of the latter is the catalytic determination of sulphide in copper metal with the iodine/sodium azide system [8]. The sample is dissolved in a suitable acid and the H_2S generated is swept out by a nitrogen stream and collected in the reaction vessel.

The joint use of chromatographic techniques and kinetic methods is no doubt the best solution to the problems arising from the low selectivity of some catalytic and chemiluminescence methods. Thus, once separated in the chromatographic system, the different species can be determined kinetically with high sensitivity. Chemiluminescence reactions, on account of their greater selectivity, have been more frequently employed than catalytic methods as indicators in chromatographic procedures.

As stated in Chapter 2, the oxidation of luminol by hydrogen peroxide is catalysed by a score of metal ions. Dulumyea and Hartkopf [9] have used this reaction as the indicator system in ion-exchange chromatography. One of the potential drawbacks of this combination is that the metal ions are usually eluted by an acid medium, but their reaction with luminol takes place in alkaline medium, and thus involves an abrupt change in pH. To decrease the acidity without affecting the chromatographic resolution, these authors use mixtures of hydrochloric acid, water and ethanol in different proportions, or sodium or lithium chloride solutions as eluents, none of which alter the pH of the indicator reaction. Figure 9.1 shows the manifold of this chemiluminescence system. A peristaltic pump mixes the sample, alkaline solution of luminol and alkaline hydrogen peroxide, the chemiluminescence yielded being sensed in the detector chamber.

In a recent paper, Nyarady *et al.* [10] reported a sensing system for use in gas chromatography, that involved a redox catalysed reaction and a chemiluminescent reaction, which they called a 'redox–chemiluminescent detector' (RCD). This was based on the gold-catalysed reduction of nitrogen dioxide by the analytes in a post-column chromatographic detector according to:

$$\text{Analyte} + NO_2 \xrightarrow[150-400^\circ C]{Au} \text{Oxidized analyte} + NO$$

followed by the chemiluminescence reaction

$$NO + O_3 \rightarrow NO_2^* + O_2$$

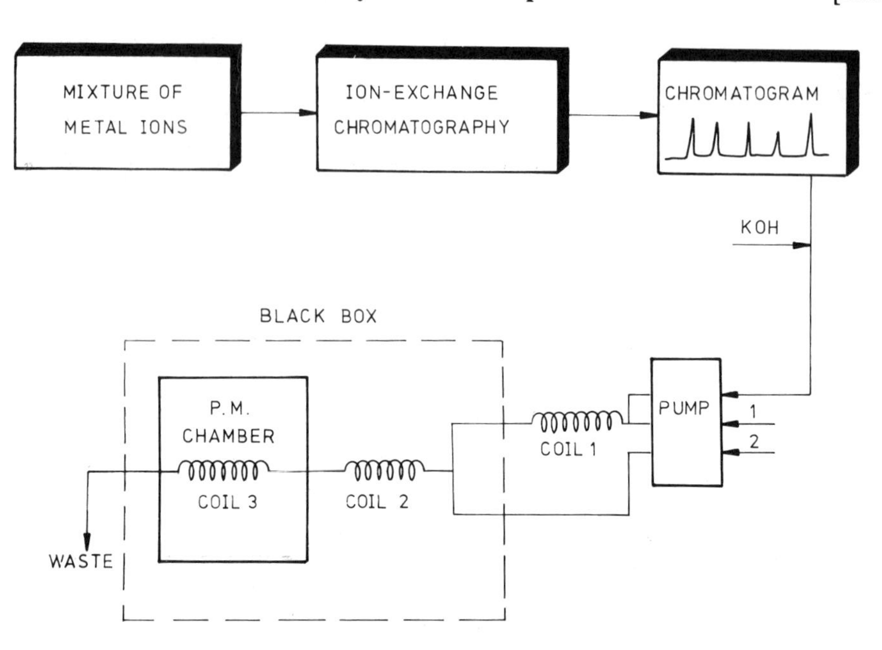

Fig. 9.1 — Scheme of a luminescence detector: (1) luminol; (2) H_2O_2.

$$NO_2^* \rightarrow NO_2 + h\nu$$

The combination of these reactions endowed this detection system with high sensitivity. The operational scheme is shown in Fig. 9.2a. The detector is highly responsive to reductants such as H_2S, CS_2, SO_2, H_2O_2, H_2, CO, alcohols, thiols, etc., but completely insensitive to most of the compounds normally occurring in environmental or industrial sample matrices — hence its great potential in these fields. In addition, it allows the direct determination of traces of such compounds, thereby avoiding separation steps and saving time. Finally, this system responds to some species that are undetectable by the conventional flame ionization detector (FID), to which it is a suitable alternative; for example, the FID senses only butanoic acid in mixtures with formic acid and hydrogen peroxide, whereas the RCD allows the determination of all three species (Fig. 9.2b).

It is not only chemiluminescent reactions which have been used as sensing systems in chromatographic techniques. Nachtmann *et al.* [11] have utilized an HPLC system and the iodide-catalysed reaction between Ce(IV) and As(III) for the determination of thyroid hormones in biological samples (serum thyroxins) at the subnanogram level.

Finally, it is worth mentioning that even uncatalysed reactions have been applied successfully in conjunction with this type of system. Kobayashi and Imai [12] determined dansylamino-acids through their reaction with bis(2,4,6-trichlorophenyl) oxalate (TCPO) and hydrogen peroxide on the basis of a chemiluminescence

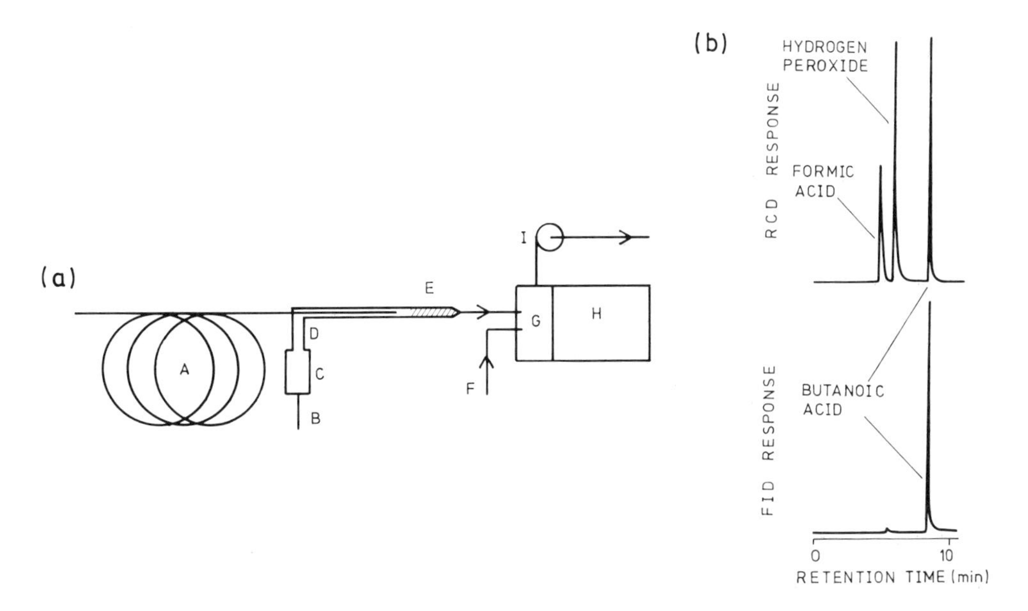

Fig. 9.2 — (a) Scheme of redox chemiluminescence detector (RCD) used with a gas chromatograph: A, capillary GC column; B, inlet for purified helium as carrier gas; C, NO_2 permeation device; D, inlet for NO_2/He reagent gas; E, heated catalyst zone for redox reaction; F, inlet for O_3; G, chemiluminescence reaction chamber; H, photomultiplier tube and electronics; I, vacuum pump. (b) Simultaneous detection of the split gas-chromatographic column effluent with an FID and an RCD in parallel. (Reproduced with permission from S. A. Nyarady, R. M. Barkley and R. E. Sievers, *Anal. Chem.*, 1985, **57**, 2074. Copyright 1985, American Chemical Society).

process. Figure 9.3 depicts the scheme of the system used. The eluate from the chromatographic column is mixed with the TCPO and H_2O_2 solutions in a reactor to yield the chemiluminescence signal. The procedure is useful for the determination of *N*-terminal amino-acids in proteins, as well as that of drugs [13], catecholamines [14] or aryl oxalates [15].

9.2 SCOPE OF APPLICATION

The applications of kinetic methods to the analysis of real samples can be divided into the six areas of interest considered in Table 9.1, of which those of environmental and clinical chemistry have received most attention, for reasons discussed below.

Investigation of the applicability of kinetic methods in a given area is strongly dependent on (among other factors) the availability of suitable samples. There are three ways of testing the performance of a kinetic method, namely by applying it to real samples, spiked real samples or simulated (synthetic) samples.

Real samples. These are analysed for the analyte of interest. If the samples used are not reference materials, the method must be compared with an accepted non-kinetic technique. Thus, the determination of nitrite in meat by a chemiluminescence reaction is compared with the classical colorimetric procedure based on the Griess

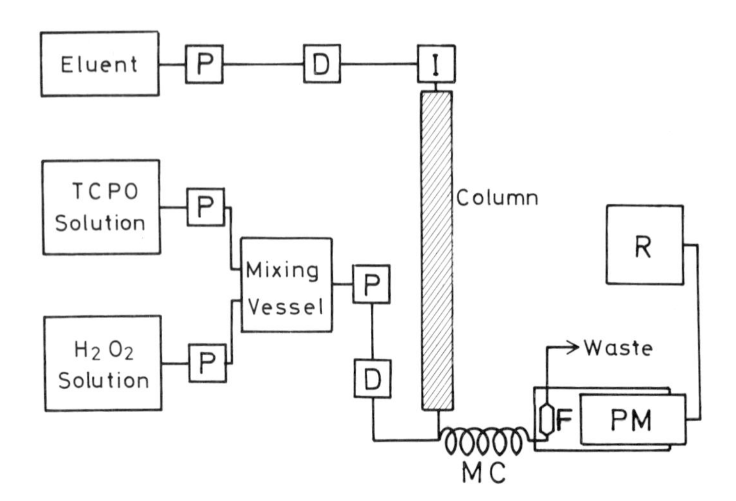

Fig. 9.3 — Schematic diagram of the apparatus used in the flow system. TCPO = bis(2,4,6-trichlorophenyl) oxalate; D, damper; F, flow-cell; I, injector; MC, mixing coil; P, pump; PM, photomultiplier; R, recorder. (Reproduced with permission, from S. Kobayashi and K. Imai, *Anal. Chem.*, 1980, **52**, 424. Copyright, 1980, American Chemical Society).

Table 9.1 — Areas of application of kinetic methods

Environmental chemistry	Waters Industrial effluents Environmental air
Clinical and pharmaceutical chemistry	Biological fluids Pharmaceuticals
Industrial products	Alloys Petroleum derivatives Other products
Geochemistry and agricultural chemistry	Minerals and rocks Oligoelements in plants and soils Fertilizers
Food analysis	Natural foods Manufactured foods
Analytical-grade reagents	High-purity metals Mineral acids Salts

reaction [16]. Atomic-absorption spectrometric techniques, spectrophotometry and neutron-activation analyses are used for comparison with kinetic determinations of metal ions. The standard-addition method is useful for this purpose, since it facilitates the detection of multiplicative interferences.

Various organizations, such as the National Bureau of Standards (NBS), the Bureau of Analysed Samples (BAS), British Chemical Standards (BCS) and the Bundesanstalt für Materialprüfung (BAM), market standard samples (standard reference materials, RSMs) of alloys, minerals and even vegetables, with contents certified as the averages of results found by expert analysts using different analytical techniques.

Analytes added as spikes to real matrices. Sometimes the sample of interest, though available, does not contain the analyte, a known amount of which (a 'spike') is therefore added to it and compared with the result obtained by the kinetic method to calculate the so-called *recovery*. Recoveries between 95 and 105% are taken as evidence of the usefulness of a given kinetic method, and of the absence of interferences in the determination of the analyte.

Simulated samples. Samples such as nuclear fuels, waste waters, etc., are often replaced in practice by simulated (synthetic) samples for purposes of comparison of analytical techniques. These should match as faithfully as possible their real counterparts, both in analyte content and treatment. Some representative examples are the kinetic catalytic determination of technetium in nuclear fuels [17], of iodide in simulated rainwater matrices by the Sandell–Kolthoff reaction [19] and of cyanide in electroplating baths [19]. As for spiked samples, the success of the kinetic method concerned is evaluated from the recovery.

Real and simulated samples are more commonly used than spiked samples for evaluating the efficiency of kinetic methods.

There follows a detailed description of the most interesting applications of kinetic methods in the six areas of chemical analysis listed above.

9.2.1 Environmental chemistry

Environmental chemistry is the most extensive field of application of kinetic methods to real samples, for several reasons.

(a) The matrix is usually quite simple, which allows direct determinations — only the analysis of domestic or industrial effluents is likely to pose problems arising from the nature of the matrix.
(b) The trace elements present in these samples are normally found at similar concentration levels, so prior separation steps for elimination of interferences are usually not needed.
(c) Environmental samples are inexpensive, abundant and readily available. Atmospheric air samples can be collected on filters for as long as dictated by the sensitivity of the kinetic method to be applied.
(d) The characteristics of these samples make them ideal for automated analysis.

The determination of metal ions in environmental samples is of great ecological significance, as these species cannot be degraded biologically or chemically; nature's attempt to decompose them occasionally results in even more toxic forms, as in the case of mercury species. Kinetic methods, on account of their high sensitivity, are thus a useful tool for determination of heavy metals in environmental water and air, where these toxic species are found at low concentration levels.

Table 9.2 gathers most of the applications of kinetic methods to environmental analytical chemistry reported over the last ten years. As can be seen, most of the entries refer to water analysis.

Table 9.2 — Applications of kinetic methods to environmental chemistry

Sample	Species	Reaction	Observations	References
Drinking water	Manganese	Catalysed, redox	Compared with NAA	[20]
		Catalysed, redox	Automatic photometric procedure	[21]
		Catalysed, redox	Standard-addition method	[22]
		Uncatalysed	Fluorimetric detection	[23]
		Catalysed, redox	Segmented-flow system (30/hr)	[24]
	Vanadium	Catalysed, redox	Prior ion-exchange separation	[25]
	Zinc, magnesium, copper and nickel	Ligand-exchange	Mixture resolution	[26]
	Iodide	Catalysed, redox	Detn. range 0.01–2 mg/l.	[27]
	Iodate, iodide	Catalysed, redox	Segmented-flow analyser (20/hr)	[28]
	Fluoride	Catalysed, redox	FIA (40/hr)	[29]
	Sulphide	Chemiluminescent		[30]
	Phosphate, silicate	Catalysed, redox	Miniature centrifugal analyser	[31]
Tap water	Copper	Catalysed, redox		[32]
		Catalysed, redox	Compared with non-flame AAS	[33]
	Cobalt	Chemiluminescent	Removal of interfering Fe and Mg	[34]
	Nitrite	Uncatalysed	Stopped-flow (360/hr)	[35]
	Fluoride	Catalysed, redox	Content $\leqslant 0.1\ \mu g/ml$	[36]
	Chlorinated compounds	Chemiluminescent	Compared with colorimetry	[37]
River water	Copper	Catalysed, redox	Compared with non-flame AAS	[38]
		Catalysed, redox	Compared with liquid–liquid extraction and AAS	[39]
	Vanadium	Catalysed, redox	Also applied to tap water	[40]
	Iron	Catalysed, redox	Also applied to sea and tap water	[41]
	Fluoride	Catalysed, redox	Interference by sulphate and phosphate	[42]
	Iodate, iodide	Catalysed, redox	Uses Technicon I AutoAnalyzer	[3]
Sea-water	Vanadium	Catalysed, redox	Filtering + concentrated HCl	[43]
	Copper	Catalysed, redox	No pretreatment	[44]
	Calcium + magnesium	Ligand-exchange	Stopped-flow mixture analysis	[45]
	Molybdenum	Catalysed, redox	Preconcentration with chelating resin	[46]
	Chlorinated compounds	Chemiluminescent	Also applied to other types of water	[47]

Continued on next page

Rain water	Vanadium	Catalysed, redox	Ion-exchange preconcentration (FIA)	[48]
	Iodide	Catalysed, redox	Simulated sample, matrix	[18]
Waste water	Mercury	Ligand-exchange, redox	Stagnant and electric power plant waters	[49]
		Uncatalysed	Industrial effluents	[50]
	Cadmium + manganese	Ligand-exchange	Stopped-flow	[51]
	Tungsten	Catalysed, redox	Mining water	[52]
	Gold	Catalysed, redox	Washing sewage	[53]
	Cyanide	Catalysed, redox	Electrolytic plating sewage	[54]
		Catalysed, redox	With and without cyanide distillation	[19]
	Bromide	Catalysed, redox	Mine water	[55]
	Chlorinated compounds	Chemiluminescent	Analysis for ClO^- and ClO_2 in disinfected water	[56]
	Nitrate	Photochemical	Organic-free effluents	[57]
	Benzaldehyde, formaldehyde	Uncatalysed	Used in the manufacture of synthetic fibres	[58]
Environmental air	Mercury	Ligand-exchange		[59]
	Manganese	Uncatalysed	Sample collected on cellulose filter (0.8 μm)	[60]
	Sulphide	Chemiluminescent	Stopped-flow/ microdistillation system	[61]
	Nitrogen oxides	Uncatalysed	Determination of NO–NO_2 mixtures	[2]
	Carbon monoxide	Uncatalysed		[62]
Environmental dust	Vanadium	Catalysed, redox	Ion-exchange prior separation	[63]
	Fluoride	Inhibition	Isolation from matrix by microdiffusion	[64]

NAA: neutron activation analysis, AAS: atomic-absorption spectroscopy, FIA: flow-injection analysis.

Kinetic methods have so far been applied to the determination of a host of metal ions, from calcium and magnesium to heavy metal ions or even noble metals such as gold and mercury, as well as to anions from halogens, sulphur, phosphorus and nitrogen. As far as the chemistry is concerned, catalysed and chemiluminescence reactions are the most frequently used; uncatalysed reactions, though far less common, are suitable for the determination of organic compounds such as benzaldehyde and formaldehyde in effluents from synthetic fibre factories [58].

Direct kinetic determination is often applied to this type of sample, and is sometimes automated, as in the automatic procedure developed by Isacsson and Wettermark [65] for the determination of chlorine, chlorine dioxide and hypochlorite in water (to which the halogen is added to make it suitable for human consumption) at the micro- or submicromolar level, on the basis of the chemiluminescent reaction between luminol and hydrogen peroxide. The continuous-flow assembly used by these authors is shown schematically in Fig. 9.4. The sample and reagents are pumped into the mixing chamber (A), the chemiluminescence signal yielded being collected by an optical fibre attached to the flow-cell. The system affords sample flow-rates of the order of 150–200 ml/min.

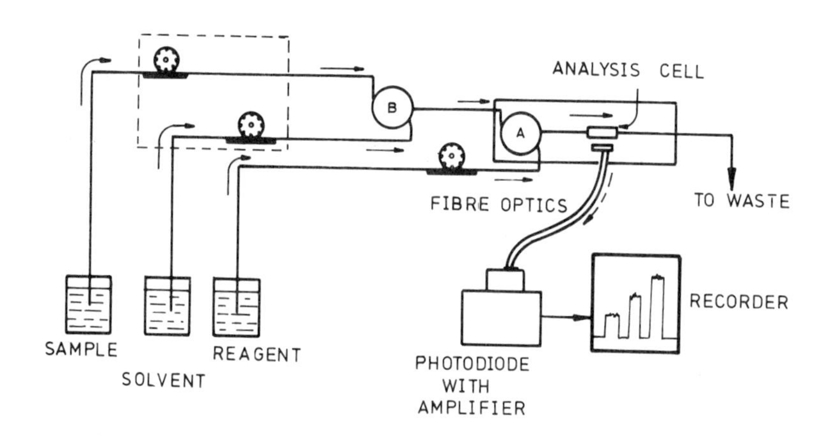

Fig. 9.4 — Equipment for continuous flow technique (Reproduced from U. Isacsson and M. Tanaka, *Anal. Chim. Acta*, 1976, **83**, 227, by permission. Copyright 1976, Elsevier Science Publishers).

It is interesting to note that direct kinetic determinations do not always allow the overall analyte concentration in the sample to be calculated. Thus, the kinetic results obtained by Nakano *et al.* [41] in the determination of iron in river water differ from those found by the AAS procedure used for comparison purposes. This discrepancy is not the result of matrix inference, but of speciation: only the free metal ion is measured by the kinetic procedure, whereas AAS determines the total iron content, including that of any organic complexes present. If the kinetic determination is carried out after appropriate pretreatment, the same results for total iron content are obtained by both procedures. Hence kinetic determination with and without pretreatment allows speciation of iron in water.

Analyte separation is sometimes essential if acceptable results are to be obtained, and a wide variety of procedures can be used. Fukasawa and Yamane [25] use a combination of cation-and anion-exchange chromatography to separate vanadium (in $HCl–H_2O_2$ medium) from other metals such as iron, copper, titanium, molybdenum and tungsten, all of which interfere with its kinetic catalytic determination in river and lake water.

As is widely known, mercury is a highly toxic metal, inhibiting the activity of some enzymes. Since its concentration in natural waters ranges between 0.1 and 5.0 ng/ml, the sensitivity of catalytic methods makes them ideal for routine monitoring of this metal ion in water. Thus, Rychkova and Dolmanova [49] determined mercury in stagnant water from electric power plants by the joint use of a ligand-exchange and a redox reaction. Tabata and Tanaka [66] have developed a distillation unit for mercury which allows its selective separation and simultaneous preconcentration from the sample matrix even at concentrations as low as a few ng/ml.

Distillation has also been used in the chemiluminescence determination of sulphide in atmospheric air [61]. The sulphide is collected as Ag_2S on filters, which are then dissolved in an acid solution of Sn(II), and sulphur is isolated from the matrix as H_2S by microdistillation in a flow of inert gas which transports it to a flow system where its chemiluminescence yield is sensed.

The determination of fluoride in airborne dust proposed by Auffarth and Klockow [64] includes an interesting procedure for the selective isolation of the halide from the sample matrix, based on microdiffusion with hexamethyldisiloxane and its subsequent detection in a flow system by virtue of its inhibitory effect on the reaction between perborate and iodide, catalysed by traces of Zr(IV). The procedure is quite simple and has been applied both to routine analyses and micro samples.

Finally, it is worth mentioning the attempts at automation of these determinations by use of different flow techniques involving segmented-flow analysers, continuous-flow systems, flow-injection analysis and even the stopped-flow technique when dealing with especially fast reactions or mixtures of species. These systems afford sampling rates between 20 and 360 samples/hr.

9.2.2 Clinical and pharmaceutical chemistry

Kinetic methods of analysis are of special relevance to clinical and pharmaceutical analysis, particularly enzymatic methods. So much so, that it is in these two areas that automatic methods have gained their present significance, partly thanks to the commercialization of segmented-flow analysers by companies such as Technicon. The basic operational scheme of these analysers is illustrated in Fig. 9.5 [99]. Though designed for enzymatic reactions, they are equally applicable to non-enzymatic processes.

Table 9.3 lists the most significant contributions in the field of non-enzymatic determinations reported over the last few years. As can be seen, most of the clinical determinations were carried out on blood and/or urine matrices and applied to inorganic and organic analytes alike, whereas the pharmaceutical determinations dealt mainly with organic substances.

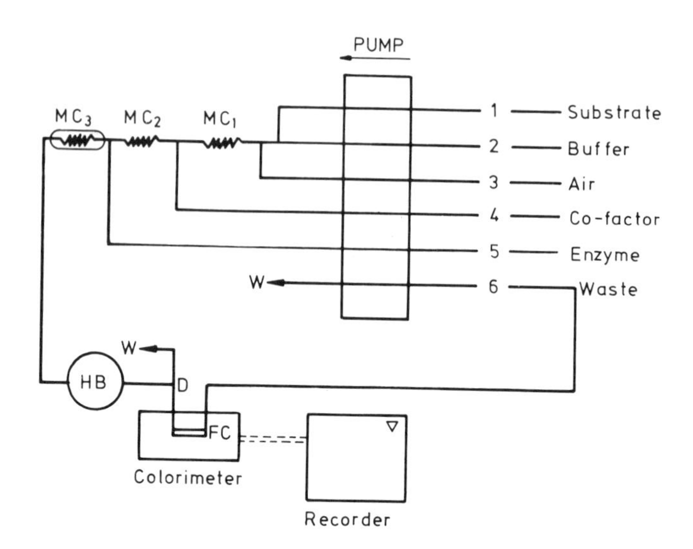

Fig. 9.5 — Straightforward flow system for enzymatic assays. D, debubbler; FC, flow-cell; HB, heating bath; MC, mixing coils (MC_3 is jacketed to operate at the temperature of the heating bath); W, waste. (Reproduced from D. B. Roodyn, *Automated Enzyme Assays*, by permission. Copyright 1970, Elsevier Science Publishers).

It is interesting to note the wider application of uncatalysed reactions in these areas, compared to others considered in Table 9.1. This partly results from the greater occurrence and importance of organic compounds in clinical and pharmaceutical chemistry on the one hand, and their inertness as catalysts on the other. Nevertheless, as these compounds are not normally found at very low levels, the sensitivity of these uncatalysed reactions is generally sufficient to allow the analyst to dispense with the use of a preconcentration step.

The advantages provided by the use of kinetic methods of analysis in these areas can be summed up as (a) sample consumption in uncatalysed reactions is rather low, which is of greater significance in the analysis of biological fluids, and (b) measurements on uncatalysed reactions are faster to make than those in equilibrium methods.

Most of these determinations call for sample pretreatment, either wet or dry in the case of inorganic species and wet for organic compounds. Wet pretreatment usually involves digestion with acids for inorganic compounds and sodium hydroxide for organic substances. Dry pretreatment is usually in the form of ashing at 600°C, dissolution of the residue in dilute nitric acid, evaporation to dryness and extraction with water.

The Sandell–Kolthoff reaction is no doubt the most commonly employed in the analysis of biological samples; so much so that some clinical handbooks recommend its use for the determination of thyroid hormones such as tri- (T_3) or tetra-iodothyronine (T_4) [100]. The normal content of these species in serum ranges between 4 and 8 μg/100 ml; concentrations below the lower or above the upper limit give rise to hypothyroidism and thyrotoxicosis, respectively. There are a number of procedures available for the determination of these species. Thus, Knapp and

Table 9.3 — Applications of kinetic methods to clinical and pharmaceutical chemistry

Sample	Species	Reaction	Observations	References
Serum	Copper	Catalysed, redox	Prior separation of albumin	[67]
		Catalysed, redox	Automatic analyser (20/hr)	[68]
		Chemiluminescent	Ashing + HNO_3	[69]
	Cobalt	Catalysed, redox		[70]
	Molybdenum	Catalysed, redox	Ashing and extraction of Mo	[71]
	Chloride	Chemiluminescent	Prior electrolysis	[72]
	Iodide	Catalysed, redox	Prior mineralization	[73]
		Catalysed, redox	Serum volumes of 12.3 μl	[74]
	Carbon monoxide	Uncatalysed	Range 0.05–20 mg/100 ml	[75]
	Iodoproteins	Catalysed, redox	Acidification + bromine	[76,77]
	Uric acid	Promotion	Fast analysis	[78,79]
	Uric and ascorbic acid	Chemiluminescent	Sensitivity increased by use of oxidoreductase	[80]
	Creatinine	Uncatalysed	Ion-exchange separation (Dowex 50W-X8)	[81]
Urine	Copper	Catalysed, redox	Statistical study on the population	[82]
	Iodide	Catalysed, redox	Dialysis at 37°C; AutoAnalyzer	[83]
		Catalysed, redox	With and without pretreatment	[84]
	Uric acid	Promotion	Simple, selective, sensitive procedure	[78,79]
	Sulphonamides	Uncatalysed	Stopped-flow	[85]
	Morphine	Decomposition	Compared with HPLC	[86]
Liver tissue	Copper	Catalysed, redox	Digestion with HNO_3 + Cu extraction	[87]
Human hair	Selenium	Catalysed, redox		[88]
Pig viscera	Copper	Catalysed, redox	Ashing at 600°C + HNO_3 + boiling to dryness + H_2O	[89]
Rat brain	Dopamine	Uncatalysed	Range 0.76–6.83 μg/ml	[90]
Pharmaceuticals	Bismuth	C = N − group exchange	Indirect determination	[91]
	Mercury	Catalytic titration	Potentiometric detection	[92]
	Nitroglycerine	Hydrolysis	Treatment with NaOH in methanol	[93]
	Paracetamol	Bromination	Mean recovery 100.5–102.6%	[94]
	Thiamine	Oxidation	Fluorimetric detection	[95]
	Thyroxine	Catalysed, redox	Treatment with NaOH + filtration	[96]
	Tetracyclines	Degradation	Formation of anhydride (0.8–4.0%)	[97]
	Catecholamines, sulphonamides and barbiturates	Catalytic titration	Excipient influences titration	[98]

Leopold [77] have developed an automatic digital system capable of detecting down to 1 ng/ml in serum samples. As thyroid hormones are accompanied by other iodine compounds in these samples, they have to be isolated on a chromatographic column prior to their determination. Recently, it was shown that these hormones occur largely bound to blood proteins, so the amount of protein-bound iodine, normally determined by the Sandell–Kolthoff reaction, is a measure of the thyroxine content in the serum.

Grases *et al.* [84] recently developed a determination for iodine in urine, with thermometric detection. By carrying out the analysis with and without mineralization of the sample, these authors determined organically bound iodine by difference.

Another species of interest commonly found in serum and urine is oxalic acid, occasionally involved in metabolic disfunctions. The range of methods (precipitation, liquid–liquid extraction, isotope dilution, enzymatic reactions, etc.) available for its determination has been broadened by Dutt and Mottola [79], who have developed a fast and selective procedure affording detection limits of a few μg. After isolation from the matrix, the acid is determined (by the standard-addition method) through its promoting effect on a redox reaction.

Analyses of pharmaceutical preparations are somewhat less frequent and usually involve organic compounds such as paracetamol, thiamine, thyroxine, tetracyclines, etc., which are generally determined by uncatalysed reactions. Hadjiioannou *et al.* [96] determined thyroxine in tablets by its catalytic effect on the Ce(IV)/As(III) system, with recoveries ranging from 95 to 108%. Among the few references to the determination of organic compounds by catalytic titrations are those of catecholamines, sulphonamides and barbiturates proposed by Greenhow and Spencer [98].

As regards the determination of inorganic compounds in pharmaceutical preparations, it is worth noting that of bismuth in suppositories by a reaction involving the exchange of $C=N-$ groups [91], and that of mercury in various pharmaceuticals by catalytic titration, on the basis of the inhibitory effect of this metal on the iodide-catalysed reaction between Ce(IV) and As(III) [92].

Finally, it is interesting to note that current trends in the application of kinetic methods to clinical and pharmaceutical determinations seem to point to the resolution of mixtures by differential kinetic analysis.

9.2.3 Industrial chemistry

The application of kinetic methods to the monitoring of traces of species in industrial products is focused on metal ions which are normally determined by use of catalysed reactions. This field of application embraces three areas, namely: (a) metals and alloys; (b) petroleum products and (c) other products (Table 9.4).

The examples of application to metallurgical analysis are all exploratory, since the speed required in production control is best obtained by purely instrumental techniques such as spectrography.

An area of greater interest is the analysis of crude oils and their derivatives, which can contain trace metals arising from their geological source, or from contamination during the refining process, or as additives designed to improve the performance of the refined products. Determination of these metals is of vital importance, as it may help to clarify the mechanism by which oil is formed within the earth and because

Table 9.4 — Applications of kinetic methods to the analysis of industrial products

Sample	Species	Reaction	Observations	References
Aluminium alloy	Silicon	Catalysed, redox	Decomposition with 30% NaOH; prior separation of Mg, Mn and Fe	[101]
	Iron, nickel, copper and manganese	Catalytic titration	Treatment with conc. HCl and a few drops of H_2O_2	[102]
	Iron + manganese	Catalysed, redox	Simultaneous determination	[103]
Aluminium bronze	Iron	Catalysed, redox	Treatment with HNO_3 (1:1 v/v)	[104]
Copper alloy	Ruthenium	Catalysed, redox	Fusion with Na_2O_2 at 650°C for 30 min and dissolution in HCl	[105]
	Iron + manganese	Catalysed, redox	Fluorimetric detection	[103]
Zinc alloy	Silver	Catalysed, redox	Prior extraction of Ag with dithizone	[106]
Nickel alloy	Copper	Catalysed, redox	Direct determination	[107]
Ni–Cr alloy	Iron	Catalysed, redox	Prior separation of Ni by ion-exchange (Dowex 1–X8)	[104]
Stainless steel	Nickel	Catalytic titration	Also determines Fe content	[102]
Crude oils	Vanadium	Uncatalysed	Evaporation + ashing at 600–900°C + dissolution in HNO_3	[108]
		Catalysed, redox	Combustion in oxygen flask	[109]
		Catalysed, redox	Acidification with H_2SO_4 + ashing at 525°C + dissolution in HNO_3–H_2SO_4	[110]
Nuclear fuel	Technetium	Catalysed, redox	Simulated HCl matrix	[17,111]
Catalysts	Ruthenium	Catalysed, redox	Decomposition with Na_2O_2	[112]
Optical fibre	Copper, iron	Catalysed, redox	Decomposition with HF; masking of fluoride with Al(III)	[113]
Borosilicate glass	Borate	Hydrolysis	Attack with Na_2CO_3 + ashing at 850–900°C + dissolution in dilute HCl	[114]
Semiconductors	Silver	Catalysed, redox	Sample content: CdS + $CdSe_x$ $Te_{(1-x)}$; direct determination	[115]
Portland cement	Iron	Catalysed, redox	Separation of silica by precipitation	[104]
Plastic food containers	Lead	Catalysed, redox	Ashing at 500°C (2 hr) + 0.2*M* HNO_3	[116]
Various materials	Ruthenium	Catalysed, redox	Fusion with Na_2O_2 at 700°C and dissolution in dilute HCl	[117]

their presence may considerably affect the performance of any catalysts used in the refining process.

One of the elements most frequently determined kinetically in crude oils is vanadium. Prior ashing of the crude is inadvisable in this case because vanadium occurs as volatile porphyrins in oil. However, pretreatment with sulphuric acid destroys these complexes and allows subsequent ashing without loss. All things considered, the best technique is combustion in oxygen [109].

A recent paper on the determination of technetium in nuclear fuels [17] is of interest. The method is the first kinetic catalytic determination of this element and is a suitable alternative to instrumental techniques such as radiochemistry or mass spectrometry. The simulated matrix used is shown in Table 9.5, and contains mainly

Table 9.5 — Composition of the simulated HAO (high-activity oxide) nuclear fuel solution (2*M* HCl medium). (Reproduced with permission, from F. Grases and J. G. March, *Anal. Chem.*, 1985, **57**, 1419. Copyright 1985, American Chemical Society)

Species	Concentration (g/l.)
Cu(III)	0.002
Ag(I), Al(III), Sb(III)	0.016
Cd(II), Ni(II), Sn(II)	0.025
Cr(VI), Ce(III)	0.50
Mo(VI), Y(III), Fe(III), Na(I)	0.200
La(III), Sr(II)	0.400
Cs(I), Nd(III), Zr(IV)	1.000
U(VI)	200

U(VI). Technetium can be determined in it (with a photometric detector) at the 0.1–1.5 μg/ml level in as small a sample volume as 10 μl.

The remainder of the applications of kinetic methods to analysis of industrial products are varied in nature. Among them are the determination of borate in borosilicate glass [113], the direct determination of silver in semiconductors of the $A^{II}B^{IV}$ type because of its influence on their conductivity [114], and that of lead in plastic food containers [116]. In the last case, owing to the cumulative nature of the toxic effects of lead, the sensitivity of the kinetic methods used (enhanced by the fluorimetric sensing system employed) is a significant asset in monitoring the quality of the container in order to avoid contamination of foods.

9.2.4 Geochemistry and agricultural chemistry

Catalysed reactions are again by far the most frequently used of the kinetic systems and hence metal ions are the commonest subjects for analysis (see Table 9.6). The minerals are usually attacked with acids or acid mixtures, and rocks by alkaline fusion. In some instances, either the major component or the analyte is isolated from the matrix prior to the determination, e.g. the kinetic catalytic determination of copper in galena [5] is preceded by separation of the lead as $PbSO_4$, and lead in moss is kinetically determined after its extraction with dithizone [125].

It is worth mentioning here the determination of iron in the NBS fruit-tree foliage RSM, developed by Ditzler and Gutknecht [124], because of the procedure used to

Table 9.6 — Applications of kinetic methods to the analysis of geological materials and agricultural products

Sample	Species	Reaction	Observations	References
Zinc concentrate	Iron	Catalysed, redox	Treatment with HCl (1:1 v/v) + concentrated HNO_3	[104]
Blende	Zinc	C = N − group exchange	Treatment with HCl–HNO_3 (1:1 v/v)	[91]
Galena	Copper	Catalysed, redox	Separation of lead as $PbSO_4$	[5]
Dolomite	Iron	Catalysed, redox	Ion-exchange separation of Ca and Mg	[104]
Sulphurized minerals	Rhenium	Catalysed, redox		[118]
Molybdenite	Molybdenum	Catalysed, redox	W masked with H_3PO_4; laborious treatment	[119]
Silicate rocks	Chromium	Catalysed, redox		[120]
	Bromide	Catalysed, redox	Fusion with KOH; oxidation to Br_2 and extraction into benzene	[121]
Various rocks	Iridium, rhodium and palladium	Catalysed, redox	Range 0.01–0.002 g/ton	[122]
	Chloride	Catalysed, redox	Fusion with NaOH and dissolution in H_2O	[123]
Soils	Iron + manganese	Catalysed, redox	Simultaneous mixture resolution	[103]
Fruit-tree foliage	Iron	Catalysed, redox	NBS standards; attack with various acids	[124]
Moss	Lead	Catalysed, redox	Extraction of Pb with dithizone	[125]
Vegetables	Bismuth	Catalysed, redox	Ashing at 500°C + 6M HNO_3 + boiling to dryness + 1M HNO_3	[126]
Plants	Iodide	Catalysed, redox	Simulation of a nuclear accident; assimilation of radioactive iodide	[127]
			AutoAnalyzer	[128]
Fertilizers	Phosphorus	Catalysed, redox	Digestion by AOAC method	[129]

monitor the course of the reaction. This involves the combination of a kinetic catalytic method and gas chromatography (GC) for detection purposes. The amount of product detected by the chromatograph will be proportional to the catalyst concentration.

The procedure relies on the catalytic action of Fe(III) on the oxidation of anisole by hydrogen peroxide in a slightly acidic medium in the presence of hydroquinone, which yields *o*-hydroxyanisole according to:

$$\underset{\text{anisole}}{C_6H_5OCH_3} + H_2O_2 \xrightarrow[\text{hydroquinone}]{Fe(III)} \underset{o\text{-hydroxyanisole}}{C_6H_4(OH)OCH_3} + H_2O$$

Once the reaction has developed for the preset time, the reaction mixture is extracted into chloroform (so that the reaction is immediately stopped) and a 2.0-μl aliquot of the extract is injected into the gas chromatograph.

Figure 9.6a displays a typical chromatogram obtained for an injection of 2.0 μl corresponding to 300 ng/ml iron and a reaction time of 10 min. The heights of the peaks corresponding to the reaction product and obtained for different reaction times are used to produce the calibration graphs shown in Fig. 9.6b, from which the iron concentration can be readily determined.

The procedure is highly selective, as iron(III) is the only effective catalyst for this reaction, and it is also rather sensitive (detection limit 1.5 ng/ml). The method has also been exploited in the determination of iron in river water and vitamin preparations with excellent results.

One of the most important species determined kinetically in agricultural chemistry is phosphate. Phosphorus is one of the essential nutrients influencing the metabolism, structure and reproduction of plant cells. Most of the methods available for its determination rely on formation of molybdophosphoric acid, which is the basis for the oficially recommended AOAC procedure. These equilibrium methods have drawbacks such as low sampling rates or low selectivity arising from side-reactions and/or matrix effects. These shortcomings are easily circumvented by the use of kinetic methods.

McCracken and Malmstadt [129] have developed a novel reaction-rate stopped-flow method for the determination of phosphorus in cereals, feed and fertilizers, based on the well-known production of molybdenum blue with ascorbic acid. The reaction time is always less than 10 sec.

Table 9.7 compares the results found in some of the kinetic analyses of feeds and pet food, with those obtained in analyses performed in twenty laboratories by the AOAC official method. Agreement between the results is extremely good, as the rsd is only 0.2–0.7%.

9.2.5 Food analysis

The application of kinetic methods in the field of food analysis has grown enormously in the last few years, so much so that no mention of it is made in Müller's 1982 review

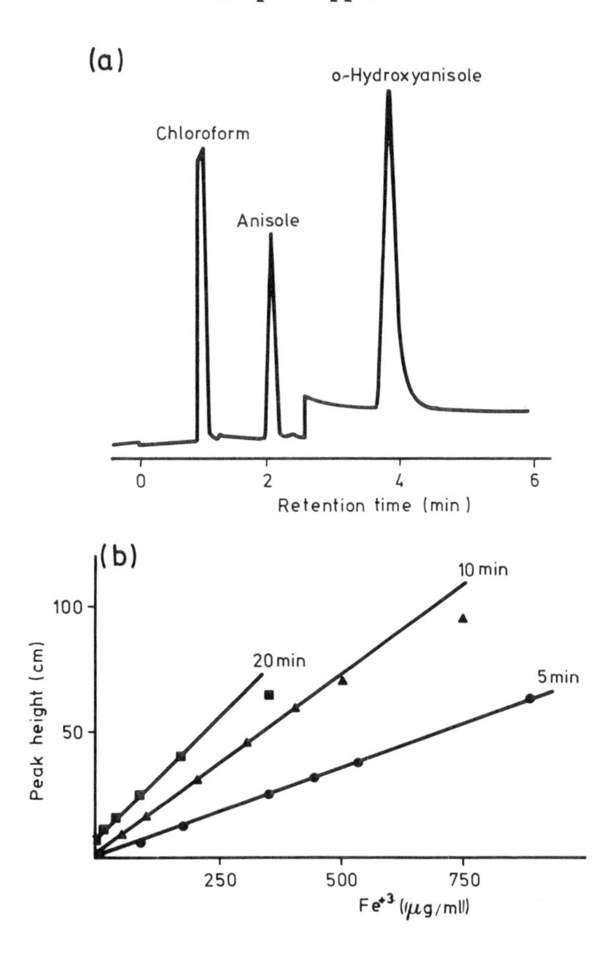

Fig. 9.6 — (a) Chromatogram representing the measurement of 300 ng/ml Fe^{3+}. Injected volume, 2 μl; reaction time, 10 min. Solvent (chloroform), reactant (anisole) and product (*o*-hydroxyanisole) measured at relative sensitivities of 1, 6.25 and 312.5, respectively. (b) GC response to *o*-hydroxyanisole *vs*. Fe(III) concentration at various reaction times. The signal height was normalized to an instrumental attenuation factor of 80. Reaction times: 5 min (circles), 10 min (triangles) and 20 min (squares). (Reproduced with permission, from M. A. Ditzler and W. F. Gutknecht, *Anal. Chem.*, 1980, **52**, 614. Copyright 1980, American Chemical Society).

of catalytic methods of analysis [130].

As can be seen from Table 9.8, most of the applications of kinetic methods to the analysis of foods deal with inorganic species.

Manganese and copper are the elements most commonly determined in trace amounts in foods. It is worth noting the determination of manganese in wines and brandies [1], since its concentration in them is one of the features of the *appellation d'origine*. The procedure is direct and is better than the usual AAS method, which calls for sample pretreatment and preconcentration. Table 9.9 lists some results obtained in the kinetic catalytic determination of manganese in spirits of different

Table 9.7 — Comparison of the stopped-flow method with the official AOAC method for phosphorus determination in feeds and pet food. (Reproduced by permission, from M. S. McCracken and H. V. Malmstadt, *Talanta*, 1979, **26**, 467. Copyright 1979, Pergamon Press)

Sample	Stopped-flow, %	AOAC method, %[b]	RSD (%)
7726 Swine ration	0.600 ± 0.002[a,c]	0.598 ± 0.042[a,c]	0.3
7727 Expanded pet food	1.027 ± 0.006	1.037 ± 0.115	0.6
7729 Pig feed	0.588 ± 0.001	0.580 ± 0.025	0.2
7730 Broiler finisher	0.571 ± 0.004	0.560 ± 0.026	0.7

[a]Average of 3 digestions with 4 determinations per digestion.
[b]Reported in the AAFCO Check Sample Program by 20 laboratories during 1977.
[c]Standard deviation.

characteristics, with and without mineralization, by the standard-addition method, compared with those found by the AAS procedure. The consistency between both groups of results is the best proof of the value of the kinetic method.

There are very few literature reports of analysis of real samples by modified catalytic effects. One such determination is that of zinc in various types of milk by virtue of its activating effect on a manganese-catalysed reaction [132]. Mineralization of the samples at 500°C for 2 hr is required, however.

As in other cases of application, iodide is the anion most frequently determined. Lauber [139] has developed a procedure for the determination of this halide in feeds with low iodine contents, by use of the reaction between Ce(IV) and As(III). The possible loss of iodine during the ashing of the samples at 630°C for 4 hr was studied with ^{125}I-thyroxine; according to the results, the losses were always insignificant (about 2.5%). Moxon and Dixon [133] have devised a procedure for the determination of total iodide in foods, based on its catalytic effect on the reaction between thiocyanate and nitrite. After a comprehensive study to select the mineralization procedure best suited to the case, and establish the effect of storage time on the iodine content of the foods, these authors used the flow manifold shown in Fig. 9.7 to perform the semi-automatic determination of the halide at a sampling rate of 20 samples/hr. The manifold is quite simple and involves the colorimetric determination of the iodine by a fixed-time (17 min) method.

The methods available for the determination of nitrite in foods are of outstanding interest as this ion is a major precursor in the formation of nitrosamines. Doerr *et al.* [16] have developed a chemiluminescence method for this ion in cured meat, based on the formation of nitric oxide with tartaric acid and sodium ascorbate and injection of small volumes (8 ml) into a chemiluminescence detector with the aid of an injection valve. A triplicate determination takes only 10 min.

There are few literature references to the analysis of organic compounds by kinetic methods. One of the most interesting is the determination of tryptophan in various foods, proposed by Steinhart [138]. The samples are hydrolysed in alkaline medium and the tryptophan present reacts specifically with formaldehyde to yield a highly fluorescent product. The kinetic procedure possesses three basic advantages over its equilibrium counterpart, namely: (a) it is faster, as the analysis is direct; (b) the risk of analyte losses is much lower as a result of the smaller number of steps involved; and (c) it is more sensitive.

Table 9.8 — Applications of kinetic methods to food analysis

Sample	Species	Reaction	Observations	References
Milk	Manganese	Catalysed, redox	Treatment with acids + a few drops of H_2O_2	[131]
	Zinc	Activation	Six types of milk; compared with AAS	[132]
	Iodide	Catalysed, redox	Automatic analyser	[133]
Cheese	Iron + manganese	Catalysed, redox	Acid digestion with H_2SO_4 (1:1 v/v) + a few drops of H_2O_2	[103]
Dairy products	Manganese	Catalysed, redox	Ashing at 500°C (2 hr) + acids	[134]
Fruits	Copper	Catalysed, redox	FIA stopped-flow	[135]
Rice	Copper	Catalysed, redox	Compared with AAS with prior liquid–liquid extraction	[135]
Corn flour	Iodide	Catalysed, redox	Decomposition with $HClO_4$/HNO_3/HCl at 170°C for 30 min	[136]
Cereals	Phosphorus	Catalysed, redox	Stopped-flow	[129]
Wines and brandies	Manganese	Catalysed, redox	Direct determination; compared with AAS	[1]
Beer	Manganese	Catalysed, redox	Treatment with Na_2CO_3 + ashing at 500°C (1 hr) + dilute HCl	[131]
	Iron + manganese	Catalysed, redox	Simultaneous determination	[103]
Eggs	Iodide	Catalysed, redox	Study of the effect of storage in refrigerator	[133]
Fish	Iodide	Catalysed, redox	150 mg sample; detn. limit 0.3 ng	[136]
Cod liver oil	Nitrite	Chemiluminescent	Conversion of nitrite into NO	[137]
Cured meat	Nitrite	Chemiluminescent	Compared with Griess method and differential pulse polarography	[16]
Various foods	Tryptophan	Uncatalysed	Vegetables, animal foods, fish and other foods	[138]
Pet food	Iodide	Catalysed, redox	Biological studies on the influence of thyroid hormones	[139]
		Catalysed, redox	Mineralization with KOH–$ZnSO_4$ + ashing at 450°C to avoid I^- losses	[140]
Feed	Phosphorus	Catalysed, redox	AOAC treatment	[129]

Table 9.9 — Determination of manganese in wines and brandies. (Reproduced from D. Pérez-Bendito, J. Peinado and F. Toribio, *Analyst*, 1984, **109**, 1297, by permission. Copyright 1984, Royal Society of Chemistry)

Sample	Manganese found (μg/ml)[a]		
	Kinetic method[b]		
	Ashed	Untreated	AAS method
WINES			
Oloroso (Montilla)[c]	257 ± 4.7	255 ± 2.9	256
Rama (Montilla)[c]	297 ± 6.3	298 ± 3.5	301
Manzanilla (Jerez)[c]		130 ± 1.7	131
White (Valdepeñas)[c]	—	265 ± 3.2	263
White (Jumilla)[c]	—	544 ± 1.5	545
Rosé (Rioja)[c]	—	556 ± 3.4	551
BRANDIES			
Jerez[c]	47 ± 1.3	42 ± 1.7	44
Villafranca del Penedés[c]	—	144 ± 2.8	146

[a]Average of six individual determinations and their standard deviations.
[b]By a standard-addition procedure.
[c]Spanish wine-growing areas.

9.2.6 Quality control of analytical purity

The kinetic determination of traces in pure reagents is one of the areas of application where the inherent sensitivity of kinetic methods of analysis is best displayed. Table 9.10 gathers some of these applications, involving high-purity metals, mineral acids and salts (both of alkali metals and transition metals).

No doubt, the simplest determinations are those involving mineral acids and alkali metal salts on account of the kinetic inertness of the matrix components, their sole potential interference arising from their high ionic content, which influences the reaction rate through the ionic strength.

The analysis of high-purity metals calls for extremely selective methods if the analysis is to be applied in a direct fashion, hence the scarcity of such methods, since analysis of high-purity metals usually involves separation stages. A typical example of direct determination of this kind is that of traces of Pt in palladium metal and its salts by a chemiluminescence reaction [144].

Among the commoner determinations involving prior separations are the enthalpimetric method for sulphide in copper metal, based on the reaction between iodide and sodium azide, in which the hydrogen sulphide generated on acidification is swept by a nitrogen stream into the reaction vessel [8], the determination of manganese in elemental selenium after separation of the latter as the water-soluble dioxide [141], and the determination of phosphate and silicate [4] in $FeCl_3$, involving prior

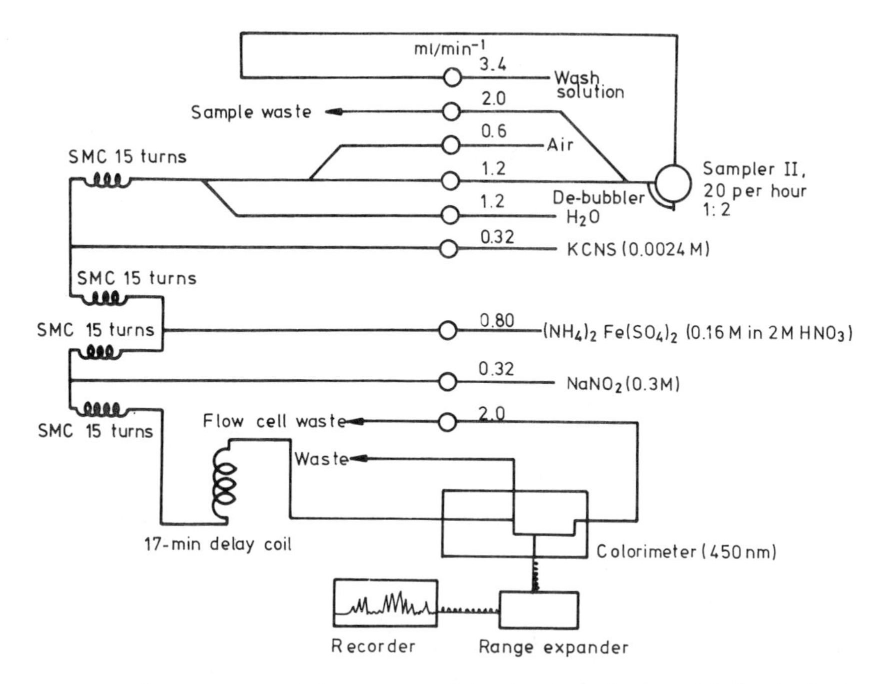

Fig. 9.7 — Flow diagram of the set-up used for the determination of iodine in foods. (Reproduced from R. E. D. Moxon and E. J. Dixon, *Analyst*, 1980, **105**, 344, by permission. Copyright 1980, Royal Society of Chemistry).

extraction of the iron(III) into methyl isobutyl ketone from 8*M* hydrochloric acid medium.

9.3 PRESENT AND FUTURE OF KINETIC METHODS OF ANALYSIS

Figures 9.8–9.10 summarize the developments in the field of kinetic methods applied to the analysis of real samples over the past ten years. Environmental, clinical and pharmaceutical chemistry have gained most from the advantages provided by kinetic methodology (Fig. 9.8).

Catalysed reactions are the most commonly used, followed by chemiluminescence and uncatalysed reactions, the last-named being of especial relevance to clinical and pharmaceutical chemistry.

Logically, the species for which the largest number of kinetic methods are available are also the most frequently determined in real samples. The relative frequency of assay of such species (generally inorganic) in determinations on real samples (in terms of published reports) is shown in Fig. 9.9. As can be seen, manganese and iodide are the cation and anion most frequently involved, the latter being usually determined by difference on the basis of the Sandell–Kolthoff reaction. Generally one particular method will be predominant for determination of an analyte, an exception being the case of nickel, most of the kinetic methods for its determination having been exploited.

Table 9.10 — Applications of kinetic methods to the determination of impurity traces in analytically pure reagents

Sample	Species	Reaction	Observations	References
High-purity metals	Manganese	Catalysed, redox	Determination in Se, isolated as water-soluble SeO_2	[141]
	Tin	Chemiluminescent	Determination in Ni and Cd metal	[142]
	Ruthenium	Catalysed, redox	Determination in Pt metal	[143]
	Platinum	Chemiluminescent	Determination in Pd metal	[144]
	Sulphide	Catalysed, redox	Determination in Cu metal; treatment with acids and collection of H_2S in 0.01*M* NaOH	[8]
Mineral acids	Manganese	Catalysed, redox	Determination in HF and HNO_3; evaporation with 1 ml of 1*M* H_2SO_4	[145]
	Rhodium	Catalysed, redox	Determination in HCl, $HClO_4$ and H_2SO_4	[146]
	Iodide	Uncatalysed	Determination in HNO_3	[147]
Alkali metal salts	Cobalt	Catalysed, redox	Determination in NaCl and KCl	[148–150]
	Iron	Chemiluminescent	Determination in LiCl, KCl, RbCl and CsCl	[151]
	Nickel	Chemiluminescent	Determination in NaCl and Na_2SO_4	[152]
	Chromium	Chemiluminescent	Determination in $(NH_4)_2MoO_4$, $(NH_4)_2WO_4$ and MoO_3	[153]
	Iodide	Uncatalysed	Determination in K_2SO_4, Na_2SO_4 and KH_2PO_4	[147]
	Carbonate	Complex-formation	Determination in KCl, KBr and CsI	[154,155]
Transition metal salts	Iron	Catalysed, redox	Determination in $MgSO_4$ and other salts	[156–158]
	Cobalt	Catalysed, redox	Determination in zinc and cadmium selenide; prior ion-exchange separation	[159]
	Copper, iron, titanium, vanadium and niobium	Catalysed, redox	Determination of nitrates, carbonates and sulphates of various metal ions; prior extraction of ions	[160,161]
	Ruthenium	Catalysed, redox	Determination of $PdCl_2$, $PtCl_4$ and $RuCl_3$	[143]
	Platinum	Catalysed, redox	Determination in $PdCl_2$ and Pd oxide	[144]
	Phosphate, silicate	Catalysed, redox	Determination in $FeCl_3$; prior extraction of iron into MIBK	[4]
Organic solvents	Vanadium	Catalysed, redox	Determination of toluene and acetone	[162]

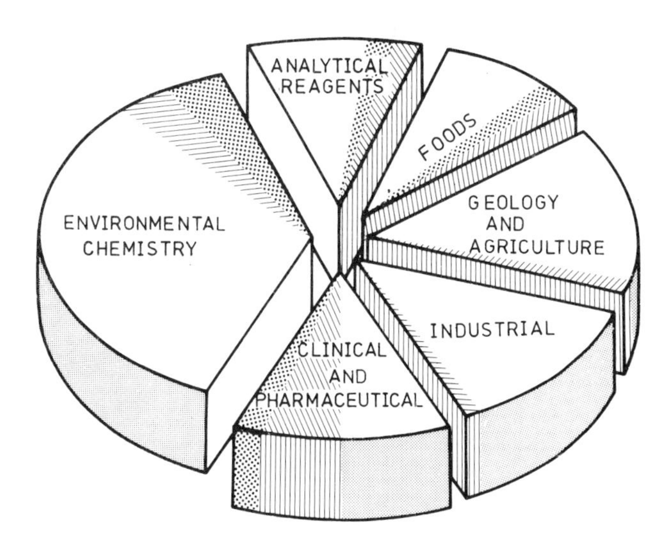

Fig. 9.8 — Significance of kinetic methods in their different areas of application: catalytic reactions (clear zones), uncatalysed reactions (shaded zones) and chemiluminescent reactions (dotted zones).

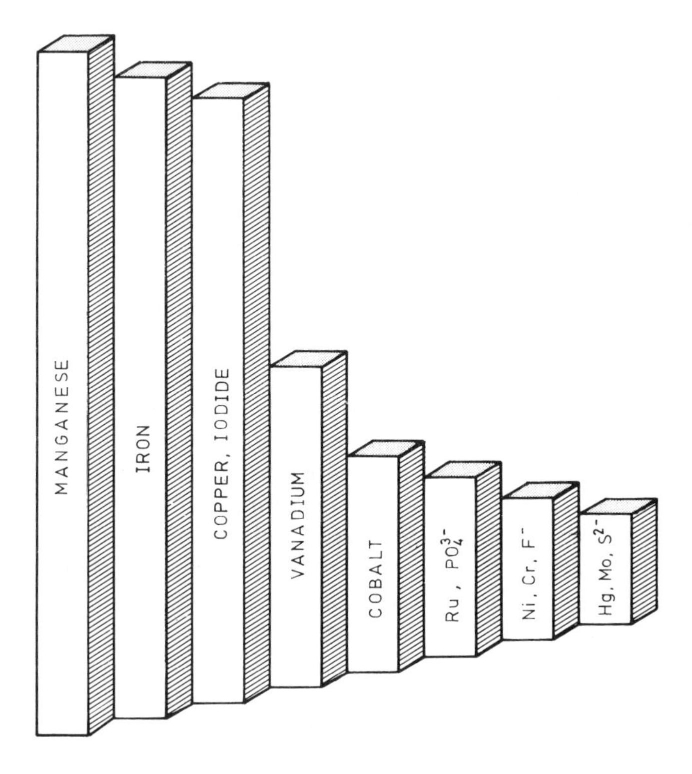

Fig. 9.9. — Species most frequently assayed in real samples by kinetic methods.

Figure 9.10 illustrates the evolution of applications of these methods over the period 1975–84. As can be seen, the number of applications reported reached a peak in 1979 and another in 1984 after levelling off for five years. This evolution is almost parallel to that of kinetic methods themselves, but we believe that the ratio of methods applied to methods developed has increased enormously in the last few years. This is logical to a certain extent, since newly developed methods are aimed at the solution of specific problems in the various areas of application rather than to the theoretical exposition of the method. Some even simply apply already described reactions to particular practical problems.

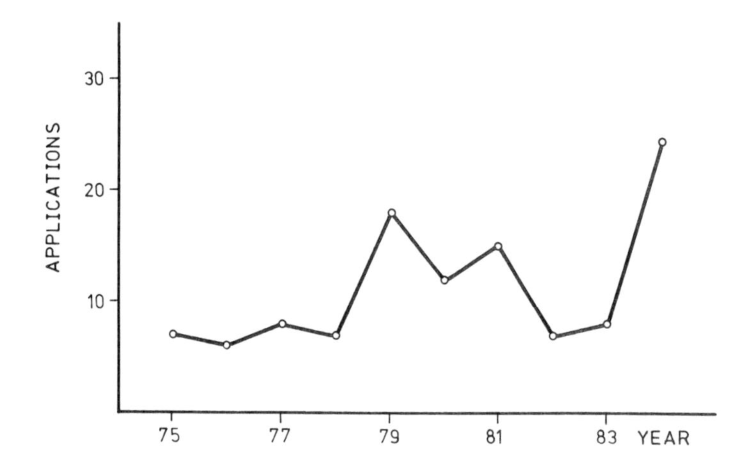

Fig. 9.10 — Evolution of the number of applications of kinetic methods reported in the literature in the last few years.

As can be seen, the current degree of application of kinetic methods to the analysis of a variety of species reflects to a certain extent the degree of development of these methods.

The last biennial reviews published in *Analytical Chemistry*, and the topics presented at the two International Symposia on Kinetics in Analytical Chemistry held so far, have set the guidelines for future developments in this field.

As regards methodology, it is foreseeable that methods developed in the near future will rely on more comprehensive use of information — linear and non-linear regression methods are superseding methods based on the use of a few data from the kinetic curve. No doubt, these methods will provide more accurate and reproducible results. Their present degree of development and future expansion rely heavily on the use of microcomputers in analytical chemistry.

Other potential developments will arise from the joint use of kinetic methodology and other techniques such as direct or indirect photoactivation applied to the photosensitization of catalysed reactions as an alternative to classical activation.

Kinetic methods will also be coupled to pre- and post-column chromatographic techniques. Pre-column devices will extend the use of gas chromatography to the analysis of inorganic species without the need for a prior liquid–liquid extraction to produce volatile complexes. In short, the combined technique involves stopping the catalysed reaction at a preset time by extracting its product (organic substance) into an appropriate solvent, and subsequent detection of this by the chromatograph. The coupling of post-column devices will result in improved selectivity of the kinetic methods and increased sensitivity of the chromatographic detection technique, e.g. mixtures of thyroid hormones are first separated by liquid chromatography and the individual species sensed by a kinetic catalytic method.

As far as instrumentation is concerned, the on-line use of different computer systems with existing configurations will foster the automation of a large number of manual methods. This in turn will facilitate routine analysis in control laboratories.

Another field of development of kinetic methods is that of fast kinetics instrumentation. Thus, stopped-flow techniques will probably be used much more extensively, especially if the instrumentation involved becomes less expensive as a result of the incorporation of straightforward mixing chambers into conventional spectrofluorimeters, as the required graphical recording and data collection pose no problems today, thanks to the use of suitable microcomputers. The development of optical multidetection systems (e.g. the diode-array detector) used in conjunction with the stopped-flow technique offers promising advantages for the simultaneous kinetic analysis of mixtures of species of interest (some substances and their degradation products), some applications of which have already been described in the literature.

REFERENCES

[1] D. Pérez-Bendito, J. Peinado and F. Toribio, *Analyst*, 1984, **109**, 1297.
[2] J. F. Coetzee, D. R. Bayla and P. K. Chattopadhyay, *Anal. Chem.*, 1973, **45**, 2266.
[3] V. W. Truesdale and P. J. Smith, *Analyst*, 1975, **100**, 111.
[4] K. Ohashi, H. Kawaguchi and K. Yamamoto, *Anal. Chim. Acta*, 1979, **111**, 301.
[5] F. Salinas-López, J. J. Berzas-Nevado and A. Espinosa-Mansilla, *Talanta*, 1984, **31**, 325.
[6] T. Fukasawa and T. Yamane, *Bunseki Kagaku*, 1973, **22**, 168.
[7] S. U. Kreingol'd, A. N. Kasnem and G. B. Sevebryakova, *Zavodsk. Lab.*, 1974, **40**, 6.
[8] N. Kiba, M. Nishijima and M. Furusawa, *Talanta*, 1980, **27**, 1090.
[9] R. Delumyea and A. V. Hartkopf, *Anal. Chem.*, 1976, **48**, 1402.
[10] S. A. Nyarady, R. M. Barkley and R. E. Sievers, *Anal. Chem.*, 1985, **57**, 2074.
[11] F. Nachtmann, G. Knapp and H. Spitzy, *J. Chromatog.*, 1978, **149**, 693.
[12] S. Kobayashi and K. Imai, *Anal. Chem.*, 1980, **52**, 424.
[13] G. J. de Jong, N. Lammers, F. J. Spruit, U. A. Th. Brinkman and R. W. Frei, *Chromatographia*, 1984, **18**, 1299.
[14] S. Kobayashi, J. Sekino, K. Honda and K. Imai, *Anal. Biochem.*, 1981, **112**, 99.
[15] K. Honda, K. Miyaguchi and K. Imai, *Anal. Chim. Acta*, 1985, **177**, 103.
[16] R. C. Doerr, J. B. Fox Jr., L. Lakritz and W. Fiddler, *Anal. Chem.*, 1981, **53**, 381.
[17] F. Grases and J. G. March, *Anal. Chem.*, 1985, **57**, 1419.
[18] Q. Wang and D. Chen, *Fenxi Huaxue*, 1981, **9**, 686.
[19] S. Rubio, A. Gómez-Hens and M. Valcárcel, *Talanta*, 1984, **31**, 783.
[20] D. P. Nikolelis and T. P. Hadjiioannou, *Analyst*, 1977, **102**, 591.
[21] D. P. Nikolelis and T. P. Hadjiioannou, *Anal. Chim. Acta*, 1978, **97**, 111.
[22] Y. R. Zeng and R. Line, *Huaxue Xuebao*, 1983, **41**, 960.
[23] E. A. Morgan, N. A. Vlasov and L. A. Kozhemyakina, *Zh. Analit. Khim.*, 1972, **27**, 2064.
[24] H. Schurnig and H. Müller, *Acta Hydrochim. Hydrobiol.*, 1979, **7**, 281.
[25] T. Fukasawa and T. Yamane, *Anal. Chim. Acta*, 1977, **88**, 147.
[26] E. Mentasti, V. Dlask and J. S. Coe, *Analyst*, 1985, **110**, 1451.
[27] S. U. Kreingol'd, L. I. Sosenkova, A. A. Panteleimonova and L. V. Lavrelashvili, *Zh. Analit. Khim.*, 1978, **33**, 2168.

[28] S. D. Jones, C. P. Spencer and V. W. Truesdale, *Analyst*, 1982, **107**, 1417.
[29] K. Toda, I. Sanemasa and T. Deguchi, *Bunseki Kagaku*, 1985, **34**, 31.
[30] N. M. Lukovskaya and N. L. Anatienko, *Ukr. Khim. Zh.*, 1978, **44**, 199.
[31] W. D. Bostick, C. A. Burtis and C. D. Scott, *Anal. Lett.*, 1976, **9**, 65.
[32] S. Nakano, K. Kuramoto and T. Kawashima, *Nippon Kagaku Kaishi*, 1981, 91.
[33] S. Nakano, K. Kuramoto and T. Kawashima, *Chem. Lett.*, 1980, 849.
[34] L. A. Montano and J. D. Ingle, Jr., *Anal. Chem.*, 1979, **51**, 926.
[35] M. A. Koupparis, K. M. Walczak and H. V. Malmstadt, *Analyst*, 1982, **107**, 1309.
[36] A. Sakuragawa, M. Tsukada, T. Okutani and S. Utsumi, *Bunseki Kagaku*, 1982, **31**, 224.
[37] D. F. Marino and J. D. Ingle, Jr., *Anal. Chim. Acta*, 1981, **123**, 247.
[38] S. Nakano, H. Enoki and T. Kawashima, *Chem. Lett.*, 1980, 1173.
[39] S. Nakano, M. Tanaka, M. Fushihara and T. Kawashima, *Mikrochim. Acta*, 1983 **I**, 457.
[40] K. Hirayama and N. Unohara, *Bunseki Kagaku*, 1980, **29**, 733.
[41] S. Nakano, M. Odzu, M. Tanaka and T. Kawashima, *Mikrochim. Acta*, 1983 **I**, 403.
[42] T. Okutani, A. Sakuragawa and R. Uematsu, *Bunseki Kagaku*, 1984, **33**, 1.
[43] S. Nakano, E. Kasahara, M. Tanaka and T. Kawashima, *Chem. Lett.*, 1981, 597.
[44] J. L. Ferrer-Herranz and D. Pérez-Bendito, *Anal. Chim. Acta*, 1981, **132**, 157.
[45] J. B. Pausch and D. W. Margerum, *Anal. Chem.*, 1969, **41**, 226.
[46] M. Kataoka, S. Tahara and K. Ohzeki, *Z. Anal. Chem.*, 1985, **321**, 146.
[47] D. J. Saksa and R. B. Smart, *Environ. Sci. Technol.*, 1985, **19**, 450.
[48] T. Fukasawa, S. Kawakubo, T. Okabe and A. Mizuike, *Bunseki Kagaku*, 1984, **33**, 609.
[49] V. I. Rychkova and I. F. Dolmanova, *Zh. Analit. Khim.*, 1979, **34**, 1094.
[50] N. Xie, M. Xu, Z. Pan and J. Miao, *Fenxi Huaxue*, 1984, **12**, 281.
[51] K. Haraguchi, K. Nakagawa, T. Ogata and S. Ito, *Bunseki Kagaku*, 1981, **30**, 149.
[52] R. N. Voevutskaya, V. K. Pavlova and A. T. Pilipenko, *Zh. Analit. Khim.*, 1979, **34**, 1299.
[53] O. A. Bilenko, N. B. Potekhina and S. P. Mushtakova, *Zh. Analit. Khim.*, 1984, **39**, 804.
[54] V. G. Badding and J. L. Durney, *Plat. Surf. Finish.*, 1980, **67**, 49.
[55] M. P. Babkin, *Zavodsk. Lab.*, 1971, **37**, 524.
[56] U. Isacsson and G. Wettermark, *Anal. Lett.*, 1978, **11**, 13.
[57] E. I. Dodin, L. S. Makarenko, V. F. Tsvetkov, I. P. Kharlamov and A. M. Pavlova, *Zavodsk. Lab.*, 1973, **39**, 1050.
[58] G. A. Zolotova, I. D. Strel'tsova, T. A. Gorchkova, Ba Yasuf, M. A. Volodina and I. F. Dolmanova, *Zh. Analit. Khim.*, 1984, **39**, 1886.
[59] S. Raman, *Indian J. Chem.*, 1975, **13**, 1229.
[60] A. Navas and F. Sánchez Rojas, *Talanta*, 1984, **31**, 437.
[61] J. Teckentrup and D. Klockow, *Talanta*, 1981, **28**, 663.
[62] T. Simonescu, V. Rusu and L. Kiss, *Revta. Chim.*, 1975, **26**, 75.
[63] T. Fukasawa and T. Yamane, *Bunseki Kagaku*, 1977, **26**, 692.
[64] J. Auffarth and D. Klockow, *Anal. Chim. Acta*, 1979, **111**, 89.
[65] U. Isacsson and G. Wettermark, *Anal. Chim. Acta*, 1976, **83**, 227.
[66] M. Tabata and M. Tanaka, *Anal. Lett.*, 1980, **13**, 427.
[67] I. F. Dolmanova, O. I. Melnikova, G. I. Tsizin and T. N. Shekhovtsova, *Zh. Analit. Khim.*, 1980, **35**, 728.
[68] S. Gantcheva and P. R. Bontchev, *Talanta*, 1980, **27**, 893.
[69] W. Dong and Z. Zhang, *Fenxi Huaxue*, 1984, **12**, 186.
[70] A. I. Merkulov and R. I. Skvortsova, *Zh. Analit. Khim.*, 1981, **36**, 1778.
[71] G. D. Christian and G. J. Patriarche, *Anal. Lett.*, 1979, **12**, 11.
[72] V. N. Kachibaya, I. L. Siamashvili and M. V. Mamukashvili, *Zh. Analit. Khim.*, 1971, **26**, 1848.
[73] J. O. Peyrin and Y. Barbier, *Chem. Abstr.*, 1973, **78**, 81712u.
[74] H. Hoch and C. G. Lewallen, *Clin. Chem.*, 1969, **15**, 204.
[75] T. Simonescu and V. Rusu, *Revta. Chim.*, 1976, **27**, 164.
[76] G. Palumbo, M. F. Tecce and G. Ambrosio, *Anal. Biochem.*, 1982, **123**, 183.
[77] G. Knapp and H. Leopold, *Anal. Chem.*, 1974, **46**, 719.
[78] V. V. S. Dutt and H. A. Mottola, *Anal. Chem.*, 1974, **46**, 1777.
[79] V. V. S. Dutt and H. A. Mottola, *Biochem. Med.*, 1974, **9**, 148.
[80] J. E. Frew and P. Jones, *Anal. Lett.*, 1985, **18**, 1579.
[81] E. P. Diamandis and T. P. Hadjiioannou, *Clin. Chem.*, 1981, **2**, 455.
[82] D. P. Nikolelis and T. P. Hadjiioannou, *Mikrochim. Acta*, 1977 **I**, 125
[83] P. J. Garry, D. Lashley and G. M. Owen, *Clin. Chem.*, 1973, **19**, 950
[84] F. Grases, R. Forteza, J. G. March and V. Cerdá, *Talanta*, 1985, **32**, 123.
[85] A. G. Xenakis and M. I. Karayannis, *Anal. Chim. Acta*, 1984, **159**, 343.
[86] G. A. Milovanović and M. A. Sekheta, *Mikrochim. Acta*, 1984 **III**, 477.

[87] M. Otto, H. Müller and W. Werner, *Talanta*, 1979, **26**, 781.
[88] S. Chen, S. Peng and D. Yuan, *Fenxi Huaxue*, 1984, **12**, 913.
[89] F. Hao, L. Cai, J. Pan, S. Chen and F. Bian, *Fenxi Huaxue*, 1983, **11**, 857.
[90] G. A. Milovanović and Lj. Stefanović-Ristić, *Microchem. J.*, 1985, **31**, 293.
[91] A. Ríos and M. Valcárcel, *Analyst*, 1984, **109**, 1147.
[92] F. C. Gaál and B. F. Abramović, *Mikrochim. Acta*, 1982 **I**, 465.
[93] H. Fung, P. Dalecki, E. Tse and T. C. Rhodes, *J. Pharm. Sci.*, 1973, **25**, 3378.
[94] M. Elsayed, *Pharmazie*, 1979, **34**, 569.
[95] M. A. Ryan and J. D. Ingle, Jr., *Anal. Chem.*, 1980, **52**, 2177.
[96] M. Timotheou-Potamia, E. G. Sarantonis, A. C. Calokerinos and T. P. Hadjiioannou, *Anal. Chim. Acta*, 1985, **171**, 363.
[97] M. A. H. Elsayed, M. H. Barary and H. Mahgoub, *Anal. Lett.*, 1985, **18**, 1357.
[98] E. J. Greenhow and L. E. Spencer, *Analyst*, 1973, **98**, 485.
[99] D. B. Roodyn, *Automated Enzyme Assays*, Chapter 3. Elsevier, Amsterdam, 1970.
[100] J. A. Loraine and E. T. Bell, *Hormone Assays and Their Clinical Application*, 4th Ed., Chapter 12, Churchill-Livingstone, Edinburgh, 1976.
[101] R. P. Morozova and L. V. Il'enko, *Zh. Analit. Khim.*, 1973, **28**, 1835.
[102] T. Rayo-Sara and D. Pérez-Bendito, *Anal. Chim. Acta*, 1985, **172**, 273.
[103] A. Moreno, M. Silva and D. Pérez-Bendito, *Anal. Chim. Acta*, 1984, **159**, 319.
[104] A. Moreno, M. Silva, D. Pérez-Bendito and M. Valcárcel, *Anal. Chim. Acta*, 1984, **157**, 333.
[105] A. P. Rysev and L. P. Zhitenko, *Zh. Analit. Khim.*, 1981, **36**, 126.
[106] R. L. Wilson and J. D. Ingle, Jr., *Anal. Chem.*, 1977, **49**, 1066.
[107] V. F. Toropova, M. G. Gadbullin, A. R. Garifzyanov and R. A. Cherkasov, *Zh. Analit. Khim.*, 1984, **39**, 267.
[108] F. Salinas, F. García-Sánchez, F. Grases and C. Genestar, *Anal. Lett.*, 1980, **13**, 473.
[109] T. Fukasawa and T. Yamane, *Anal. Chim. Acta*, 1980, **113**, 123.
[110] M. Hernández-Córdoba, P. Viñas and C. Sánchez-Pedreño, *Analyst*, 1985, **110**, 1343.
[111] F. Grases and J. G. March, *Analyst*, 1985, **110**, 795.
[112] N. N. Gusakova and S. P. Mustakova, *Zh. Analit. Khim.*, 1981, **36**, 317.
[113] P. I. Kuznetsov, B. D. Luft, V. V. Shemet, V. N. Antonov, S. U. Kreingol'd, A. A. Panteleimonova and M. S. Chupakhin, *Zavodsk. Lab.*, 1976, **42**, 657.
[114] J. C. Gijsbers and J. G. Kloosterboer, *Anal. Chem.*, 1978, **50**, 455.
[115] L. M. Matat, I. B. Mizetskaya, V. K. Pavlova and A. T. Pilipenko, *Zh. Analit. Khim.*, 1982, **37**, 2165.
[116] S. Rubio, A. Gómez-Hens and M. Valcárcel, *Analyst*, 1984, **109**, 597.
[117] A. P. Rysev, L. P. Zhitenko and V. A. Nadezhdina, *Zavodsk. Lab.*, 1981, **47**, No. 6, 20.
[118] L. G. Pavlova and T. V. Gurkina, *Zh. Analit. Khim.*, 1979, **34**, 1787.
[119] K. Tan, *Fenxi Huaxue*, 1983, **11**, 433.
[120] A. T. Tashkhodzhaev, L. E. Zel'tser, S. T. Talipov and K. Khikmatov, *Zh. Analit. Khim.*, 1976, **31**, 485.
[121] K. Takahashi, M. Yoshida, T. Ozawa and I. Iwasaki, *Bull. Chem. Soc. Japan*, 1970, **43**, 3159.
[122] N. N. Nikol'skaya, L. P. Tikhonova, Z. A. Ezhkova and I. Yu. Davydova, *Zh. Analit. Khim.*, 1979, **34**, 171.
[123] G. Zhang and Z. Pan, *Ti Ch'iu Hua Hsueh*, 1979, 353.
[124] M. A. Ditzler and W. F. Gutknecht, *Anal. Chem.*, 1980, **52**, 614.
[125] R. G. Anderson and B. C. Brown, *Talanta*, 1981, **28**, 365.
[126] L. Lan and Y. Hu, *Fenxi Huaxue*, 1984, **12**, 118.
[127] W. Matthes, T. Kis and M. Stoeppler, *Zh. Analit. Chem.*, 1973, **267**, 89.
[128] H. van Vliet, W. D. Basson and R. G. Boehmer, *Analyst*, 1975, **100**, 405.
[129] M. S. McCracken and H. V. Malmstadt, *Talanta*, 1979, **26**, 467.
[130] H. Müller, *CRC Crit. Rev. Anal Chem.*, 1982, **13**, 313.
[131] A. Moreno, M. Silva, D. Pérez-Bendito and M. Valcárcel, *Talanta*, 1983, **30**, 107.
[132] A. Moreno, M. Silva, D. Pérez-Bendito and M. Valcárcel, *Analyst*, 1983, **108**, 85.
[133] R. E. D. Moxon and E. J. Dixon, *Analyst*, 1980, **105**, 344.
[134] S. Rubio, A. Gómez-Hens and M. Valcárcel, *Analyst*, 1984, **109**, 717.
[135] F. Lázaro, M. D. Luque de Castro and M. Valcárcel, *Anal. Chim. Acta*, 1984, **165**, 177.
[136] H. Gstrein, B. Maichin, P. Eustecehio and G. Knapp, *Mikrochim. Acta*, 1979 **I**, 291.
[137] C. L. Walters, M. J. Downes, R. J. Hart, S. Perse and P. L. R. Smith, *Z. Lebensm. Unters. Forsch.*, 1978, **167**, 229.
[138] H. Steinhart, *Anal. Chem.*, 1979, **51**, 1012.
[139] K. Lauber, *Anal. Chem.*, 1975, **47**, 769.
[140] P. Fioravanti and M. Halmi, *Mitt. Geb. Lebensmittelunters. Hyg.*, 1971, **62**, 388.

[141] T. Fukasawa, T. Yamane and T. Yamazaki, *Bunseki Kagaku*, 1977, **26**, 202.
[142] N. M. Lukovskaya and N. F. Kushchevskaya, *Ukr. Khim. Zh.*, 1985, **51**, 511.
[143] T. Suwinska, Z. Gregorowizc and M. D. Matysek, *Chem. Abst.*, 1982, **96**, 115090d.
[144] V. I. Rigin, A. S. Bakhmurov and A. I. Blokhin, *Zh. Analit. Khim.*, 1975, **30**, 2413.
[145] Y. Fukasawa and T. Yamane, *Bunseki Kagaku*, 1973, **22**, 168.
[146] A. T. Pilipenko, T. S. Maksimenko and N. M. Lukoskaya, *Zh. Analit. Khim.*, 1979, **34**, 523.
[147] T. Pérez-Ruiz, C. Martínez-Lozano and J. Ochotorena, *Talanta*, 1982, **29**, 479.
[148] S. U. Kreingol'd, L. I. Sosenkova and I. F. Vzorova, *Metody Anal. Kontrolya Proizvod Khim. Promsti.*, 1976, No. 2, 38.
[149] V. K. Zinchuk, V. S. Besidka, Ya. P. Skorobogatyi and R. F. Markovskaya, *Zh. Analit. Khim.*, 1981, **36**, 701.
[150] R. P. Pantaler, L. D. Alfimova, A. M. Balgakova and I. V. Pulyaeva, *Zh. Analit. Khim.*, 1975, **30**, 946.
[151] V. I. Rigin and A. I. Blokhin, *Zh. Analit. Khim.*, 1977, **32**, 312.
[152] V. K. Zinchuk and R. N. Gal'chun, *Zh. Analit. Khim.*, 1984, **39**, 56.
[153] V. I. Rigin and A. S. Bakhmurov, *Zh. Analit. Khim.*, 1976, **31**, 93.
[154] A. B. Blauk, R. P. Pantaler, I. V. and L. P. Eksperiandova, *Zh. Analit. Khim.*, 1978, **33**, 1771.
[155] R. P. Panteler and I. V. Pulyaeva, *Zh. Analit. Khim.*, 1979, **34**, 287.
[156] M. Otto and H. Müller, *Anal. Chim. Acta*, 1977, **90**, 159.
[157] T. Pérez-Ruiz, C. Martínez-Lozano and V. Tomas, *Analyst*, 1984, **109**, 1401.
[158] A. P. Gumenyuk and S. P. Mushtakova, *Zh. Analit. Khim.*, 1984, **39**, 1278.
[159] Z. Gregorowicz, D. Matysek-Mejewska and T. Suwinska, *Mikrochim. Acta*, 1985 **I**, 237.
[160] S. U. Kreingol'd and E. D. Loginova, *Zh. Analit. Khim.*, 1983, **38**, 1397.
[161] R. Baranowski, I. Baranowska and Z. Gregorowicz, *Microchem. J.*, 1979, **24**, 367.
[162] M. Otto, J. Stach and R. Kirme, *Anal. Chim. Acta*, 1983, **147**, 277.

Index